Gene Regulation

A eukaryotic perspective

Fourth Edition

David S. Latchman
Professor of Human Genetics and Dean
Institute of Child Health
University College London

First edition published in 1990
Second edition published in 1995 by:
Chapman & Hall
Third edition published in 1998 by:
Stanley Thornes (Publishers) Ltd

Fourth edition published in 2002
Nelson Thornes Ltd
Delta Place
27 Bath Road
CHELTENHAM
GL53 7TH
United Kingdom

02 03 04 05 06 / 10 9 8 7 6 5 4 3 2 1

A catalogue record for this book is available from the British Library

ISBN 0 7487 6530 1

Typeset by Saxon Graphics Ltd, Derby
Printed in Croatia by Zrinski

To my mother and in memory of my father

CONTENTS

Comparative

Preface to the Fourth Edition

Once again, considerable information on the process of Gene Regulation has accumulated since the third edition of this book was published. Interestingly, much of this information involves the two topics of modulation of chromatin structure by histone modifications/remodelling complexes and the role of co-activators/co-repressors which were discussed in detail for the first time in the third edition. To reflect this, the topic of histone modifications now warrants a separate section (Section 6.6) and the discussion of chromatin remodelling complexes (Section 6.7) and of co-activators/co-repressors (Section 8.3) has been considerably expanded.

As well as a general updating of the work, enhanced consideration has also been given to the methods used to analyse gene expression such as transcriptomic and proteomic techniques (Chapter 1) with a new section added on methods for examining DNA binding by transcription factors (Section 7.2.4). Similarly, new sections have been added on topics which have become of increasing importance such as the coupling of transcription with post-transcriptional processes (Section 3.3.4) and negatively acting sequence elements (Section 7.4). In addition, the more extensive discussion of the aberrations in gene regulation which result in diseases other than cancer in this new edition has led to the title of Chapter 9 being changed to "Gene regulation and human disease" (from "Gene regulation and cancer").

Overall, therefore, the book provides an up-to-date and comprehensive account of the processes involved in gene expression and the mechanisms by which such expression is regulated, as well as an introduction to the methods used to study these processes.

Finally, I would like to thank Maruschka Malacos who has taken over the difficult task of extensively modifying the existing word processed text and inserting much new material. I would also like to thank the staff of Nelson Thornes Ltd, and particularly Catherine Shaw, for their enthusiasm to commission a new edition of this book and their efficiency in producing it.

David S. Latchman

PREFACE TO THE FIRST EDITION

An understanding of how genes are regulated in humans and higher eukaryotes is essential for the understanding of normal development and disease. For these reasons this process is discussed extensively in many current texts. Unfortunately, however, many of these works devote most of their space to a consideration of gene regulation in bacteria and then discuss the complexities of the eukaryotic situation more briefly. Although such an approach was the only viable one when only the simpler systems were understood, it is clear that sufficient information is now available to discuss eukaryotic gene regulation as a subject in its own right.

This work aims to provide a text which focuses on how cellular genes are regulated in eukaryotes. For this reason bacterial examples are included only to illustrate basic control processes, while examples from the eukaryotic viruses are used only when no comparable cellular gene example of a type of regulation is available.

It is hoped that this approach will appeal to a wide variety of students and others interested in cellular gene regulation. Thus, the first four chapters provide an introduction to the nature of transcriptional regulation as well as a discussion of the other levels at which genes can be regulated, which is suitable for second-year undergraduates in biology or medicine. A more extensive discussion of the details of transcriptional control, at a level suitable for final-year undergraduates, is contained in Chapters 5–7. The entire work should also provide an introduction to the topic for starting researchers, or those moving into this area from other fields and wishing to know how the gene which they are studying might be regulated. Clinicians moving into molecular research should find the concentration on the mechanisms of human cellular gene regulation of interest, and for these readers in particular, Chapter 8 focuses on the malregulation of gene expression which occurs in human diseases and, specifically, in cancer.

I would like to thank Professor Martin Raff and Dr Paul Brickell for their critical reading of the text, and Dr Joan Heaysman who suggested that I should write this book. I am also extremely grateful to Mrs Rose Lang for typing the text and tolerating my continual changes, and to Mrs Jane Templeman who prepared the illustrations.

David S. Latchman

PREFACE TO THE SECOND EDITION

In the four years since the first edition of *Gene Regulation* was published, considerable advances have been made in understanding the details of gene regulation. Nonetheless, the basic principles outlined in the first edition have remained unchanged and this second edition therefore adopts the same general approach as its predecessor. The opportunity has been taken, however, of updating the discussion of most topics to reflect recent advances. This has required the complete rewriting of some sections where progress has been particularly rapid such as those dealing with alternative splicing factors (Chapter 4) or the mechanism of transcriptional activation (Chapter 7). In addition, several new sections have been added reflecting topics which are now known to be of sufficient importance in gene regulation to merit inclusion, as well as others suggested by comments on the first edition. These include regulation of transcription by polymerase pausing (Chapter 3), X chromosome inactivation and genomic imprinting (Chapter 5), locus control regions (Chapter 6), negative regulation of transcription (Chapter 7) and anti-oncogenes (Chapter 8). It is hoped that these changes and additions will allow the new edition of *Gene Regulation* to build on the success of the first edition and provide a comprehensive overview of our current understanding in this area.

Finally, I would like to thank Mrs Rose Lang who has coped most efficiently with the need to add, delete or amend large sections of the original word processed text of the first edition. I am also most grateful to Rachel Young and the staff at Chapman & Hall for deciding to commission a new edition of this work and their efficiency in producing it.

David S. Latchman

Preface to the Third Edition

As before, considerable progress has been made in the study of gene regulation in the three years since the second edition of this book was published. Once again, this has necessitated the re-writing of many sections of the book including, as previously, those on alternative splicing factors (Chapter 5) and transcriptional activation (Chapter 8) where progress has again been rapid. In addition, several new topics are discussed in detail for the first time to reflect their increasing importance: these include regulation of RNA transport (Chapter 4), the SWI-SNF complex (Chapter 6), the effect of specific transcription factors on histone acetylation (Chapters 6 and 8), as well as the role of co-repressors/co-activators and the mediator complex by which the inhibitory or activating effects of transcription factors are often mediated (Chapters 8 and 9).

The most significant change in this edition, however, is the addition of an extra chapter (Chapter 3) which describes the basic processes (transcription, capping, polyadenylation, splicing, transport and translation) by which DNA is converted into RNA and then into protein. Since it is this process which is regulated by the mechanisms described in the remainder of the book, this chapter renders the work complete in itself as a textbook dealing with gene expression and its regulation. In addition, this extra chapter allows the extensive discussion of the RNA polymerases themselves and their associated factors which carry out the basic process of transcription and are hence the ultimate target for the transcriptional regulatory processes which are the major regulatory mechanisms operating to control gene expression.

Lastly, I would like to thank Sarah Chinn who has taken over the difficult tasks of typing the new material and extensively amending the text of the previous edition, as well as Jane Templeman who has once again prepared the illustrations. I am also most grateful to Nigel Balmforth and the staff at Chapman & Hall for commissioning another edition and for their continued efficiency in its production.

David S. Latchman

ACKNOWLEDGEMENTS

I would like to thank all those colleagues who have given permission for material from their papers to be reproduced in this book and have provided prints suitable for reproduction. Other than those listed below, all figures have been drawn especially for this work. Figures 1.1 and 1.2 photographs kindly provided by Dr A. Moore, from Moore *et al.*, *Eur. J. Biochem.* **152**, 729 (1985), by permission of Springer-Verlag; Figure 1.6 from Latchman *et al.*, *Biochim. Biophys. Acta* **783**, 130 (1984), by permission of Elsevier Science Publishers; Figure 1.7, photograph kindly provided by Dr P. Brickell, from Devlin *et al.*, *Development* **103**, 111 (1988), by permission of the Company of Biologists Ltd; Figure 1.9 redrawn from Bishop *et al.*, *Nature* **250**, 199 (1974), by permission of Dr J. Bishop and Macmillan Magazines Ltd; Figure 1.10 redrawn from Hastie and Bishop, *Cell* **9**, 761 (1976), by permission of Dr J. Bishop and Cell Press.

Figure 2.2, photograph kindly provided by Dr A.T. Sumner; Figure 2.4, redrawn from Cove, D.J., *Genetics* (1971), by permission of Professor D.J. Cove and Cambridge University Press; Figure 2.5, redrawn from Steward *et al.*, *Science* **143**, 20 (1964), by permission of Professor F.C. Steward and the American Association for the Advancement of Science (AAAS); Figure 2.6, redrawn from Gurdon, *Gene Expression* (1974), by permission of Professor J.B. Gurdon and Oxford University Press; Figure 2.10, photograph kindly provided by Dr M. Ashburner; Figure 2.11, photograph kindly provided by Dr R. Manning, from Manning and Gage, *J. Biol. Chem.* **253**, 2044 (1978); Figure 2.12, redrawn from Spradling and Mahowald, *PNAS* **77**, 1096 (1980), by permission of Dr A. Spradling.

Figures 3.4 and 3.5, photographs kindly provided by Professor D.D. Brown, 3.4 from Sakonju *et al.*, *Cell* **19**, 13 (1980), 3.5 from Sakonju and Brown, *Cell* **31**, 395 (1982), by permission of Cell Press; Figure 3.22 photograph kindly provided by Professor B. Daneholt from Visa *et al.*, *Cell* **84**, 253–254 by permission of Cell Press.

Figures 4.2 and 4.3, photographs kindly provided by Professor B.W. O'Malley, from Roop *et al.*, *Cell* **15**, 671 (1978), by permission of Cell Press; Figure 4.9, photograph kindly provided by Dr M. Ashburner; Figure 4.11, photograph kindly provided by Dr R.S. Hill, from Hill and Macgregor, *J. Cell Sci.* **44**, 87 (1980), by permission of the Company of Biologists Ltd.

Figure 5.9, redrawn from Breitbart and Nadal-Ginard, *Cell* **49**, 793 (1987), by permission of Professor B. Nadal-Ginard and Cell Press; Figure 5.24, redrawn from Guyette *et al.*, *Cell* **17**, 1013 (1979), by permission of Professor J. Rosen and Cell Press, Figure 5.25, redrawn from Casey *et al.*, *Science* **240**, 924 (1988), by permission of Dr J.B. Harford and AAAS; Figure 5.30 photograph kindly provided by Dr N. Standart and Dr T. Hunt.

Figure 6.3, redrawn from Gurdon, J.B., *Gene Expression* (1974), by permission of Professor J.B. Gurdon and Oxford University Press; Figure

6.4, redrawn from Cove D.J., *Genetics* (1971), by permission of Professor D.J. Cove and Cambridge University Press; Figure 6.6, photographs kindly provided by Dr J.T. Finch, from Finch *et al.*, *PNAS* **72**, 3320 (1975); Figure 6.7, photograph kindly provided by Dr F. Thoma, from Thoma *et al.*, *J. Cell. Biol.* **83**, 403 (1979), by permission of Rockefeller Press; Figure 6.8, photograph kindly provided by Dr J. McGhee, from McGhee *et al.*, *Cell* **33**, 831 (1983), by permission of Cell Press; Figure 6.10, photograph kindly provided by Professor O.L. Miller, from McKnight *et al.*, *Cold Spring Harbor Symposium* **42**, 741 (1978), by permission of Cold Spring Harbor Laboratory; Figure 6.15, redrawn from Weintraub *et al.*, *Cell* **24**, 333 (1981), by permission of Professor H. Weintraub and Cell Press; Figure 6.16, redrawn from Keshet *et al.*, *Cell* **44**, 535 (1986), by permission of Professor H. Cedar and Cell Press; Figure 6.28, photograph kindly provided by Professor P. Chambon, from Kaye *et al.*, *EMBO Journal* **3**, 1137 (1984), by permission of Oxford University Press; Figure 6.30, photograph kindly provided by Professor M. Yaniv, from Saragosti *et al.*, *Cell* **20**, 65 (1980), by permission of Cell Press.

Figures 7.11 and 7.12, photographs kindly provided by Professor D. Hanahan, from Hanahan, *Nature* **315**, 115 (1985), by permission of Macmillan Magazines Ltd.

Figure 8.4, photograph kindly provided by Professor W.J. Gehring, from Gehring, *Science* **236**, 1245 (1987), by permission of AAAS; Figure 8.11, redrawn from Schleif, *Science* **241**, 1182 (1988), by permission of Dr R. Schleif and AAAS; Figure 8.14, redraw from Evans and Hollenberg, *Cell* **52**, (1988), by permission of Professor R.M. Evans and Cell Press; Figure 8.23, redrawn from Turner and Tjian, *Science* **243**, 1189 (1989), by permission of Professor R. Tjian and AAAS; Figure 9.5, redrawn from Takeya and Hanafusa, *Cell* **32**, 881 (1985), by permission of Professor H. Hanafusa and Cell Press.

I am also especially grateful to those who have provided the colour plates which are included for the first time in this edition. Colour plate 1 kindly provided by Kevin Mills and Professor Brian Winchester. Colour plate 2 kindly provided by Dr Rachel Chambers and Dr Mike Hubank. Colour plate 3 kindly provided by the Roslin Institute. Colour plate 4 kindly provided by Dr Stefan Moisyadi and Professor Ryuzo Yanagimachi from Wakayama *et al.*, *Nature* **394**, 369 (1998) by kind permission of Macmillan Magazines Ltd. Colour plate 5 kindly provided by Dr Patrick Schultz from Brand *et al.*, *Science* **286**, 2151 (1999) by kind permission of the American Association for the Advancement of Science. Colour plate 6 kindly provided by Dr Eva Nogales from Andell *et al.*, *Science* **286**, 2153 (1999) by kind permission of the American Association for the Advancement of Science. Colour plate 7, kindly provided by Dr J.H. Geiger. Colour plate 8, kindly provided by Professor Sir Aaron Klug from the front cover of *Science* 8 June 2001, by kind permission of the American Association for the Advancement of Science. Colour plate 9, kindly provided by Dr Y. Muto and Professor S. Yokoyama, from Handa *et al.*, *Nature* **398**, 579 (1999) by kind permission of Macmillan Magazines Ltd. Colour plate 10, panel a kindly provided by

Professor Tim Richmond, from Luger *et al.*, *Nature* **389**, 251 (1997), panel b kindly provided by Dr Daniela Rhodes, from Rhodes, *Nature* **389**, 231 (1997) both by kind permission of Macmillan Magazines Ltd. Colour plate 11, kindly provided by Dr P. Wright, from Lee *et al.*, *Science* **245**, 635 (1989) by kind permission of the American Association for the Advancement of Science. Colour plate 12, kindly provided by Professor R. Kaptein from Hard *et al.*, *Science* **249**, 157 (1990) by kind permission of the American Association for the Advancement of Science. Colour plate 13, kindly provided by Dr D. Rhodes from Schwabe *et al.*, *Cell* **75**, 567 (1993) by kind permission of Elsevier Science.

TISSUE-SPECIFIC EXPRESSION OF PROTEINS AND MESSENGER RNAs

1

1.1 INTRODUCTION

The evidence that eukaryotic gene expression must be a highly regulated process is available to anyone visiting a butcher's shop. The various parts of the mammalian body on display differ dramatically in appearance, ranging from the muscular legs and hind quarters to the soft tissues, of the kidneys and liver. However, all these diverse types of tissues arose from a single cell, the fertilised egg or zygote, raising the question of how this diversity is achieved. It is the aim of this book to consider the processes regulating tissue-specific gene expression in mammals and the manner in which they produce these differences in the nature and function of different tissues.

1.2 TISSUE-SPECIFIC EXPRESSION OF PROTEINS

The fundamental dogma of molecular biology is that DNA produces RNA which in turn produces proteins. Thus the genetic information in the DNA specifying particular functions is converted into an RNA copy, which is then translated into protein. The action of the protein then produces the phenotype, be it the presence of a functional globin protein transporting oxygen in the blood or the activity of a proteinaceous enzyme capable of producing the pigment causing the appearance of brown, rather than blue, eyes. Hence, if the differences in the appearance of mammalian tissues described above are indeed caused by differences in gene expression in different tissues, they should be produced by differences in the proteins present in these tissues. Such differences can be detected both by specific methods, which study the expression of one particular protein, and by general methods aimed at studying the expression of all proteins in a given tissue.

1.2.1 Specific methods for studying the protein composition of tissues

A simple means of determining whether differences in protein composition exist in different tissues involves investigating the expression of individual known proteins in such tissues. These methods might involve the isolation of widely differing amounts of an individual protein (or none of the protein at all) from different tissues, using an established purification procedure, or the detection of an enzymatic activity associated with the protein in extracts of only one particular tissue. We shall consider in detail, however, only methods where the expression of a specific protein is monitored by the use of a specific antibody to it. Normally, such an antibody is produced by injecting the protein into an animal such as a rabbit or mouse. The resulting immune response results in the presence in the animal's blood of antibodies which specifically recognise the protein and can be used to monitor its expression in particular tissues.

Such antibodies can be used in conjunction with the one-dimensional polyacrylamide gel electrophoresis technique to investigate the expression of a particular protein in different tissues. In this technique, proteins are denatured by treatment with the detergent sodium dodecyl sulfate (SDS) and then subjected to electrophoresis in a polyacrylamide gel which separates them according to their size (Laemmli, 1970). Subsequently, in a technique known as Western blotting (Gershoni and Palade, 1983), the gel-separated proteins are transferred to a nitrocellulose filter which is incubated with the antibody. The antibody reacts specifically with the protein against which it is directed and which will be present at a particular position on the filter, dependent on how far it moved in the electrophoresis step and hence on its size. The binding of the antibody is then visualised by a radioactive or enzymatic detection procedure. If a tissue contains the protein of interest, a band will be observed in the track containing total protein from that tissue and the intensity of the band observed will provide a measure of the amount of protein present in the tissue. If none of the particular protein is present in a given tissue, no band will form (Fig. 1.1). Hence this method allows the presence or absence of a specific protein in a particular tissue to be assessed using one-dimensional gel electrophoresis without the complicating effect of other unrelated proteins of similar size, since these will fail to bind the antibody. Similarly, quantitative differences in the expression of a particular protein in different tissues can be detected on the basis of differences in the intensity of the band obtained when extracts prepared from the different tissues are used in this procedure.

The specific reaction of a protein with an antibody can also be used directly to investigate its expression within a particular tissue. In this method, known as immunofluorescence, thin sections of the tissue of interest are reacted with the antibody, which binds to those cell types expressing the protein. As before, the position of the antibody is visualised by an enzymatic detection procedure or, more usually, by the use of a fluorescent dye which can be seen in a microscope when appropriate filters are used (Fig. 1.2). Therefore this method can be used not only to provide information about the tissues expressing a particular protein but, since individual cells can be examined, it also allows detection of the specific cell types within the tissue which are expressing the protein.

The results of experiments using these specific methods to study the expression of particular proteins indicate that, while some proteins are present at similar abundances in all tissues, others are present at widely different abundances in different tissues, and a large number are specific to one or a few tissues or cell types. Hence the differences between different tissues in appearance and function are correlated with qualitative and quantitative differences in protein composition, as assayed using antibodies to individual proteins.

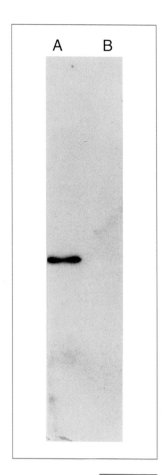

Figure 1.1

Western blot with an antibody to guinea-pig casein kinase showing the presence of the protein in lactating mammary gland (A) but not in liver (B).

Figure 1.2

Use of the antibody to casein kinase to show the presence of the protein (bright areas) in frozen sections of lactating mammary gland (a) but not in liver (b). (c) A phase-contrast photomicrograph of the liver section, confirming that the lack of staining with the antibody in (b) is not due to the absence of liver cells in the sample.

1.2.2 General methods for studying the protein composition of tissues

As well as examining the expression of individual proteins in different tissues, it is also possible to use more general methods to compare the overall protein composition of different tissues. All the proteins present in a tissue can be subjected to one-dimensional gel electrophoresis (see Section 1.2.1) and examined by using a stain which reacts with all proteins. Hence this method allows the visualisation of all cellular proteins, separated according to their size on the gel rather than using Western blotting with a particular antibody to focus on a specific protein. However, this method is of relatively limited use to investigate variations in the overall protein content of different tissues. This is because of the very large number of proteins in the cell and the limited resolution of the technique, which

Figure 1.3

Two-dimensional gel electrophoresis.

3

separates proteins solely on the basis of their size. Thus, for example, two entirely different proteins in two tissues may be scored as being the same protein simply on the basis of a similarity in size.

A more detailed investigation of the protein composition of different tissues can be achieved by two-dimensional gel electrophoresis (O'Farrell, 1975). In this procedure (Fig. 1.3), proteins are first separated on the basis of differences in their charge, in a technique known as isoelectric focusing, and the separated proteins, still in the first gel, are layered on top of an SDS–polyacrylamide gel. In the subsequent electrophoresis the proteins are separated by their size. Hence a protein moves to a position determined both by its size and its charge. The much greater resolution of this method allows a number of differences in the protein composition of particular tissues to be identified. Thus some spots or proteins are found in only one or a few tissues and not in many others, while others are found at dramatically different abundances in different tissues (Fig. 1.4).

Hence, the different appearances of different tissues are indeed paralleled by both qualitative and quantitative differences in the proteins present in each tissue. It should be noted, however, that some proteins can be shown by two-dimensional gel electrophoresis to be present at similar levels in virtu- ally all tissues. Presumably such so-called housekeeping proteins are involved in basic metabolic processes common to all cell types.

Initial experiments using two-dimensional gel electrophoresis simply involved examining the patterns of spots generated by different tissues and describing which spots were present or absent in specific tissues or showed different intensities in different tissues. More recently, however, methods have been developed which allow the protein responsible for forming a spot of particular interest to be identified and characterised. These methods have allowed the development of a new field of study, known as proteomics, in which the power of two-dimensional gels to separate a wide range of proteins is combined with the ability to study specific proteins individually.

Figure 1.4

Schematic results of two-dimensional gel electrophoresis allowing the detection of proteins specific to one tissue or expressed at different levels in different tissues.

In these methods (for reviews see Blackstock and Weir, 1999; Pandey and Mann, 2000) individual spots of interest are excised from the two-dimensional gel and digested into their constituent peptides using the proteolytic enzyme, trypsin. The individual peptides from a single protein spot are then analysed by mass spectrometry. This can involve both matrix-assisted laser desorption/ionisation (MALDI), which allows the molecular weight of the peptide to be determined, and nanoelectrospray mass spectrometry, which allows the amino acid sequence of a particular peptide to be obtained (Fig. 1.5). Often these two techniques are performed sequentially in a tandem mass spectrometer which first determines the molecular weight of a peptide produced by trypsin digestion and then fragments it further, allowing its sequence to be determined.

This peptide molecular weight and sequence information can then be used to search databases of known protein sequences which have been obtained either by direct protein sequencing or, more frequently, predicted from the ever-expanding amount of DNA sequence information. As trypsin always cleaves protein after a lysine or arginine residue, it is possible to align the peptide sequence against that of a known protein and hence identify the protein responsible for the original spot (Fig. 1.5; colour plate 1) (for further discussion of the techniques of mass spectrometry see Yates, 2000).

Figure 1.5

Use of mass spectrometry to determine the molecular weight and amino acid sequence of a particular protein spot on a two-dimensional gel. The information obtained in this way can be used to search databases of known protein sequences and hence identify the protein producing the spot.

This method can thus be used to identify an individual protein which shows an interesting pattern of expression in different tissues or whose expression is altered in a specific disease. Hence it combines the power of two-dimensional gel electrophoresis to look at a wide range of proteins with the ability to identify an individual protein. It is thus being widely used, for example, to identify the proteins associated with individual cellular organelles when these are isolated and subjected to two-dimensional gel electrophoresis. Similarly, it can be used to identify other proteins associated with a known protein in the cell. This can be achieved by carrying out an immunoprecipitation with an antibody to the known protein and then analysing the immunoprecipitate by two-dimensional gel electrophoresis to identify the other proteins which have been isolated by virtue of their association with the immunoprecipitated protein.

In terms of gene regulation, proteomic methods involving two-dimensional gel electrophoresis and mass spectrometry allow us to identify specific proteins whose expression varies between different tissues. It thus combines the methods for looking at specific proteins (e.g. with antibodies) or looking at the general population of proteins (using two-dimensional gel electrophoresis alone) which were described previously and reinforces our conclusion that the protein content of different tissues is both qualitatively and quantitatively different.

1.3 TISSUE-SPECIFIC EXPRESSION OF MESSENGER RNAs

Proteins are produced by the translation of specific messenger RNA molecules on the ribosome (see Section 3.3.6). Hence, having established that quantitative and qualitative differences exist in the protein composition of different tissues, it is necessary to ask whether such differences are paralleled by tissue-specific differences in the abundance of their corresponding mRNAs. Thus, although it seems likely that differences in the mRNA populations of different tissues do indeed underlie the observed differences in proteins, it is possible that all tissues have the same mRNA species and that production of different proteins is controlled by regulating which of these are selected by the ribosome to be translated into protein.

As with the study of proteins, both specific and general techniques exist for studying the mRNAs expressed in a given tissue.

1.3.1 Specific methods for studying the mRNAs expressed in different tissues

A number of different specific methods exist which can be used to detect and quantify one specific mRNA using a cloned DNA probe derived from its corresponding gene. In the most commonly used of such methods, Northern blotting (Thomas, 1980), the RNA extracted from a particular tissue is electrophoresed on an agarose gel, transferred to a nitrocellulose filter and hybridised to a radioactive probe derived from the gene encoding the mRNA of interest. The presence of the RNA in a particular tissue will result in binding of the radioactive probe and the visualisation of a band on autoradiography, the intensity of the band being dependent on the amount of RNA present (Fig. 1.6). Although such experiments do detect the RNA encoding some proteins, such as actin or tubulin, in all tissues, very many others are found only in one particular tissue. Thus the RNA for the globin protein is found only in reticulocytes, that for myosin only in muscle, while (in the example shown in Fig. 1.6) the mRNA encoding the foetal protein, α-foeto-protein, is shown to be present only in the embryonic yolk sac and not in the adult liver. Similar results can also be obtained using other methods for detecting specific RNAs in material isolated from different tissues, such as S1 nuclease analysis, RNAse protection and primer extension (for a review of these methods see Mason *et al.*, 1993).

As with protein studies, methods studying expression in RNA isolated from particular tissues can be supplemented by methods allowing direct visualisation of the RNA in particular cell types. In such a method, known as *in situ* hybridisation, a radioactive probe specific for the RNA to be detected is hybridised to a section of the tissue of interest in which cellular morphology has been maintained. Visualisation of the position at which the radioactive probe has bound (Fig. 1.7) not only allows an assessment of whether the particular tissue is expressing the RNA of interest but also of which individual cell types within the tissue are responsible for such expression. This technique has been used to show that the expression of some mRNAs is confined to only one cell type, paralleling the expression of the corresponding proteins by that cell type.

1.3.2 General methods for studying the mRNAs expressed in different tissues

As in the case of proteins, methods for studying individual RNAs can be supplemented by methods for studying RNA populations. Early experiments in this area made use of the $R_o t$ curve (Bishop *et al.*, 1974). In this method a complementary DNA (cDNA copy) of the RNA is first made, using radioactive precursors so that the cDNA is made radioactively labelled. After dissociating the cDNA and the RNA, the rate at which the radioactive cDNA reanneals or hybridises to the RNA is followed. In this process, more molecules of an abundant RNA will be present than of an RNA which is less abundant. The cDNA from an abundant RNA will find a partner more rapidly than one derived from a rare one, simply because more potential partners are available. Hence, when used with the RNA of one tissue, the rate of hybridisation or reannealing of any cDNA molecule provides a measure of the abundance of its corresponding RNA in that tissue. The higher the abundance of a particular RNA species, the more rapidly it anneals to its complementary cDNA (Fig. 1.8). Typical $R_o t$ curves, showing three predominant abundance classes of RNA in any given tissue, can readily be constructed (Fig. 1.9).

This method can be extended to provide a comparison of the RNA populations of two different tissues (Hastie and Bishop, 1976). In this method the RNA of one tissue is mixed with labelled cDNA prepared from the RNA of another tissue and the mixture annealed. Clearly, the rate of hybridisation

Figure 1.6

Northern blot hybridisation using a probe specific for the α-foetoprotein mRNA. The RNA is detectable in the embryonic yolk sac sample (track A) but not in the adult liver sample (track B).

Figure 1.7

Localisation of the RNA for type I collagen (a) and type II collagen (b) in the 10-day chick embryo leg by *in situ* hybridisation. Note the different distributions of the bright areas produced by binding of each probe to its specific mRNA. c, cartilage; t, tendons; p, perichondrium.

of the radiolabelled cDNA will be determined by the abundance of its corresponding RNA in the tissue from which the RNA is derived. If the RNA is absent altogether in this tissue (being specific to that from which the cDNA is derived), no reannealing will occur. If the corresponding RNA is present, the rate at which the cDNA hybridises will be determined by its abundance in the tissue from which the RNA is derived. Hence, by carrying out an inter-tissue $R_o t$ curve, both qualitative and quantitative differences between the RNA populations of different tissues can be observed. Such experiments (see Fig. 1.10) lead to the conclusion that, although sharing of rare RNAs (presumably encoding housekeeping proteins) does occur, both qualitative and quantitative differences between the RNA populations of different tissues are observed, with RNA species that are abundant in one cell type being rare or absent in other tissues. Thus, in a comparison of kidney and liver messenger RNAs, Hastie and Bishop (1976) found that, while between 9500 and 10 500 of the 11 000 rare messenger RNA species found in kidney were also found in liver, the six highly abundant RNA species found in the kidney were either absent or present at very low levels in the liver (Fig. 1.10).

More recently, methods have been developed that can combine the ability to look at variation in the total RNA population of different tissues with the ability to look specifically at the variation in a specific mRNA. By analogy with the proteomic methods which achieve this at the protein level (see Section 1.2.2), these methods are often referred to as transcriptomic methods (for reviews see Lockhart and Winzeler, 2000; Young, 2000).

In these methods, so-called "gene chips" are prepared in which DNA sequences derived from thousands of different mRNAs are laid out in a regular array at a density of over 10 000 different DNA sequences per square centimetre (Fig. 1.11). This is achieved either by spotting out cDNA clones derived from the different mRNAs at high density on a glass slide or

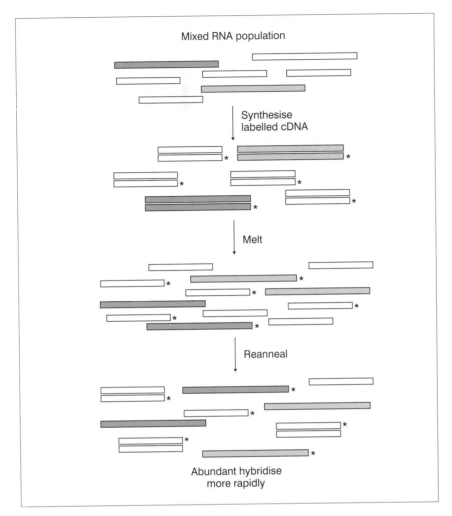

Mixed RNA population

Synthesise
labelled cDNA

Melt

Reanneal

Abundant hybridise
more rapidly

Figure 1.8

Hybridisation of messenger RNA and its radioactive complementary DNA copy (indicated by the stars) in a $R_o t$ curve experiment. The abundant RNA finds its partner and hybridises more rapidly.

by synthesising short oligonucleotides derived from each mRNA *in situ* on the gene chip.

Once this has been achieved the gene chips can be hybridised with radioactive probes prepared from the total mRNA present in different tissues and the pattern of hybridisation observed (Fig. 1.11). Thus, where a specific mRNA is present in a tissue, it will be present in the radioactive probe and a signal will be obtained on the gene chip, with the strength of the signal being proportional to the amount of the RNA which is present. Similarly, if the mRNA is absent then it will be absent from the probe and no signal will be obtained. As more and more DNA sequence information is obtained, it is possible to include increasing numbers of DNA sequences derived from individual mRNAs on the chips, taking advantage of the ability to spot out many different DNA sequences onto a very small chip. Hence, the expression of virtually all the mRNAs in a cell can now be compared in different tissues and cell types (Fig. 1.11; colour plate 2).

Figure 1.9

Typical $R_o t$ curve, produced using the mRNA of a single tissue. Note the three phases of the curve, produced by hybridisation of the highly abundant, moderately abundant, and rare RNA species.

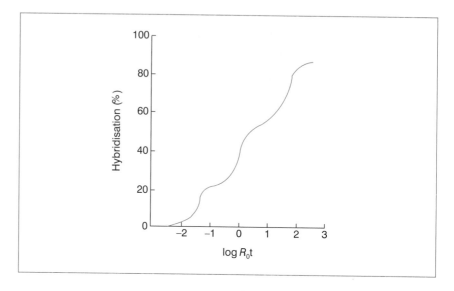

Figure 1.10

Inter-tissue $R_o t$ curve, in which mRNA from the liver has been hybridised to cDNA prepared from liver, kidney or brain mRNA. Note the lower extent of hybridisation, with the kidney and brain samples indicating the absence or lower abundance of some liver RNAs in these tissues.

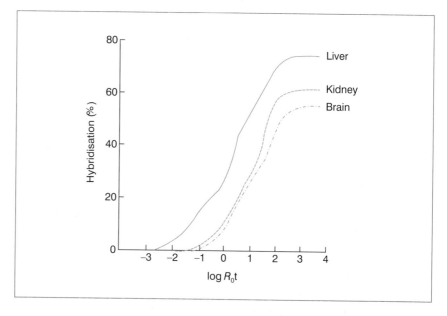

This very powerful method thus allows the expression profile of individual tissues and cell types to be obtained and compared in a manner which combines the ability of Northern blotting or *in situ* hybridisation to study individual mRNAs with the ability of $R_o t$ curve analysis to look generally at the entire mRNA population. The use of this method has provided further support for the conclusion that the mRNA populations of different tissues show both qualitative and quantitative differences. Because of its great power in combining both general and specific methods of looking at RNA populations, methods of this type have also been used to probe for subtle

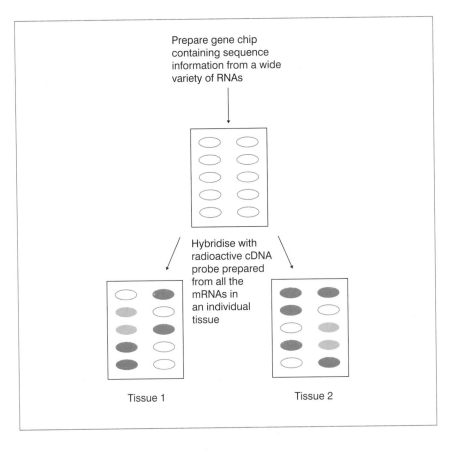

Prepare gene chip
containing sequence
information from a wide
variety of RNAs

Hybridise with
radioactive cDNA
probe prepared
from all the
mRNAs in
an individual
tissue

Tissue 1

Tissue 2

Figure 1.11

Gene chip analysis of mRNA expression patterns in different tissues. Radioactive cDNA is prepared from the total mRNA present in each tissue. Hybridisation of this radioactive cDNA to the gene chip containing many different gene sequences will reveal the expression level of each gene in each tissue. The figure shows this for genes which are not expressed in a particular tissue (open circle), genes which are expressed at low level (shaded circle) and genes which are expressed at high level (solid circle).

differences between closely related samples. Thus, for example, it has been used to characterise the gene expression differences between different types of breast cancer with different characteristics and different outcomes for the patient (Perou *et al.*, 2000).

1.4 CONCLUSIONS

It is clear, therefore, that the qualitative and quantitative variation in protein composition between different tissues is paralleled by a similar variation in the nature of the mRNA species present in different tissues. Indeed, in a study of a wide range of yeast proteins and mRNAs the abundance of an individual protein, as determined by two-dimensional gel electrophoresis, generally correlated very well with the abundance of the corresponding mRNA as determined by a gene chip approach (Futcher *et al.*, 1999) and a similar correlation has also been observed in a survey of human liver mRNAs and proteins (Anderson and Seilhamer, 1997). Hence, the production of different proteins by different tissues is not regulated by selecting which of a common pool of RNA molecules is selected for translation by the ribosome, although some examples of such regulated translation exist (see Section 5.6).

In order to understand the basis for such tissue-specific variation in protein and mRNA composition, it is therefore necessary to move one stage further back and examine the nature of the DNA encoding these mRNAs in different tissues.

REFERENCES

Anderson, L. and Seilhamer, J. (1997). A comparison of selected mRNA and protein abundances in human liver. *Electrophoresis* **18**, 533–537.

Bishop, J.O., Morton, J.G., Rosbash M. and Richardson, M. (1974). Three abundance classes in HeLa cell messenger RNA. *Nature* **250**, 199–204.

Blackstock, W.P. and Weir, M.P. (1999). Proteomics: quantitative and physical mapping of cellular proteins. *Trends in Biotechnology* **17**, 121–127.

Futcher, B., Latter, G.I., Monardo, P., McLaughlin, C.S. and Garrels, J.I. (1999). A sampling of the yeast proteome. *Molecular and Cellular Biology* **19**, 7357–7368.

Gershoni, J.M. and Palade, G.E. (1983). Protein blotting: principles and applications. *Analytical Biochemistry* **131**, 1–15.

Hastie, N.B. and Bishop, J.O. (1976). The expression of three abundance classes of messenger RNA in mouse tissues. *Cell* **9**, 761–774.

Laemmli, U.K. (1970). Cleavage of structural proteins during the assembly of the head of the bacteriophage T4. *Nature* **227**, 680–685.

Lockhart, D.J. and Winzeler, E.A. (2000). Genomics, gene expression and DNA arrays. *Nature* **405**, 827–836.

Mason, P.J., Enver, T., Wilkinson, D. and Williams, J.G. (1993). Assay of gene transcription *in vitro*. In: *Gene Transcription: A Practical Approach*, (eds Hames, B.D. and Higgins, S.J.) IRL Press, Oxford, pp 5–63.

O'Farrell, P.H. (1975). High resolution two-dimensional electrophoresis of proteins. *Journal of Biological Chemistry* **250**, 4007–4021.

Pandey, A. and Mann, M. (2000). Proteomics to study genes and genomes. *Nature* **405**, 837–846.

Perou, C.M., Sørlie, T., Eisen, M.B., Van de Rijn, M., Jeffrey, S.S., Rees, C.A., Pollack, J.R., Ross, D.T., Johnsen, H., Akslen, L.A., Fluge, Ø.A., Pergamenschikov, A., Williams, C., Zhu, S.X., Lønning, P.E., Børresen-Dale, A-L., Brown, P.O. and Botstein, D. (2000). Molecular portraits of human breast tumours. *Nature* **406**, 747–752.

Thomas, P.S. (1980). Hybridization of denatured RNA and small DNA fragments transferred to nitrocellulose. *Proceedings of the National Academy of Sciences of the USA* **77**, 5201–5205.

Yates, J.R. (2000). Mass spectrometry from genomics to proteomics. *Trends in Genetics* **16**, 5–8.

Young, R.A. (2000). Biomedical discovery with DNA arrays. *Cell* **102**, 9–15.

THE DNA OF DIFFERENT CELL TYPES IS SIMILAR IN BOTH AMOUNT AND TYPE

2

2.1 INTRODUCTION

Having established that the RNA and protein content of different cell types shows considerable variation, the fundamental dogma of molecular biology, in which DNA makes RNA which in turn makes protein, directs us to consider whether such variation is caused by differences in the DNA present in each tissue. Thus, in theory, DNA which corresponded to an RNA required in one particular tissue only might be discarded in all other tissues. Alternatively, it might be activated in the tissue where RNA was required, by a selective increase in its copy number, in the genome via an amplification event or by some rearrangement of the DNA necessary for its activation. The possible use of each of these mechanisms (DNA loss, amplification or rearrangement) will be discussed in turn (for a review see Brown, 1981).

2.2 DNA LOSS

2.2.1 DNA loss as a mechanism of gene regulation

In theory, a possible method of gene control would involve the genes for all proteins being present in the fertilised egg, followed by a selective loss of genes which were not required in particular tissues (Fig. 2.1). Thus only the genes for housekeeping proteins and the proteins characteristic of a particular cell type would be retained in that cell type. Hence the presence of the RNA for immunoglobulin only in the antibody-producing B-cells would be paralleled by the presence of the corresponding gene only in such cells, whereas the gene for a muscle-specific protein (such as myosin) would be

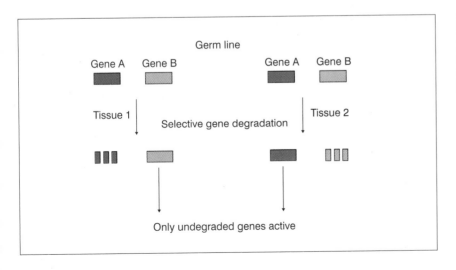

Figure 2.1

Model for gene regulation by the specific degradation of genes whose products are not required in a particular tissue.

absent in B-cell DNA but present in that of a muscle cell, where the immunoglobulin gene had been deleted. Such a mechanism would obviate the need for any other means of regulating gene activity, the tissue-specific RNA and protein populations described in Chapter 1 being produced by the constitutive activity of the genes retained in any particular cell type.

A variety of evidence suggests, however, that selective DNA loss is not a general mechanism of gene control. Such evidence is available from chromosomal, functional, and molecular studies, which will be considered in turn.

2.2.2 Chromosomal studies

The DNA-containing chromosomes, which can be visualised shortly before cell division, differ from one another in both size and the location of the centromere, by which they attach to the mitotic spindle prior to cell division. They can be further distinguished by a variety of staining techniques, which produce a specific banded appearance. The most widely used of these, known as G-banding, involves brief proteolytic or chemical treatment of the chromosome, followed by staining with Giemsa stain, and produces a characteristic pattern of bands (Fig. 2.2), allowing the cytogeneticist to identify individual chromosomes in humans and other species and to distinguish cells from different species.

In general, when such techniques are applied to chromosomes prepared from different differentiated cell types of the same species, however, the numbers and sizes of the chromosomes observed and their banding patterns are indistinguishable. Thus the patterns obtained from a liver cell or a brain cell are very similar and cannot be distinguished even by experienced cytogeneticists. Hence such karyotypic studies provide strong evidence against the loss of whole chromosomes or specific parts of chromosomes (visualised as G-bands) in particular differentiated cell types, and argue against selective DNA loss as a general mechanism for gene regulation.

None the less, a small number of cases of such a loss of DNA have been observed in particular cell types and these will be considered in turn.

Loss of the nucleus in mammalian red blood cells

The best known case occurs in the development of the mammalian red blood cell, or erythrocyte (for a review see Bessis and Brickal, 1952), which is a highly specialised cell containing large amounts of the blood pigment haemoglobin and functions in the transport of oxygen in the blood. Such cells are produced from progenitors, known as erythroblasts, which divide to produce two daughter cells, one of which maintains the stem cell population and later divides again, while the other undergoes differentiation. In this differentiation, the region of the cell containing the nucleus is pinched off, surrounded by a region of the cell membrane, and is eventually destroyed (Fig. 2.3). The resulting cell, known as the reticulocyte, is thus entirely anucleate and completely lacks DNA. However, it continues synthesis of large amounts of globin (the protein part of the haemoglobin molecule), as well as small amounts of other reticulocyte proteins, by

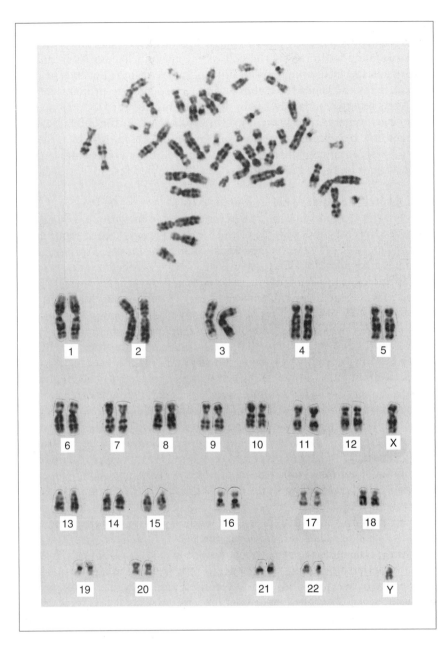

Figure 2.2

Human chromosomes stained with Giemsa stain. The upper panel shows a chromosome spread; in the lower panel the chromosomes have been arrayed in order of their chromosome number.

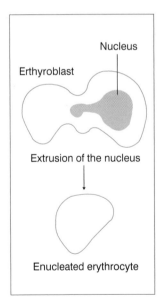

Figure 2.3

Extrusion of the nucleus from an erythroblast, resulting in an anucleate erythrocyte.

repeated translation of messenger RNA molecules produced prior to the loss of the nucleus. Such RNA molecules must therefore be highly stable and resistant to degradation (see Section 5.5). Eventually, however, other cytoplasmic components (including ribosomes) are lost, protein synthesis ceases and the cell assumes the characteristic structure of the erythrocyte, which is essentially a bag full of oxygen-transporting haemoglobin molecules.

Although this process offers us an example of DNA loss during differentiation, it is clearly a highly specialised case in which the loss of the nucleus is primarily intended to allow the cell to fill up with haemoglobin and to assume a shape facilitating oxygen uptake, rather than as a means of gene control. Thus the genes for globin and other reticulocyte proteins are not selectively retained but are lost with the rest of the nuclear DNA. Hence the tissue-specific pattern of protein synthesis observed in the reticulocyte is achieved not by selective gene loss but by the processes which, earlier in development, resulted in high-level transcription of the globin genes and the production of long-lived stable messenger RNA molecules.

Loss of specific chromosomes in somatic cells

The reticulocyte case does not, therefore, provide us with an example of selective DNA loss and, indeed, no such case has, as yet, been reported in mammalian cells. However, a number of such cases, in which specific chromosomes are lost during development, have been described in other eukaryotes, including some nematodes, crustaceans and insects (Wilson, 1928). In these cases a clear distinction is made early in embryonic development between the cells which will produce the body of the organism (the somatic cells) and those which will eventually form its gonads and allow the organism to reproduce (the germ cells). Early on in the development of the embryo, during the cleavage stage, the somatic cells lose certain chromosomes, with only the germ cells, which will give rise to subsequent generations, retaining the entire genome. Such losses can be quite extensive, for example in the gall midge *Miastor* only 12 of the total 48 chromosomes are retained in the somatic cells. It is assumed that the lost DNA is only required for some aspects of germ cell development. It is known to be rich in the highly repeated (or satellite) type of DNA, which is thought to be involved in homologous chromosome pairing during the meiotic division that occurs in germ cell development (John and Miklos, 1979). Hence it is more economical for the organism to dispense with such DNA in the development of somatic tissue, rather than to waste energy replicating it and passing it on to all the somatic cells where it has no function.

Although the mechanism by which this selective loss of DNA occurs is not yet fully understood, there is evidence that it is mediated by cytoplasmic differences between regions of the fertilised egg or zygote. Thus, in the nematode worm *Parascaris* the distinction between somatic and germ cells is established by the first division of the zygote, the smaller of the two cells produced losing chromosomes and giving rise to the soma while the larger retains a full chromosome complement and gives rise to the germ line (Fig. 2.4). If, however, the zygote is centrifuged, the plane of the first division can be altered, resulting in the production of two similar-sized cells, neither of which loses chromosomes. Hence the centrifugation process has abolished some heterogeneity in the cytoplasm responsible for this effect. Interestingly, although both daughter cells retain a full complement of chromosomes, their inability to lose chromosomes prevents normal embryonic development from occurring.

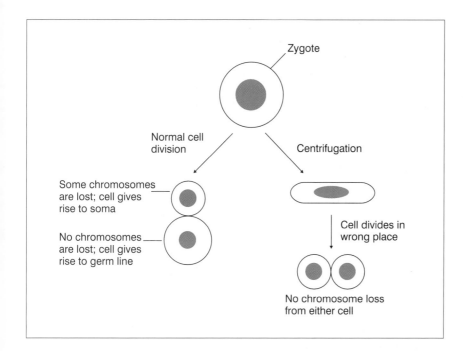

Figure 2.4

Early development in the nematode, and the consequences of changing the plane of the first cell division by centrifugation of the zygote.

Although cases of DNA loss observable by examination of the chromosomes have been observed in some organisms, it must be emphasised that in the vast majority of cases no such loss of DNA is observed by this means. However, because the amount of DNA in a Giemsa-stained band is very large (about 10 million base pairs), it is possible that losses of short regulatory regions necessary for gene expression, or losses of individual genes, could occur and not be observed by this technique. The large-scale losses seen in chromosome studies would then represent the tip of the iceberg, in which varying degrees of gene loss occur in different situations. A consideration of this possibility requires the use of more sensitive methods to search for gene deletions. These involve both functional and molecular studies.

2.2.3 Functional studies

A model in which selective gene loss controls differentiation must view differentiation as essentially an irreversible process. Thus, for example, once the gene for myosin has been eliminated from an antibody-producing B-lymphocyte it will not be possible for such a cell, whatever the circumstances, to give rise to a myosin-containing muscle cell. Although the requirement for such a change from one differentiated cell type to another is not likely to occur in normal development, such a phenomenon, known as transdifferentiation, has been achieved experimentally in Amphibia (Yamada, 1967). Thus, if the lens is surgically removed from an eye of one of these organisms, some of the neighbouring cells in the iris epithelium lose their differentiated phenotype and begin to proliferate. Eventually these cells differentiate into typical lens cells and produce large quantities of the lens-specific proteins, such as the crystallins, which are readily detectable with appropriate antibodies. The genes for the lens proteins must therefore be intact within the iris cells, even though in normal development such genes would never be required in these cells.

The case against selective loss of DNA is strengthened still further by the very dramatic experiments in which a whole new organism has been produced from a single differentiated cell. This was achieved a number of years ago in both plants (Steward, 1970) and Amphibia (Gurdon, 1968).

In plants such regeneration has been achieved in several species, including both the carrot and tobacco plants. In the carrot, for example, regeneration can occur from a single differentiated phloem cell, which forms part of the tubing system by which nutrients are transported in the plant. Thus, if a piece of root tissue is placed in culture (Fig. 2.5), single quiescent cells of the phloem can be stimulated to grow and divide, and an undifferentiated callus-type tissue forms. The disorganised cell mass can be maintained in culture indefinitely but, if the medium is suitably supplemented at various stages, embryonic development will occur and eventually result in a fully functional flowering plant, containing all the types of differentiated cells and tissues normally found. The plant that forms is fertile and cannot be distinguished from a plant produced by normal biological processes.

Figure 2.5

Scheme for the production of a fertile carrot plant from a single differentiated cell of the adult plant phloem by growth in culture as a free cell suspension which develops into an embryo and then into an adult plant.

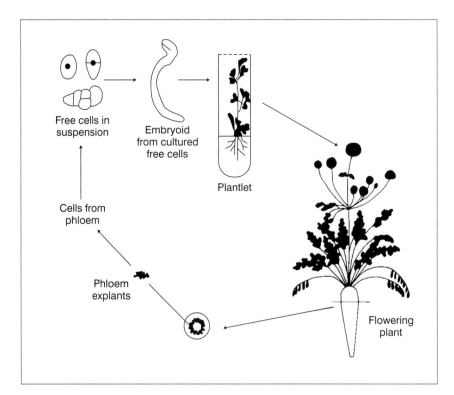

Free cells in suspension

Embryoid from cultured free cells

Plantlet

Cells from phloem

Phloem explants

Flowering plant

The ability to regenerate a functional plant from fully differentiated cells eliminates the possibility that genes required in other cell types are eliminated in the course of plant development. It is noteworthy, however, that in this case, as with lens regeneration, one differentiated cell type does not transmute directly into another; rather, a transitional state of undifferentiated proliferating cells serves as an intermediate. Hence, although differentiation does not apparently involve permanent irreversible changes in the DNA, it appears to be relatively stable and, although reversible, requires an intermediate stage for changes to occur. This semi-stability of cellular differentiation will be discussed further in Chapter 6.

Although no complex animal has been regenerated by culturing a single differentiated cell in the manner used for plants, other techniques have been

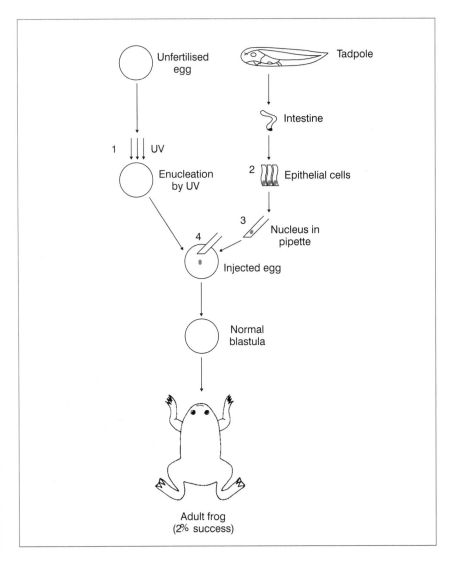

Figure 2.6

Nuclear transplantation in Amphibia. The introduction of a donor nucleus from a differentiated cell into a recipient egg whose nucleus has been destroyed by irradiation with ultraviolet light can result in an adult frog with the genetic characteristics of the donor nucleus.

used to show that differentiated cell nuclei are capable of giving rise to very many different cell types (Gurdon, 1968). These experiments involve the use of nuclear transplantation (Fig. 2.6). In this technique the nucleus of an unfertilised frog egg is destroyed, either surgically or by irradiation with ultraviolet light, and a donor nucleus from a differentiated cell of a genetically distinguishable strain of frog is implanted. Development is then allowed to proceed in order to test whether, in the environment of the egg cytoplasm, the genetic information in the nucleus of the differentiated cell can produce an adult frog. When the donor nucleus is derived, for example, from differentiated intestinal epithelial cells of a tadpole, an adult frog is indeed produced in a small proportion of cases (about 2%) and development to at least a normal swimming tadpole occurs in about 20% of cases, with the organisms having the genetic characteristics of the donor nucleus. Such successful development supported by the nucleus of a differentiated frog cell is not unique to intestinal cells and has been achieved with the nuclei of other cell types, such as the skin cells of an adult frog. Hence, although these experiments are technically difficult and have a high failure rate, it is possible for the nuclei of differentiated cells specialised for the absorption of food to produce organisms containing a range of different cell types and which are perfectly normal and fertile.

Although these experiments were carried out in Amphibia over thirty years ago, it is only within the last few years that similar experiments have been successfully performed in mammals. In 1997, however, Wilmut and co-workers reported experiments in which a donor nucleus derived from a mammary gland of a six year old sheep was introduced into an unfertilised egg whose own nucleus had been destroyed (Wilmut *et al.*, 1997; Stewart, 1997). In this manner, they produced an adult sheep (Dolly: colour plate 3) which, as in Gurdon's experiments with frogs, had the genetic characteristic of the sheep from which the mammary gland cell nucleus had been derived. Similar "cloning" experiments have now been successfully carried out in several other mammalian species, including mice (Wakayama *et al.*, 1998) (colour plate 4) and pigs (Polejaeva *et al.*, 2000). Although these studies have provoked considerable ethical discussion on the potential cloning of humans, their relevance for the area of gene regulation is that they indicate that, in mammals as in Amphibia, a nucleus from an adult differentiated cell is capable of supporting embryonic development and producing the full range of differentiated cell types present in an adult organism.

As with plants, it is therefore clear that, in animals, selective loss of DNA is not a general mechanism by which gene control is achieved.

2.2.4 Molecular studies

The conclusions of functional studies carried out in the 1960s and early 1970s were confirmed abundantly in the late 1970s and 1980s by molecular investigations involving the use of specific DNA probes derived from individual genes. Such probes can be used in Southern blotting experiments (Southern, 1975) to investigate the structure of a particular gene in different tissues. In this technique (Fig. 2.7) DNA from a particular tissue is digested

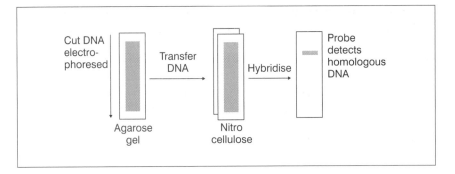

Figure 2.7

Procedure for Southern blot analysis, involving electrophoresis of DNA which has been cut with a restriction enzyme, its transfer to a nitrocellulose filter and hybridisation of the filter with a specific DNA probe for the gene of interest.

into specific fragments with restriction endonucleases which cut the DNA at particular points. Following electrophoresis on an agarose gel, the cut DNA is transferred to nitrocellulose and hybridised with a radioactive probe for the gene being studied. The structure of the gene can then be examined by determining the sizes of fragments observed when the DNA is cut with particular restriction enzymes.

As with the use of Northern blotting and Western blotting (see Chapter 1) to study, respectively, the RNA and protein content of different tissues, Southern blotting can be used to investigate the structure of the DNA in indi-

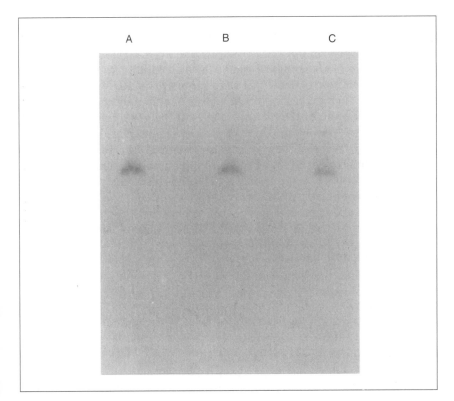

Figure 2.8

Southern blot hybridisation of a radioactive probe specific for the α-foetoprotein gene to EcoRI-cut DNA prepared from foetal liver (track A), adult liver (track B) and adult brain (track C). Note the presence of an identically sized gene fragment in each tissue, although the gene is only expressed in the foetal liver.

vidual tissues. When this is done, no losses of particular genes are observed in specific tissues. Thus, the gene for β-globin is clearly present in the DNA of brain or spleen where it is never expressed, as well as in the DNA of erythroid tissues where expression occurs (Jeffreys and Flavell, 1977) and numerous other examples have been reported (Fig. 2.8).

These techniques, which could detect losses of less than 100 bases (as opposed to the millions of bases resolvable by chromosomal techniques), have now been supplemented by cases where different laboratories have isolated particular genes from the DNA of different tissues which either do or do not express the gene. When the DNA sequence of the gene was determined in each case, no differences were detected in the expressing or non-expressing tissues. Similarly, the recent full DNA sequence of the human genome (International Human Genome Sequencing Consortium, 2001; Venter *et al.*, 2001) did not reveal any deletions in the sequence of specific genes where the DNA had been obtained from a tissue where that gene was not expressed. Such techniques, which could detect the loss of even a single base of DNA, have therefore confirmed the conclusions of all other studies, indicating that selective DNA loss is not a general mechanism of gene regulation.

2.3 DNA AMPLIFICATION

2.3.1 DNA amplification as a mechanism of gene regulation

Having eliminated selective DNA loss as a general means of gene control, it is necessary to consider the possible role of gene amplification in the regulation of gene expression (for reviews see Stark and Wahl 1984; Kafatos *et al.*, 1985). A possible mechanism of gene control (Fig. 2.9) would involve the selective amplification of genes that were expressed at high levels in a particular tissue, the high expression of such genes simply resulting from normal rates of transcription of the multiple copies of the gene. Unlike gene-deletion models, this possibility is not incompatible with the ability of nuclei from differentiated cells, such as frog intestinal cells, to regenerate a new organism (see Section 2.2.3). Thus, although amplification of intestine-specific genes would have occurred in such cells, this would presumably not prevent the amplification of other tissue-specific genes in the individual cell types of the new organism. The evidence against this idea comes therefore from chromosomal and molecular, rather than functional, investigations.

2.3.2 Chromosomal studies

As with DNA loss, the identity of the chromosomal complement in individual cell types provides strong evidence against the occurrence of large-scale DNA amplification. None the less, some cases of such amplification have been observed by this means and these will be discussed.

Chromosome polytenisation in *Drosophila*

One such case, which has been extensively used in cytogenetic studies, occurs in the salivary gland cells of the fruit fly *Drosophila melanogaster*. In

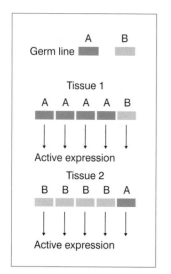

Figure 2.9

Model for gene regulation in which a gene that is expressed at high levels in a particular tissue is selectively amplified in that tissue. Gene A is amplified in tissue I and gene B in tissue 2.

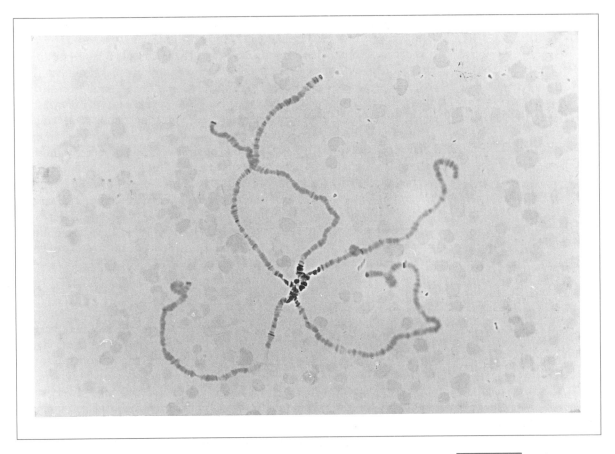

Figure 2.10

Polytene chromosomes of
Drosophila melanogaster.

these cells the DNA replicates repeatedly and the daughter molecules do not separate but remain close together, forming a single giant, or polytene, chromosome (Fig.2.10) which contains approximately 1000 DNA molecules. Although such giant chromosomes have proved very useful in experiments involving visualisation of DNA transcription (see Section 4.2.4), they cannot really be considered an example of gene control since virtually the whole of the genome (with the exception of some repeated DNA) participates in this amplification and there is no evidence for selective amplification of the genes for proteins required in the salivary gland.

Amplification of ribosomal DNA

Amplification of a specific part of the genome does occur, however, in some cases. Thus, amplification of the ribosomal DNA encoding the structural RNAs of the ribosomes is observed in the embryonic development of multi-cellular organisms, even though the germ line DNA of these organisms contains multiple copies of the ribosomal DNA prior to any amplification event. Thus, early in oogenesis the frog oocyte, like all other frog cells, contains approximately 900 chromosomal copies of the DNA encoding ribosomal RNA. During oogenesis, some of the copies are excised from the

genome and replicate extensively to produce up to 2 million extra-chromo-
somal copies of the ribosomal DNA (Brown and Dawid, 1968).

Such ribosomal DNA amplification clearly represents a response to the
intense demand for ribosomal RNA during embryonic development. This
requirement has to be met by DNA amplification since the ribosomal RNA
represents the final product of ribosomal DNA expression. By contrast, in
the case of protein-coding genes, high-level expression can be achieved both
by very efficient transcription of the DNA into RNA and also by similarly
efficient translation of the RNA into protein. Hence the observed amplification of ribosomal RNA genes does not imply that this mechanism is used in
higher organisms to amplify specific protein-coding genes. Indeed the
general similarity of the chromosome complement in different tissues argues
against large-scale amplification of such genes. As with gene deletion,
however, the sensitivity of such techniques is such that small-scale amplification of a few genes would not be detected, and it is necessary to use molecular techniques to see whether such amplification occurs.

2.3.3 Molecular studies

In addition to its use in searching for gene loss (see Section 2.2.4) the technique of Southern blotting can also be used to search for DNA amplification
in tissues expressing a particular gene. Thus, either new DNA bands hybridising to the specific probe will appear, or the same band, present in other
tissues, will be observed but it will hybridise more intensely because more
copies are present. Despite very many studies of this type only a very small
number of cases of such amplification have been reported, the vast majority
of tissue-specific genes being present in the same number of copies in all
tissues. Thus, for example, the fibroin gene is detectable at similar copy
numbers in the DNA of the posterior silk gland of the silk moth, where it is
expressed, and in the DNA of the middle silk gland, where no expression is
detectable (Fig. 2.11; Manning and Gage, 1978).

However, one exception to this general lack of amplification of protein-
coding genes is particularly noteworthy. This involves the genes encoding
the eggshell, or chorion, proteins in *Drosophila melanogaster*. As first shown
by Spradling and Mahowald (1980), the chorion genes are selectively amplified (up to 64 times) in the DNA of the cells which surround the egg follicle,
allowing the synthesis in these cells of the large amounts of chorion mRNA
and protein needed to construct the eggshell. Such amplification can be
observed readily in a Southern blot experiment using a recombinant DNA
probe derived from one of the chorion genes (Fig. 2.12). Unlike ribosomal
DNA, amplification occurs within the chromosome without excision of the
template DNA or the newly replicated copies.

This amplification event is of particular interest because it occurs in
normal cells as part of normal embryonic development, in contrast, for
example, to the amplification of cellular oncogenes that occurs in some
cancer cells (see Section 9.3) or the amplification of the gene encoding the
enzyme dihydrofolate reductase (which occurs in mammalian cells in
response to exposure to the drug methotrexate; Schimke, 1980). None the

(a) EcoRI

(b) HindIII

λ/HindIII DNA segment molecular weight (d × 10⁻⁶)

14.48

6.14

4.2

2.79

14.48

6.14

4.2

2.79

λ/HindIII

PSG

MSG

Carcass

PSG

MSG

Carcass

λ/HindIII

Figure 2.11

Southern blot of DNA prepared from the posterior silk gland (PSG), the middle silk gland (MSG) or the carcass of the silk moth, *Bombyx mori*, with a probe specific for the fibroin gene. Note the identical size and intensity of the band produced by EcoRI or HindIII digestion in each tissue, although the gene is only expressed in the posterior silk gland.

less, the amplification of the chorion genes appears to be a response to a very specialised set of circumstances, requiring a novel means of gene regulation. Thus eggshell construction in *Drosophila* occurs over a very short period (about 5 h) and probably necessitates the synthesis of mRNA for the chorion proteins at very high rates, too large to be achieved even by high-level transcription of a single unique gene. It is this combination of high-level synthesis and a very short period to achieve it which necessitates the use of gene amplification. The restricted use that is made of this mechanism is well illustrated, however, by the fact that in silk moths, in which eggshell production

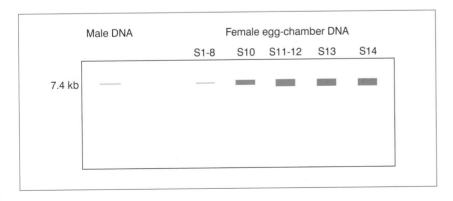

Male DNA

Female egg-chamber DNA

S1-8   S10   S11-12   S13   S14

7.4 kb

Figure 2.12

Amplification of the chorion genes in the ovarian follicle cells of *Drosophila melanogaster*. Note the dramatic increase in the intensity of hybridisation of the chorion-gene-specific probe to the DNA samples prepared from Stage 10 to Stage 14 egg chambers compared to that seen in Stages 1–8 or in male DNA.

occurs by a similar mechanism, multiple copies of the genes for the chorion proteins are present in the germ line. Thus they are present in all somatic cells, including those that do not express these genes, and are not further amplified in the ovarian follicle cells.

Such a finding serves to reinforce the conclusion of both molecular and chromosomal experiments that, as with DNA loss, DNA amplification is not a general method of gene regulation and is used only in isolated specialised cases.

2.4 DNA REARRANGEMENT

Having established that, generally, gene regulation does not occur by DNA deletion or amplification, it is necessary to consider the possibility that it might occur via specific rearrangement of genes in tissues where their protein products are required. Such activation could occur, for example, by translocation of the gene, removing it from a repressive effect of neighbouring sequences, or by bringing together the part of the gene encoding the protein and a promoter element necessary for its transcription (Fig.2.13).

Figure 2.13

Model for activation of a gene by a DNA rearrangement involving either the removal of a repressor element (A) or the joining of the gene and its promoter (B).

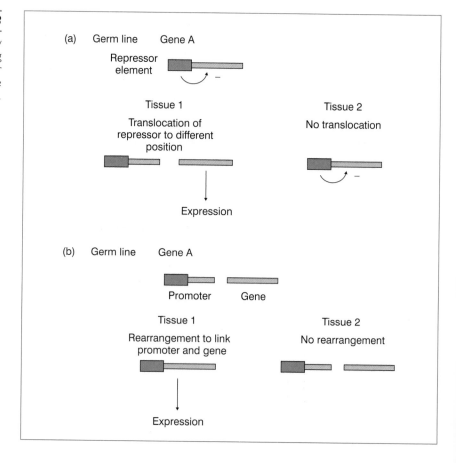

As with DNA amplification, the occurrence of such rearrangements in one particular cell type might not preclude the subsequent rearrangement of genes required in other tissues. Hence the ability of a single differentiated cell type to give rise to a variety of other differentiated cells in functional studies (see Section 2.2.3.) cannot be used to argue against the occurrence of such rearrangements. Similarly, these rearrangements might be on too small a scale to be detectable in chromosomal studies.

The evidence from molecular studies, however, shows unequivocally that these types of rearrangements are not involved in the selective activation of the vast majority of genes in particular tissues. Thus such rearrangements would produce novel bands when the DNA from a tissue in which the gene was active was cut with restriction enzymes and used in Southern blot experiments (Fig. 2.14). In fact, no such novel bands are observed in the vast majority of cases when the DNA of a tissue in which the gene is active is compared with tissues where it is inactive (Figs. 2.8 and 2.11).

The evidence against DNA rearrangement as a general means of gene control, obtained in such Southern blot experiments, has been abundantly confirmed by experiments in which the DNA sequence of a particular gene has been obtained both from a tissue where it is inactive and from one where it is active. Such experiments have failed to reveal even a single base difference in these two situations, and indicate that the gene is potentially fully functional in tissues where it is inactive. Despite these clear findings that, as with deletion and amplification, DNA rearrangement is not a general means

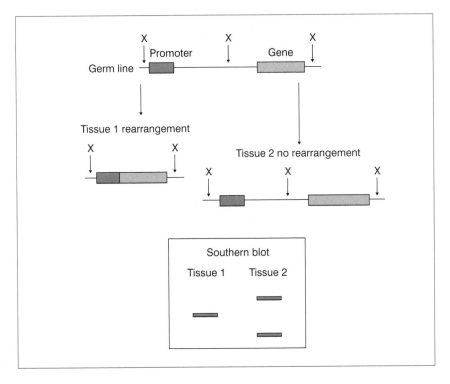

Figure 2.14

Generation of new restriction fragments detectable on a Southern blot following gene rearrangement. Sites for a particular restriction enzyme are indicated by an X and the consequences of digestion with this enzyme of unrearranged or rearranged DNA are indicated.

of gene control, it is used in a few situations such as in the mammalian immunoglobulin genes.

Antibody production in mammalian B-cells
When mammals are exposed to foreign bacteria or viruses, their immune systems respond by synthesising specific antibodies directed against the proteins of these organisms, with the aim of neutralising the harmful effects of the infection. The potential requirement for the synthesis of diverse kinds of antibody by the mammalian immune system is obviously vast. Thus, the body must be able to defend itself, for example, against a bewildering variety of challenges from infectious organisms by the production of specific antibodies against their proteins (for recent reviews of the immune system, emphasising the molecular aspects, see Parham, 2000; Roitt *et al.*, 2001). These antibodies, or immunoglobulins, are produced by the covalent association of two identical heavy chains and two identical light chains to produce a functional molecule (Fig. 2.15). The combination of specific heavy and light chains produces the specificity of the antibody molecule. In particular, each chain contains, in addition to a constant region (which is relatively similar in different antibody molecules), a highly variable region which differs widely in amino-acid sequence in different antibodies. It is this variable region that actually interacts with the antigen and hence determines the specificity of the antibody molecule.

As any heavy chain can associate with any light chain, and as approximately 1 million types of antibody specificities can be produced, at least 1000 genes encoding different heavy chains and a similar number encoding the light chains would be required. Copy-number studies have shown, however, that the germ line does not contain anything like this number of intact immunoglobulin genes, and no specific amplification events have been detected in the DNA of antibody-producing B-cells.

Rather, as first suggested by Dreyer and Bennett (1965), functional immunoglobulin genes are created by DNA rearrangements in the B-cell

Figure 2.15

Association of two heavy (H) and two light (L) chains to form a functional antibody molecule. The disulfide bonds linking the chains and the region that binds antigen are indicated.

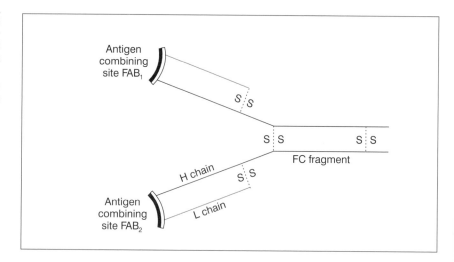

lineage. Thus the germ-line DNA contains a large number of tandemly repeated DNA segments encoding different variable regions, which are separated by over 100 kbp from a much smaller number of DNA segments encoding the constant region of the molecule (Fig. 2.16). This organisation is maintained in most somatic cell types, but in each individual B-cell a unique DNA rearrangement event occurs by which one specific variable region is brought together with one constant region by deletion of the intervening DNA. In this manner, a different functional gene, containing specific variable and constant regions, is produced in each individual B-cell.

These rearrangements can be readily detected by the appearance of novel bands in Southern blot analyses of the structure of the immunoglobulin genes in the DNA of antibody-producing B-cells. In the example shown in Fig. 2.17 (based on the data of Hozumi and Tonegawa, 1976) the digestion of DNA from non-B-cells with the restriction enzyme BamH1 produces fragments of 6.5 kb and 9.5 kb containing the variable and constant regions, respectively, whereas digestion of rearranged B-cell DNA generates a single 3.5 kb band containing both parts of the gene. This combination of different variable (V) and constant (C) regions in the production of a functional immunoglobulin gene obviously creates a wide range of different potential antibody specificities by the joining of particular V and C segments.

Despite this, more and more complex rearrangements are used in different types of immunoglobulin gene to generate still more potential variety. Thus, in the simplest case that we have discussed so far, that of the γ light chain (Fig. 2.16), one of many V regions is brought into association with a constant region, consisting of a single J segment (encoding the junction between the

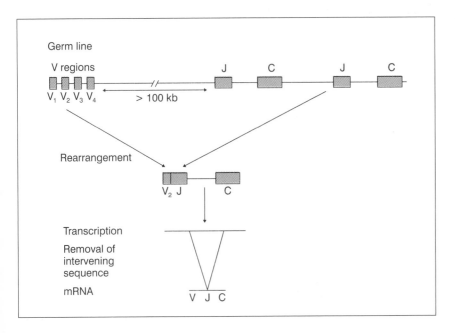

Figure 2.16

Rearrangement of the gene encoding the γ light chain of immunoglobulin results in the linkage of one specific variable region (V) to a specific joining region (J) with removal of the intervening DNA. The gene is then transcribed into RNA and the region between the joining (J) and constant regions (C) is removed by RNA splicing to create a functional mRNA.

Figure 2.17

The rearrangement of the immunoglobulin gene locus creates a different BamHI restriction fragment hybridising to immunoglobulin DNA-specific probes in the DNA of B-cells, compared to those that are present in other cell types where no rearrangement has occurred.

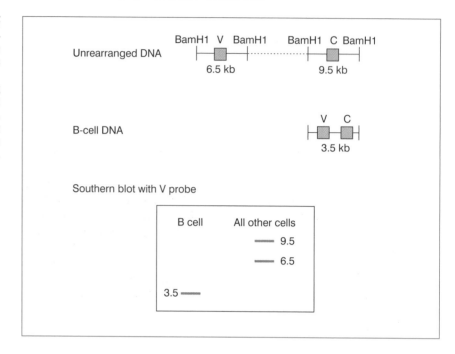

variable and constant regions) which is separated by a short region from the DNA encoding the C segment itself. The rearranged gene is then transcribed and the intervening sequence between the J and C segments removed by RNA splicing to produce a functional messenger RNA (see Section 5.2). In the genes for another type of light chain, known as κ, the situation is more complex. Thus these genes contain multiple J segments linked to individual constant segments and hence further diversity is created by the recombination of one of many V regions with one of several different J regions.

The case of the genes encoding the heavy-chain molecule is yet more complex (Fig. 2.18) in that these genes also contain a D, or diversity, segment. A functional heavy-chain gene is thus created by two successive rearrangements. In the first of these, one of several D segments is brought together with one of the J segments and the associated C region. Subsequently, a V region segment is recombined with this DJC element exactly as occurs in the VJ joining found in the light-chain genes. These two combinatorial events allow the selection of one particular copy of each of four regions to produce the functional gene, generating many different possible heavy-chain molecules.

Interestingly, the generation of diversity by DNA rearrangement is supplemented by another DNA alteration, so far unique in mammalian cells, to the genes of the immune system. It appears that immunoglobulin genes can undergo point mutations within particular B-cells, resulting in changes in the amino acids they encode and hence in their specificity. Such somatic mutation (for a review see Tian and Alt, 2000) produces forms of immunoglobulin not encoded by the DNA present in other tissues or in the

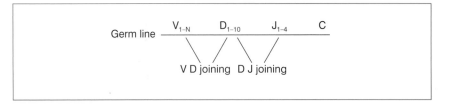

germ line, and represents another form of tissue-specific gene alteration occurring within the immunoglobulin genes.

It is clear that the primary role of the rearrangements and alterations which occur in the immunoglobulin genes in B-cells is the generation of the diversity of antibodies required to cope with the vast number of different possible antigens, rather than to produce the high-level expression of the immunoglobulin genes which occurs in such cells. None the less, high-level immunoglobulin-gene expression does occur as a consequence of such rearrangements. Thus the DNA encoding the variable regions of the antibody molecule is closely linked to the promoter elements that direct the transcription of the gene. Such elements show no activity in the unrearranged DNA of non-B-cells, but become fully active only when the variable region is joined to the constant region. In the case of the heavy-chain genes (Banerji *et al.*, 1983; Gilles *et al.*, 1983), this is because the intervening sequence between the J and C regions of these genes contains an enhancer element (see Section 7.3) which, although not a promoter itself, greatly increases the activity of the promoter adjacent to the V-region element. Hence the rearrangement not only brings the promoter element into a position where it can produce a functional mRNA containing VDJ and C regions but also facilitates the high-level transcription of this gene by juxtaposing promoter and enhancer elements that are separated by over 100 kb of DNA before the rearrangement event (Fig. 2.19).

The element of gene control in this process is, however, entirely secondary to the need to produce diversity by rearrangement. Thus, if the immunoglobulin enhancer is linked artificially close to the immunoglobulin promoter and introduced into a variety of cells, it will activate transcription

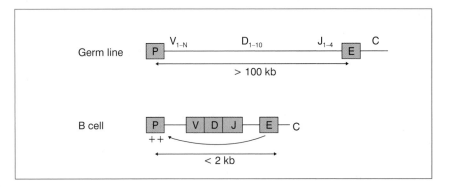

from the promoter only in B-cells and not in other cell types. Hence the enhancer is active only in B-cells and would not activate immunoglobulin transcription even if functional immunoglobulin genes containing closely linked VDJ and C segments were present in all other tissues. The immunoglobulin genes are therefore regulated by tissue-specific activator sequences in much the same way as other genes (see Chapter 7), but such activation has been made to depend on the occurrence of a DNA rearrangement whose primary role lies elsewhere.

Thus, although similar DNA rearrangements are used in other situations where diversity must be generated, for example in the genes encoding the T-cell receptor which is involved in the response to foreign antigens of T-lymphocytes, it is clear that they do not represent a general means of gene regulation applicable to other situations. Such a conclusion is entirely in accordance with the results of Southern blot and sequencing experiments which, as previously discussed, have failed to detect such rearrangements in the vast majority of genes active only in particular tissues.

2.5 CONCLUSIONS

Although we have discussed a number of cases where DNA loss, amplification or rearrangement regulate gene expression in this chapter, it is clear that these represent isolated special cases where the dictates of a particular situation have necessitated the use of such mechanisms. Thus the loss of DNA in red blood cells is dictated by the need to fill the cell with haemoglobin, the amplification of the chorion genes in *Drosophila* by the requirement to produce the corresponding proteins in a very short time and the rearrangement of the immunoglobulin genes by the need to generate a diverse array of antibodies. Hence these cases are not representative of a general mechanism of gene regulation by DNA alteration and, in general, the DNA of different cell types is quantitatively and qualitatively identical. Given that we have established previously that the RNA content of different cell types can vary dramatically, it is now necessary to investigate how such differences in RNA content can be produced from the similar DNA present in such cell types. As a background to this discussion we will first discuss, in the next chapter, the processes by which DNA is transcribed into RNA and the subsequent post-transcriptional events which convert that RNA into its final form.

REFERENCES

Banerji, J., Olson L. and Schaffner, W. (1983). A lymphocyte-specific cellular enhancer is located downstream of the joining region in immunoglobulin heavy chain genes. *Cell* **33**, 729–740.

Bessis, N. and Brickal, M. (1952). Aspect dynamique des cellules du sang. *Revues Hématologique* **7**, 407–435.

Brown, D.D. (1981). Gene expression in eukaryotes. *Science* **211**, 667–674.

Brown, D.D. and Dawid, B. (1968). Specific gene amplification in oocytes. *Science* **160**, 272–280.

Dreyer, W.J. and Bennett, J.C. (1965). The molecular basis of antibody formation: a paradox. *Proceedings of the National Academy of Sciences of the USA* **54**, 864–869.

Gillies, S.D., Morrison, S.L., Oi, V.T. and Tonegawa, S. (1983). A tissue specific enhancer is located downstream of the joining region in immunoglobulin heavy chain genes. *Cell* **33**, 717–728.

Gurdon, J.B. (1968). Transplanted nuclei and cell differentiation. *Scientific American* **219** (Dec.), 24–35.

Hozumi, N. and Tonegawa, S. (1976). Evidence for somatic rearrangement of immunoglobulin genes coding for variable and constant regions. *Proceedings of the National Academy of Sciences of the USA* **73**, 3628–3632.

International Human Genome Sequencing Consortium (2001). Initial sequencing and analysis of the human genome. *Nature* **409**, 860–921.

Jeffreys, A.J. and Flavell, R.A. (1977). The rabbit β-globin gene contains a large insert in the coding sequence. *Cell* **12**, 1097–1108.

John, B. and Miklos, G.L.G. (1979). Functional aspects of satellite DNA and heterochromatin. *International Review of Cytology* **58**, 1–113.

Kafatos, F.C., Orr, W. and Delidakis, C. (1985). Developmentally regulated gene amplification. *Trends in Genetics* **1**, 301–306.

Manning, R.F. and Gage, L.P. (1978). Physical map of the *Bombyx mori* DNA containing the gene for silk fibroin. *Journal of Biological Chemistry* **253**, 2044–2052.

Parham, P. (2000). *The Immune System*. Garland Publishing, New York.

Polejaeva, I.A., Chen, S-H., Vaught, T.D., Page, R.L., Mullins, J., Ball, S., Dai, Y., Boone, J., Walker, S., Ayares, D.L., Colman, A. and Campbell, K.H.S. (2000). Cloned pigs produced by nuclear transfer from adult somatic cells. *Nature* **407**, 86–90.

Roitt, I.M., Brostoff, J. and Male, D.K. (2001). *Immunology* (6th edn). Mosby International, London.

Schimke, R.T. (1980). Gene amplification and drug resistance. *Scientific American* **243**, 50–59.

Southern, E.M. (1975). Detection of specific sequences among DNA fragments separated by gel electrophoresis. *Journal of Molecular Biology* **98**, 503–517.

Spradling, A.C. and Mahowald, A.P. (1980). Amplification of genes for chorion proteins during oogenesis in *Drosophila melanogaster*. *Proceedings of the National Academy of Sciences of the USA* **77**, 1096–1100.

Stark, G.R. and Wahl, G.M. (1984). Gene amplification. *Annual Review of Biochemistry* **53**, 447–491.

Steward, F.C. (1970). From cultured cells to whole plants: the induction and control of their growth and morphogenesis. *Proceedings of the Royal Society, Series B* **175**, 1–30.

Stewart, C. (1997). An udder way of making lambs. *Nature* **385**, 769–771.

Tian, M. and Alt, F.W. (2000). RNA editing meets DNA shuffling. *Nature* **407**, 31–33.

Venter, J.C. *et al.* (2001). The sequence of the human genome. *Science* **291**, 1304–1351.

Wakayama, T., Perry, A.C.F., Zuccotti, M., Johnson, K.R. and Yanagimachi, R. (1998). Full-term development of mice from enucleated oocytes injected with cumulus cell nuclei. *Nature* **394**, 369–374.

Wilmut, I., Schnieke, A.E., McWhis, J., Kind, A.J. and Campbell, K.H.S. (1997). Viable offspring derived from fetal and adult mammalian cells. *Nature* **385**, 810–813.

Wilson, E.B. (1928). *The Cell in Development and Heredity*. New York: Macmillan.

Yamada, T. (1967). Cellular and sub-cellular events in Wolffian lens regeneration. *Current Topics in Developmental Biology* **2**, 249–283.

GENE EXPRESSION

3

3.1 LEVELS OF GENE REGULATION

The observation that differences in the RNA and protein content of different tissues are not paralleled by significant differences in their DNA content indicates that the process whereby DNA produces mRNA must be the level at which gene expression is regulated in eukaryotes.

In bacteria this process involves only a single stage, that of transcription, in which an RNA copy of the DNA is produced by the enzyme RNA polymerase. Even while this process is still occurring, ribosomes attach to the nascent RNA chain and begin to translate it into protein. Hence cases of gene regulation in bacteria, such as the switching on of the synthesis of the enzyme β-galactosidase in response to the presence of lactose (its substrate) are mediated by increased transcription of the appropriate gene (Jacob and Monod, 1961). Clearly, a similar regulation of gene transcription in different tissues, or in response to substances such as steroid hormones which induce the synthesis of new proteins, represents an attractive method of gene regulation in eukaryotes.

In contrast to the situation in bacteria, however, a number of stages intervene between the initial synthesis of the primary RNA transcript and the eventual production of mRNA (Fig. 3.1). The initial transcript is modified at its 5′ end by the addition of a cap structure containing a modified guanosine residue and is subsequently cleaved near its 3′ end, followed by the addition of up to 200 adenosine residues in a process known as polyadenylation. Subsequently, intervening sequences, or introns, which interrupt the protein-coding sequence in both the DNA and the primary transcript of many genes (Jeffreys and Flavell, 1977; Tilghman *et al.*, 1978) are removed by a process of RNA splicing (for reviews see Adams *et al.*, 1996; Staley and Guthrie, 1998). Although this produces a functional mRNA, the spliced molecule must then be transported from the nucleus, where these processes occur, to the cytoplasm where it can be translated into protein.

In theory any of these stages, all of which are essential for production of a functional mRNA, could be used to regulate the expression of specific genes in particular tissues. For example, it has been suggested that gene regulation could occur by transcribing all genes in all tissues and selecting whichever transcripts were appropriately spliced by correct removal of intervening sequences, thereby producing a functional mRNA (Davidson and Britten, 1979). The existence of such a plethora of possible regulatory stages has led, therefore, to much investigation as to whether all or any of these are used.

In order to provide the essential background for considering these studies, this chapter will consider the processes which occur at each of these stages between the DNA and the production of the functional mRNA. In addition, since some cases of regulation at the level of translation of the mRNA have been described, the final section of this chapter will consider

Figure 3.1

Stages in eukaryotic gene
expression which could be
regulated.

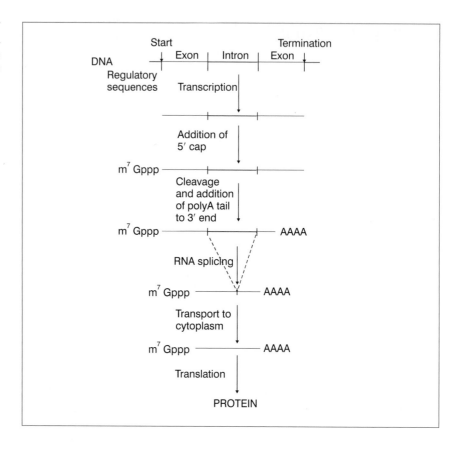

the mechanism of translation. Once this background information has been
provided, subsequent chapters will consider the evidence that the primary
control of gene expression occurs at the level of transcription (Chapter 4) as
well as a number of cases of post-transcriptional regulation (Chapter 5).

3.2 TRANSCRIPTION

Enzymes which are capable of copying the DNA so that a complementary
RNA copy is produced by the polymerisation of ribonucleotides are referred
to as RNA polymerases. In prokaryotes a single RNA polymerase enzyme is
responsible for the transcription of DNA into RNA. In eukaryotes, however,
this is not the case, and three such enzymes exist. All three RNA polymerases
are large multi-subunit enzymes (for a review see Sentenac, 1985) and
several subunits are held in common between the three enzymes. The three
enzymes can be distinguished, however, by their relative sensitivity to the
fungal toxin α-amanitin and each of them is active on a distinct set of genes
(Table 3.1). Thus, whereas all genes capable of encoding a protein, as well as
the genes for some small nuclear RNAs involved in RNA splicing (Section
3.3.3) are transcribed by RNA polymerase II, the genes encoding the 28S,
18S and 5.8S ribosomal RNAs are transcribed by RNA polymerase I, and

Genes transcribed	Sensitivity to α-amanitin
I	Ribosomal RNA (45S precursor of 28S. 18S and 5.8S rRNA)	Insensitive
II	All protein-coding genes, small nuclear RNAs U1, U2, U3, etc.	Very sensitive (inhibited 1 μg/ml)
III	Transfer RNA, 5.8S ribosomal RNA, small nuclear RNA U6, repeated DNA sequences: Alu, B1, B2, etc.; 7SK, 7SL RNA	Moderately sensitive (inhibited 10 μg/ml)

Table 3.1

Eukaryotic RNA polymerases

those encoding the transfer RNAs and the 5S ribosomal RNA are transcribed by RNA polymerase III (for a review see Sentenac, 1985).

In considering the transcriptional regulatory processes that produce tissue-specific variation in specific mRNAs and protein, our primary concern will therefore be with the regulation of transcription by RNA polymerase II. Transcription by RNA polymerases I and III is also subject to regulation, however. Moreover, the nature of the components involved in transcription by these polymerases is much simpler than those which are involved in transcription by RNA polymerase II. A prior understanding of the processes involved in basal transcription by RNA polymerases I and III therefore assists a subsequent understanding of the more complex processes involved in basal transcription by RNA polymerase II. Transcription by RNA polymerases I and III is therefore considered in Sections 3.2.1 and 3.2.2, respectively, whilst transcription by RNA polymerase II is considered in Section 3.2.3. Section 3.2.4 considers the common processes involved in transcription by each of the polymerases.

3.2.1 RNA polymerase I

RNA polymerase I is responsible for the transcription of the tandem arrays of genes encoding ribosomal RNA, such transcription constituting about one half of the total cellular transcription. As with all RNA polymerases, RNA polymerase I itself does not recognise the DNA sequences around the start site of transcription. Rather, other protein factors recognise such sequences and then recruit the RNA polymerase by a protein–protein interaction. In the case of RNA polymerase I, the essential DNA sequences which are recognised are located within the 50 bases immediately upstream of the start site of transcription. As in all genes, the sequences adjacent to the transcriptional start site, which control the expression of the gene, are known as the gene promoter. Note that in the case of all the RNA polymerases the site at which transcription begins is denoted as +1 with bases within the transcribed region being denoted as +100, +200, etc., whilst bases upstream of the start site are denoted as -100, -200, etc., as one proceeds further and further upstream (see Fig. 3.2).

In the case of RNA polymerase I, sequences around -50 are recognised by

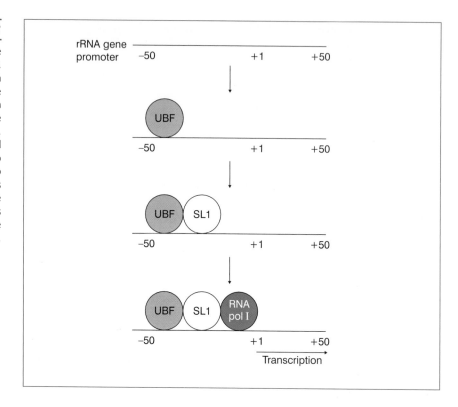

a protein transcription factor known as UBF (upstream binding factor) (for reviews see Geiduschek and Kassavetis, 1995; Jacob, 1995). Subsequently another regulatory protein known as SL1 is recruited via protein–protein interaction with UBF, and in turn SL1 recruits the RNA polymerase itself (Fig. 3.2). Hence the initiation of transcription by RNA polymerase I is relatively simple, with one essential transcription factor (SL1) being necessary to recruit the RNA polymerase itself. In turn the binding of this essential factor is facilitated by the prior binding to a specific DNA sequence of another transcription factor, UBF (for review see Geiduschek and Kassavetis, 1995; Jacob, 1995).

3.2.2 RNA polymerase III

The involvement of a specific transcription factor which acts to recruit the RNA polymerase as well as of other factors which recruit the specific factor is also illustrated by RNA polymerase III (for reviews see Gabrielsen and Sentenac, 1991; White, 1994; Geiduschek and Kassavetis, 1995).

The situation is complicated, however, by the fact that, in different genes which are transcribed by RNA polymerase III, the essential promoter DNA sequences recognised by the transcription factors which recruit the polymerase can be located either upstream or downstream of the transcribed region. Thus genes transcribed by RNA polymerase III can have either an upstream promoter (as is the case for genes transcribed by RNA polymerase

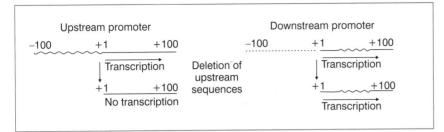

Figure 3.3

Consequences of deleting sequences upstream of the transcriptional start site in genes which have either an upstream promoter or a downstream promoter. Note that, when the promoter (wavy line) is located upstream of the transcriptional start site, deletion of upstream sequences results in an absence of transcription. In contrast, when the promoter is located within the transcribed region, deletion of upstream sequences has no effect on transcription.

I and II) or a downstream promoter located within the transcribed region (Fig. 3.3).

This type of downstream promoter, which is unique to RNA polymerase III, was first identified by detailed studies which focused on the genes that encode the 5S RNA of the ribosome. In an attempt to identify the sequences important for the expression of this gene, sequences surrounding it were deleted and the effect on the transcription of the gene in a cell-free system investigated (Sakonju *et al.*, 1980). Somewhat surprisingly, the entire upstream region of the gene could be deleted with no effect on gene expression (Fig. 3.4a,b). Indeed, deletions within the transcribed region of the 5S gene also had no effect on its expression until a boundary, 40 bases within the transcribed region, was crossed. By this means, an internal control region essential for the transcription of the 5S RNA gene was defined which is located entirely within the transcribed region.

This region of the 5S gene was shown subsequently to bind a transcription factor known as TFIIIA (for a review see Pieler and Theunissen, 1993) (Fig. 3.5). Subsequently, another transcription factor TFIIIC binds to the DNA adjacent to TFIIIA and in turn TFIIIC functions to recruit a further transcription factor TFIIIB to form a stable transcription complex (Fig. 3.6). This complex, which is stable through many cell divisions, promotes the subsequent binding of the RNA polymerase III which is recruited via a protein–protein interaction with TFIIIB and binds at the transcriptional start site (Fig. 3.6). The binding of RNA polymerase III is dependent on the presence of the stable transcription complex and not on the precise sequence of the DNA to which it binds since, as discussed above, the region to which the polymerase normally binds can be deleted and replaced by other sequences without drastically reducing transcription.

Although the assembly of transcription complexes on RNA polymerase III type genes was first defined on the 5S ribosomal RNA gene, other RNA polymerase III transcription units differ in the details of transcription complex assembly. Thus, for example, the genes encoding the transfer RNAs which play a key role in translation (see Section 3.3.6) also have an internal promoter. However, in this case, due to differences in the sequence of the promoter TFIIIA is not required. Rather, TFIIIC binds directly to sequences within the promoter and subsequently recruits TFIIIB. As in the 5S RNA promoters, TFIIIB then recruits the RNA polymerase itself. Similarly, TFIIIB plays a critical role in transcription of the RNA polymerase III genes which

have an upstream promoter such as that encoding the small nuclear RNA U6
which, unlike the other small nuclear RNAs of the splicesosome (see Section
3.3.3), is transcribed by RNA polymerase III rather than RNA polymerase II
(Table 3.1). (For a review of small nuclear RNA gene transcription see
Hernandez, 2001).

It is clear, therefore, that TFIIIB plays an essential role in the transcription

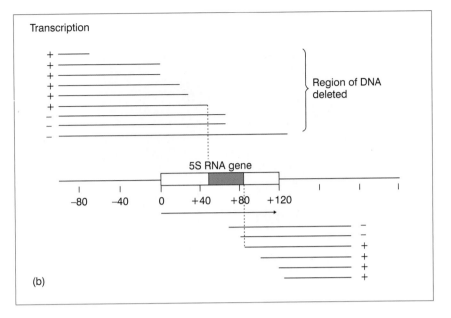

(b)

Figure 3.4b

Summary of the extent of the deletions used and their effects on transcription. The use of these deletions allows the identification of a critical control element (boxed) within the transcribed region of the 5S gene.

of the three different types of RNA polymerase III gene promoters (for a review see Hernandez, 1993) and is the functional equivalent of the RNA polymerase I SL1 factor which acts directly to recruit the RNA polymerase via protein–protein interaction (for a detailed comparison of transcriptional initiation by RNA polymerases I and III see Geiduschek and Kassavetis, 1995).

3.2.3 RNA polymerase II

The TATA box

Inspection of the region immediately upstream of the transcriptional start site reveals that a very wide variety of different genes transcribed by RNA polymerase II contain an AT-rich sequence which is found approximately 30 bases upstream of the transcription start site. This TATA box plays a critical role in promoting transcriptional initiation and in positioning the start site of transcription for RNA polymerase II. Its destruction by mutation or deletion effectively abolishes transcription of such genes (for a review see Breathnach and Chambon, 1981).

Most importantly, this TATA box acts as the initial DNA target site for the progressive assembly of the basal transcription complex for RNA polymerase II, which is considerably more complex than that for RNA polymerase I or III (for reviews see Zawel and Reinberg, 1995; Pugh, 1996; Roeder, 1996; Nikolov and Burley, 1997). Thus, initially, the TATA box is bound by the transcription factor TFIID whose binding is facilitated by the presence of another transcription factor TFIIA (Fig. 3.7a). Interestingly, structural analysis of TFIID has revealed it to have a molecular clamp structure, which consists of four globular domains around an accessible

Figure 3.5

Binding of the TFIIIA transcription factor to the internal control region of 5S DNA in a footprint assay in which binding of a protein protects the DNA from digestion by DNase I and produces a clear region lacking the ladder of bands produced by DNase I digestion of the other regions. Tracks a and e show the two DNA strands of the 5S gene in the absence of added TFIIIA while tracks b–d and f–h show the same DNA in the presence of TFIIIA.

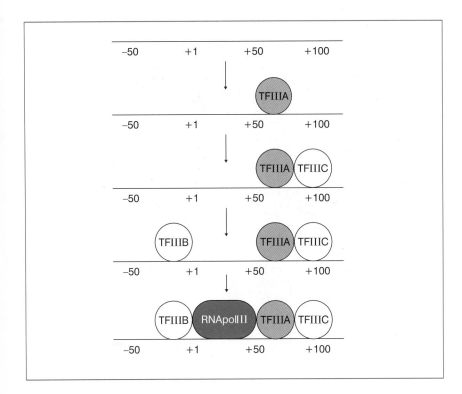

Figure 3.6

Transcription initiation by RNA polymerase III at the 5Sr RNA gene promoter. Sequences downstream of the transcription initiation site (+1) are indicated by the + signs, sequences upstream are indicated by the − signs. Following binding of the transcription factor TFIIIA to the internal control sequence, TFIIIC and TFIIIB bind with TFIIIB then acting to recruit RNA polymerase III, allowing transcription to begin. Note that within the three-dimensional structure, TFIIIB interacts directly with TFIIIC, although this is not indicated in the figure.

groove, which can accommodate the DNA to which TFIID binds (Brand *et al.*, 1999) (colour plate 5).

Subsequently, the TFIID/DNA complex is recognised by another transcription factor TFIIB (Fig. 3.7b). Structural analysis has shown that TFIIB binds on the opposite side of TFIID to that which is bound by TFIIA (Andel *et al.*, 1999) (colour plates 6 and 7). This binding of TFIIB is an essential step in the formation of the initiation complex since, as well as binding to TFIID, TFIIB can also recruit RNA polymerase II itself. Hence the binding of TFIIB allows the subsequent recruitment of RNA polymerase II to the initiation complex in association with another factor TFIIF (Fig. 3.7c). Subsequently, two other factors, TFIIE and TFIIH, associate with the complex (Fig. 3.7d).

In particular, the recruitment of TFIIH plays a critical role in allowing the RNA polymerase to initiate transcription. Thus, TFIIH is, in fact, a multicomponent complex whose molecular structure has recently been determined (Chang and Kornberg, 2000; Schultz *et al.*, 2000) and which appears to play a key role in both transcription and the repair of damaged DNA (for a review see Hoeijmakers *et al.*, 1996). In particular, TFIIH contains a kinase activity which is capable of phosphorylating the C-terminal domain of the largest subunit of RNA polymerase II. Thus this C-terminal domain contains multiple copies of the sequence Tyr–Ser–Pro–Thr–Ser–Pro–Ser. Interestingly, this repeated region serves as a site for phosphorylation by the

Figure 3.7

Transcription initiation at an RNA polymerase II promoter. Initially, the TFIID factor binds to the TATA box together with another factor TFIIA (panel a). Subsequently TFIIB is recruited by interaction with TFIID (panel b) and TFIIB then recruits RNA polymerase II and its associated factor TFIIF (panel c). TFIIE and TFIIH then bind and TFIIH phosphorylates the C-terminal domain of RNA polymerase II (panel d). This converts RNA polymerase II into a form which is capable of initiating transcription and RNA polymerase II and TFIIF then move off down the gene producing the RNA transcript, leaving TFIIA and TFIID bound at the promoter (panel e).

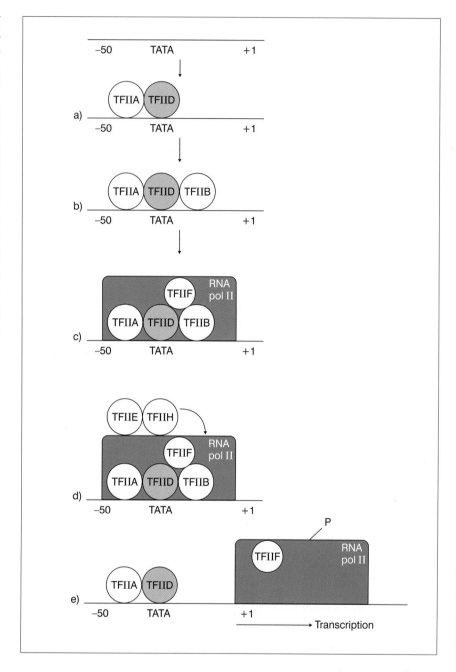

kinase activity of TFIIH and other kinases and such phosphorylation plays a critical role in allowing the polymerase to initiate transcription. Thus, it appears that the dephosphorylated form of RNA polymerase II is recruited to the DNA but that its phosphorylation triggers the start of transcriptional elongation to produce the RNA product (Fig. 3.7e).

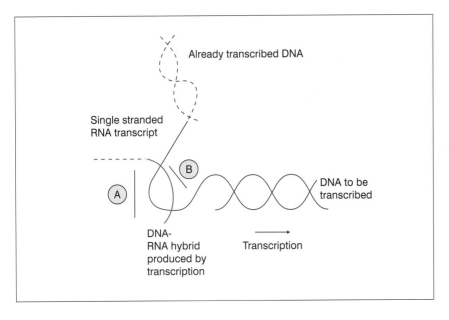

Figure 3.8

Movement of the DNA being transcribed through the RNA polymerase molecule. "A" indicates the wall within the polymerase protein which forces the DNA–RNA hybrid to make a right-angled turn, thereby allowing transcription to occur by addition of ribonucleoside triphosphates to the end of the RNA chain. "B" indicates the rudder region of the polymerase which forces the DNA–RNA hybrid to melt releasing the newly formed RNA and allowing double stranded DNA to reform. DNA which is about to be transcribed is shown by the solid lines and DNA which has already been transcribed by the dotted lines. The arrow indicates the direction of transcription.

Recently, both RNA polymerase II itself and an active transcription complex have been crystallised, allowing structural analysis (for reviews see Klug, 2001; Landick, 2001) (colour plate 8). This has revealed that, as the DNA is transcribed into RNA in the interior of the polymerase molecule, it encounters a wall of protein within the RNA polymerase (labelled A in Fig. 3.8). This forces it to make a right angled turn, exposing the end of the nascent RNA and allowing ribonucleoside triphosphates to be added to it as transcription occurs. Subsequently, the newly formed DNA–RNA hybrid produced as a consequence of transcription encounters another part of RNA polymerase, known as the rudder (labelled B in Fig. 3.8), which forces the separation of the RNA from the DNA. This allows the newly formed part of the RNA molecule to exit and double-stranded DNA to reform (Fig. 3.8).

Interestingly, as the polymerase moves off down the gene, TFIIF remains associated with it whilst TFIIA and TFIID remain bound at the promoter, allowing further cycles of recruitment of TFIIB, RNA polymerase II, etc., leading to repeated rounds of transcription. Such a role for TFIIA and TFIID in allowing repeated rounds of transcription is of particular interest in view of recent findings that some RNA polymerase is found within the cell associated with a large number of proteins, including TFIIB, TFIIF and TFIIH, to form a so-called RNA polymerase holoenzyme (for reviews see Koleskea and Young, 1995; Orphanides et al., 1996; Pugh, 1996). Hence, in addition to the stepwise pathway involving progressive recruitment of TFIIB, RNA polymerase and TFIIH individually, it appears that an alternative pathway exists in which TFIIA and TFIID can recruit a complex containing TFIIB, TFIIE, TFIIF, TFIIH and RNA polymerase itself (Fig. 3.9). In addition, the holoenzyme also contains a number of other protein components which appear to be involved either in opening up the chromatin structure, so as to

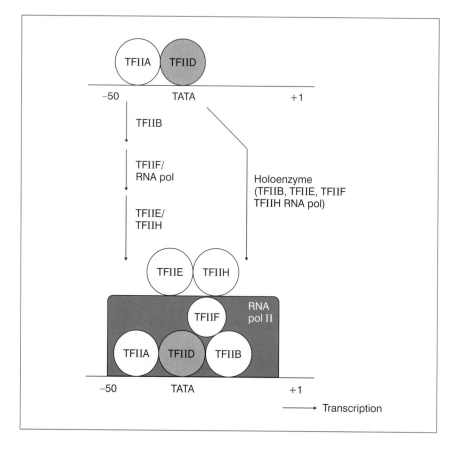

allow transcription to occur (see Section 6.7.2), or in allowing the polymerase complex to be stimulated by transcriptional activators (see Section
8.3.3).

3.2.4 Common features of transcription by the three different polymerases

Although the three different polymerases each have a different function and
different associated transcription factors, the three polymerases themselves
are all multi-subunit proteins, with several subunits being shared between
the different polymerases. Moreover, the largest subunits of each of the
three enzymes show considerable sequence homology to one another.

In addition, each of the three polymerases shows a similar pattern of
recruitment to the DNA, with one specific transcription factor binding to a
target sequence in the gene promoter followed by binding of one or more
other proteins which then recruit the polymerase itself. As discussed above
(Section 3.2.3) in the case of RNA polymerase II promoters which contain a
TATA box, the original binding to DNA is achieved by the TFIID factor
which binds to the TATA box. Interestingly, however, although TFIID was
originally identified as a single factor it is now clear that it is composed of

multiple protein components (for reviews see Burley and Roeder, 1996; Tansey and Herr, 1997). One of these proteins, known as TBP (TATA-binding protein), is responsible for binding to the TATA box whilst the other components of the complex, known as TAFs (TBP-associated factors), do not bind directly to the TATA box but appear to allow TFIID to respond to stimulation by transcriptional activators (see Section 8.3.3).

Although most of the genes transcribed by RNA polymerase II contain a TATA box, a subset of RNA polymerase II genes do not contain the TATA box (for a review see Weis and Reinberg, 1992). Paradoxically, however, TBP plays a key role also in the transcription of this class of RNA polymerase II genes. In this case, TBP does not bind to the DNA but is recruited by another DNA binding protein which binds to the initiator element overlapping the transcriptional start site. TBP then binds to this initiator binding protein, allowing the recruitment of TFIIB and the RNA polymerase itself, as occurs for promoters containing a TATA box. Hence TBP plays a central role in the assembly of the transcription complex for RNA polymerase II, joining the complex by binding to DNA in the case of TATA box containing promoters and being recruited via protein–protein interactions in the case of promoters which lack a TATA box (Fig. 3.10).

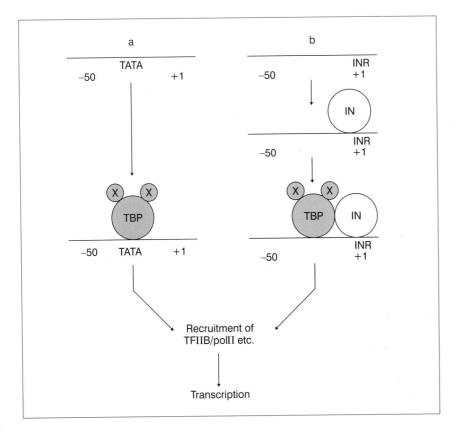

Figure 3.10

Transcription of promoters by RNA polymerase II involves the recruitment of TBP (and its associated factors (X) forming the TFIID complex) to the promoter. This may occur by direct binding of TBP to the TATA box where this is present (panel a) or by protein–protein interaction with a factor (IN) bound to the initiator element in promoters lacking a TATA box (panel b).

These findings indicate therefore that TBP may be the basic transcription factor which is essential for transcription by RNA polymerase II, paralleling the role of SL1 for RNA polymerase I and of TFIIIB for RNA polymerase III. This idea is supported by the amazing finding that TBP is actually also a component of both SL1 and TFIIIB (for a review see White and Jackson, 1992). Thus SL1 is not a single protein but is actually a complex of four factors, one of which is TBP (for a review see Sharp, 1992). Hence the recruitment of SL1 to an RNA polymerase I promoter by UBF (see Section 3.2.1) actually results in the delivery of TBP to the DNA exactly as in the case of RNA polymerase II promoters which lack a TATA box. Similarly, in the 5S and tRNA genes, TBP is delivered to the promoter as part of the TFIIIB complex following prior binding of either TFIIIA or TFIIIC (see Section 3.2.2) (for a review see Rigby, 1993). Moreover, some RNA polymerase III genes, such as that of the U6 RNA gene, contain an upstream promoter with a TATA box and hence in this case TBP can bind directly to the promoter (Fig. 3.11).

The multiple protein–protein and protein–DNA interactions which we have discussed in Sections 3.2.1, 3.2.2 and 3.2.3 therefore appear to serve merely to recruit TBP to the DNA which, in the case of all three polymerases, then leads, directly or indirectly, to the recruitment of the RNA polymerase itself. It has therefore been suggested that TBP represents an evolutionarily ancient transcription factor whose existence precedes the evolution of three independent eukaryotic RNA polymerases and which

Figure 3.11

Transcription of promoters by RNA polymerase III involves the recruitment of TBP (and its associated factors (Y) forming the TFIIIB complex) to the promoter. This may be achieved by protein–protein interactions with either TFIIIA and TFIIIC or TFIIIC alone in the case of promoters lacking a TATA box (panel a) or by direct binding to the TATA box where this is present (panel b).

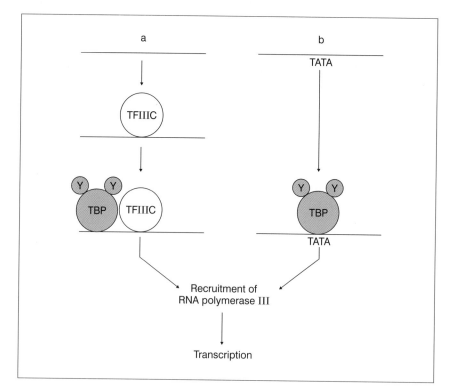

therefore plays a universal and essential role in eukaryotic transcription (for a review see Hernandez, 1993).

Although the three RNA polymerases have common features, paralleling the fact that they all copy DNA into RNA, they each have their own specific role. As RNA polymerase II which transcribes protein-coding genes is the primary target of gene regulatory processes, post-transcriptional modifications affecting the RNA transcripts produced by this enzyme will be discussed in the remainder of this chapter and its regulation is the primary subject of the rest of this book. However, the regulation of genes transcribed by RNA polymerases I and III will be described where appropriate (see in particular Sections 7.6 and 9.5.3).

3.3 POST-TRANSCRIPTIONAL EVENTS

Following transcription, a number of processes modify the primary transcript of genes which are transcribed by RNA polymerase II before it can be translated into protein (see Fig. 3.1). These modifications will now be discussed in turn.

3.3.1 Capping

Following initiation of the primary transcript, transcription proceeds by the progressive addition of ribonucleotides to the RNA chain so that an RNA molecule is produced with bases complementary to those present in the DNA being transcribed. Even before this process is complete, the 5′ end of the nascent mRNA is modified by the addition of a guanine residue which is not encoded by the DNA. The bond joining this guanine to the 5′ end of the RNA differs from the bonds linking the nucleotides in the RNA chain in two ways. Firstly, three phosphate molecules separate the G residue from the first base in the chain, whereas only a single phosphate separates each residue in the rest of the chain. Secondly, the phosphate residues are joined via a 5′ to 5′ bond rather than the standard 3′ to 5′ bond which links the nucleotides in the chain (Fig. 3.12) (for a review see Banerjee, 1980).

Figure 3.12

Structure of the cap which is present at the 5′ end of eukaryotic mRNAs. Note that the cap structure consists of a guanosine (G) residue which is linked via a 5′ to 5′ bond and three phosphate base to the first transcribed base which is normally an A or a G residue. This linkage is in contrast to the normal 3′ to 5′ link with a single phosphate residue which joins the sugar residues in the nucleotide chain.

Capping occurs only in eukaryotes and not in prokaryotes. It is believed to be necessary due to a difference in the process of translation between eukaryotic and prokaryotic mRNAs (see Section 3.3.6). Prokaryotic mRNAs are, in general, polycistronic, which means that several different proteins are translated from the same mRNA. Thus, the ribosome can initiate translation to produce protein internally within an mRNA molecule being guided by specific sequences (Shine–Dalgarno sequences) within the mRNA which are complementary to the 16S RNA of the small ribosomal subunit. In contrast, eukaryotic mRNAs are generally monocistronic, with only one protein being derived from each mRNA. These mRNAs therefore lack the guide sequences present in prokaryotic mRNAs and translation is initiated by the ribosome binding the 5′ end of the mRNA by recognising the methylated cap (Fig. 3.13).

Interestingly, in prokaryotes the so-called Shine–Dalgarno sequence which is recognised by the 16S ribosomal RNA is located within 10 bases upstream of the AUG sequence at which translation is initiated (see Section 3.3.6). In contrast, in eukaryotes the AUG initiation codon may be located several hundred bases away from the 5′ end of the RNA. Therefore, following binding at the cap, the ribosome migrates along the mRNA until it encounters the appropriate AUG initiation codon.

However, translation is not initiated at the first AUG which is encountered by the ribosome. Rather, the AUG triplet must be set within the appropriate consensus sequence which is related to GCCA/GCCAUGG. This so-called scanning hypothesis, in which the ribosome moves along the mRNA looking for this sequence, was originally propounded by Kozak and the rules which determine whether a particular sequence surrounding an AUG codon will allow the ribosome to initiate transcription are known as Kozak's rules (Kozak, 1986).

As well as its function in allowing recognition by the ribosome, the cap structure also has an important role in protecting the 5′ end of the mRNA from attack by exonuclease enzymes which would otherwise recognise a free 5′ end and would digest the RNA in a 5′ to 3′ direction. Indeed, the removal of the cap structure is one of the steps in mRNA degradation by which the mRNA molecule is ultimately broken down (for a review see Tuite, 1996).

3.3.2 Polyadenylation

As discussed above, the 5′ end of the mRNA differs from that of the primary transcript only by the addition of the single modified G residue forming the cap. In contrast, however, at the 3′ end the mRNA molecule is much shorter than the original transcript and differs from it in terminating with a run of up to 200 adenosine residues which are added post-transcriptionally in a process known as polyadenylation (Fig. 3.14) (for reviews see Barabino and Keller, 1999; Keller, 1995; Proudfoot, 1996). The process involves firstly the cleavage of the RNA transcript and then the subsequent addition of the A residues at the resulting free 3′ end.

The site at which cleavage occurs (the polyadenylation site) is flanked by two conserved sequence elements (Fig. 3.15). Upstream of the poly(A) site

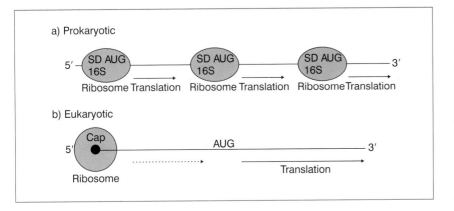

Figure 3.13

Translational initiation in prokaryotes (panel a) and eukaryotes (panel b). Note that prokaryotic mRNAs are polycistronic with several proteins been translated from a single mRNA. Initiation is achieved by the 16S RNA of the ribosome binding to a complementary sequence known as the Shine–Dalgarno (SD) sequence at various points within the mRNA molecule. Translation is then initiated at an AUG codon adjacent to the Shine–Dalgarno sequence. In contrast, eukaryotic mRNAs are monocistronic with only a single protein being translated from the mRNA. Translation is initiated by the ribosome binding to the cap structure at the 5′ end of the mRNA and moving along the mRNA to initiate translation at an AUG codon which is located in an appropriate surrounding sequence.

the RNA contains the essentially invariant sequence AAUAAA and downstream of the poly(A) site it contains a less well conserved region which is rich in G and U residues. These two sequence elements are recognised respectively by two protein complexes known as the cleavage and polyadenylation specificity factor (CPSF) and the cleavage stimulation factor CstF (see Fig. 3.15). Following the binding of these complexes they are thought to interact with one another and the cutting of the RNA occurs in the region between them. Note that this cutting is an endonucleolytic cleavage where cutting occurs within the RNA chain rather than the exonucleolytic cleavage which would occur following removal of the cap (see Section 3.3.1) where degradation begins at a free end of the RNA molecule. Subsequent to cleavage the enzyme poly (A) polymerase then adds the run of A residues to the free 3′ end (Fig. 3.15).

Thus the RNA transcript is modified at both ends, firstly by capping at the 5′ end and subsequently by cleavage and polyadenylation at the 3′ end. It is believed that the primary role of the polyadenylation process is to protect the mRNA from degradation by exonucleases which would otherwise attack its free 3′ end and rapidly degrade it. Thus, in many cases, an mRNA which is to be degraded is first deadenylated, allowing its 3′ end to be attacked, and is only then uncapped (see Section 3.3.1), allowing its 5′ end to be attacked

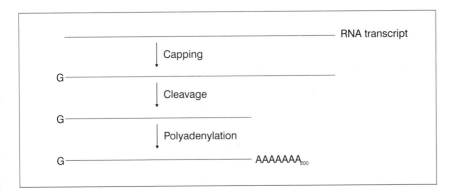

Figure 3.14

Modifications at the 5′ and 3′ end of RNA polymerase II transcripts. Note that the 5′ end of the RNA is modified by the addition of a single G residue in the capping process whereas the 3′ end is cleaved with removal of a large stretch of RNA and subsequent addition of up to 200 A residues to the free 3′ end in the process of polyadenylation.

Figure 3.15

Polyadenylation. This process involves the binding of protein factors to the AAUAAA signal upstream of the poly(A) site (CPSF) and to the G/U rich sequence downstream of the poly(A) site (CstF). CPSF and CstF then interact with one another, leading to cleavage in between them at the polyadenylation site. Subsequently the sequence downstream of the poly(A) site is degraded whilst the free 3' end of the upstream region is polyadenylated by the enzyme poly(A) polymerase.

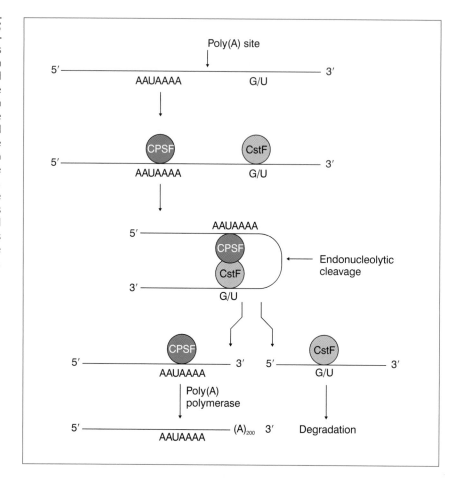

(for reviews see Beelman and Parker, 1995; Tuite, 1996). In addition, it appears that the presence of the poly A tail can also regulate the efficiency by which an mRNA is translated (see Section 5.6). Hence, like the cap, the poly A tail appears to have a dual function in regulating both translation of the mRNA into protein and its degradation.

Interestingly, not all mRNAs encoding proteins are polyadenylated. In particular, the mRNAs encoding the histone molecules. which play a critical role in chromatin structure (see Section 6.3), are not polyadenylated. In this case, however, the 3' end of the mRNA is protected from degradation since it can fold itself into a double stranded stem-loop structure which protects the mRNA from degradation. Interestingly, the folding and unfolding of this stem loop plays a critical role in regulating the stability of the histone mRNA resulting, for example, in its becoming highly stable during the S phase of the cell cycle when DNA is being synthesised and more histones are required (see Section 5.5). Such an example is, however, the exception and the vast majority of RNA species transcribed by RNA polymerase II are polyadenylated at their 3' ends following capping at the 5' end.

3.3.3 RNA splicing

Probably the most unexpected finding to have been made during the modern era of gene cloning was the discovery that the protein coding regions of genes transcribed by RNA polymerase II are interrupted by DNA sequences which do not encode parts of the protein. These intervening sequences are known as introns because, as we shall see, although they are transcribed into RNA, their RNA product remains *in* the nucleus whereas the coding regions, or exons, *exit* the nucleus.

Thus, when the DNA is transcribed, both the exons and the introns which lie between them are transcribed into a single primary RNA transcript. As well as being modified at its 5' and 3' ends by capping and polyadenylation, this primary transcript must therefore also have its introns removed before it is transported to the cytoplasm. Otherwise, if it is translated, irrelevant amino acid sequences will appear in the middle of the protein or translation will be prematurely terminated due to the presence of a stop codon (see Section 3.3.6) within the intervening sequence. The process whereby these introns are removed is known as RNA splicing (Fig. 3.16) (for reviews see Sharp, 1994; Adams *et al.*, 1996; Staley and Guthrie, 1998; Tollervey and Caceres, 2000).

Initial studies on RNA splicing concentrated on a DNA sequence analysis of the introns and the adjacent regions of the exons. Such studies revealed that virtually all eukaryotic introns for genes transcribed by RNA polymerase II begin with the bases GT and end with the bases AG (see Fig. 3.17). Although other features of the intron and the adjacent regions of the exon have been defined, such as a run of pyrimidine residues (the polypyrimidine tract) located adjacent to the AG sequence within the intron, these are much less well conserved than the GT/AG sequences.

Although this finding was of interest it did not of itself provide any functional insights into the mechanisms by which splicing occurred. These were initially provided by studies using nuclear extracts which could carry out the splicing reaction in the test tube. These experiments showed, that, when a simple RNA molecule containing two exons interrupted by an intron was added to the splicing reaction, it was possible to produce the spliced product with the two exons joined together and a free intron. Most interestingly,

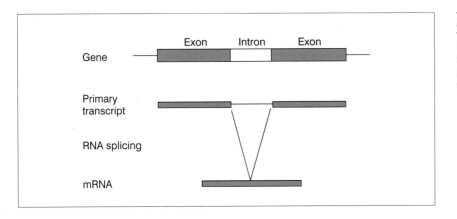

Figure 3.16

Removal of an intervening sequence (intron) from the primary RNA transcript by RNA splicing.

Figure 3.17

Exon–intron junctions in RNA polymerase II transcripts. Note that the first two bases of the intron are normally GU whilst the last two bases are normally AG.

however, the intron emerged in a lariat structure in which its 5′ end had folded back on itself and formed a 5′ to 2′ bond with an A residue located between 18–40 bases upstream of the AG sequence at the 3′ end of the intron (Fig. 3.18).

This indicated that splicing proceeds in a series of steps (Fig. 3.18) in which the transcript is first cleaved at the junction between the upstream exon and the intron (the 5′ splice site), resulting in a free upstream exon and the intron joined to exon II. In order to prevent a back reaction in which the intron rejoins with the upstream exon, the free 5′ end of the exon rapidly folds back on itself and forms the 5′ to 2′ bond with the A residue within the intron. This

Figure 3.18

Mechanism of RNA splicing. Splicing is initiated by cleavage at the 5′ splice site, generating an intron–exon 2 structure with a free 5′ end. This free 5′ end then forms a 5′ to 2′ bond with an A residue within the intron (the branch point). This is followed by cleavage at the 3′ splice site and subsequent joining of exon I to exon 2.

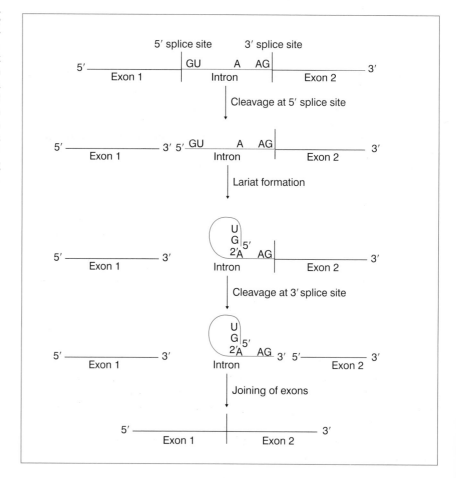

A residue is known as the branch point. Following this process, the intron is then freed from the downstream exon by cleavage immediately after the next AG sequence following the branch point. Subsequently the downstream exon then joins with the upstream exon to form the spliced product (Fig. 3.18). Hence the reaction has achieved its purpose by removing the intron from between the two exons and joining them together in a splicing reaction.

In considering this process, one might ask why the upstream exon does not simply diffuse away whilst the downstream exon is still linked to the intron and cannot be joined to it. The answer to this question is that this reaction does not take place with the RNA isolated within the nucleus. Rather, it takes place in a complex structure known as the spliceosome which consists of a number of different RNA and protein components which together catalyse this process and hold the various intermediates in the splicing process in the correct orientation relative to one another (for a review see Staley and Guthrie, 1998).

The RNA components of the spliceosome are small RNA molecules ranging from 56–217 bases in size and known as the U RNAs because they are rich in uridine residues (for a review see Neilsen, 1994). Each of these U RNAs has a specific role in the splicing process and each of them is associated with specific proteins forming small nuclear ribonucleoprotein particles (snRNP) which consist of a U RNA with its associated proteins. Some of these proteins, such as the Sm proteins, are common to all the different U snRNPs whereas others, such as the U1snRNP A protein, are associated with only one specific RNP particle (for a review see Lamond, 1999).

The splicing process is initiated by the binding of the U1 snRNP to the 5' splice site (Fig. 3.19a). Subsequently the U2 accessory factor (U2AF) binds to the polypyrimidine tract located between the branch point A and the AG at the end of the intron. This then results in the U2 snRNP binding at the adjacent branch point (Fig. 3.19b). Next, the RNA is bound by the U5 snRNP which binds to the upstream exon and by the single snRNP which contains the U4 and U6 RNAs bound to one another. This binding results in the displacement of the U1 snRNP from the 5' splice site where it is replaced by the U6 snRNP (Fig. 3.19c). This results in the release of the U1 snRNP and of the U4 RNA which is now no longer base paired to U6.

The U6 and the U2 snRNPs then interact with one another, bringing the 5' splice site close to the branch point. This is followed by the cleavage of exon 1 from intron 1, lariat formation and the joining of exon 1 to exon 2 as outlined in Fig. 3.18. The U2 snRNP, the U6 snRNP and the U5 snRNP (which has moved from the upstream exon to the intron during this process) are released with the intron (Fig. 3.19d).

Hence, the entire process of splicing takes place in the tightly ordered structure of the spliceosome and this explains, for example, why exon 1 cannot simply diffuse away after its cleavage from the intron–exon 2. Indeed, it has been demonstrated that a specific component of the spliceosome, hSLu7, binds tightly to exon 1 and holds it in close proximity to the AG of the correct 3' splice site, so allowing selection of the correct 3' splice site to occur (Chua and Reed, 1999).

Figure 3.19

Involvement of snRNP particles bearing different U RNAs in RNA splicing. Splicing is initiated by binding of the U1 particle to the 5′ splice site (panel a) followed by binding of the U2 particle to the branch point (panel b). This is followed by binding of the U5 particle to the upstream exon and of the U4/U6 particle to the 5′ splice site with release of the U1 particle and U4 RNA. U6 and U2 particles then interact with one another (panel c) bringing the 5′ splice site adjacent to the branch point. This is followed by cleavage at this site, lariat formation and subsequent cleavage at the 3′ site with exon joining. The free intron is released in association with the U2, U5 and U6 particles (panel d), with the U5 particle having moved from the upstream exon to a position within the intron.

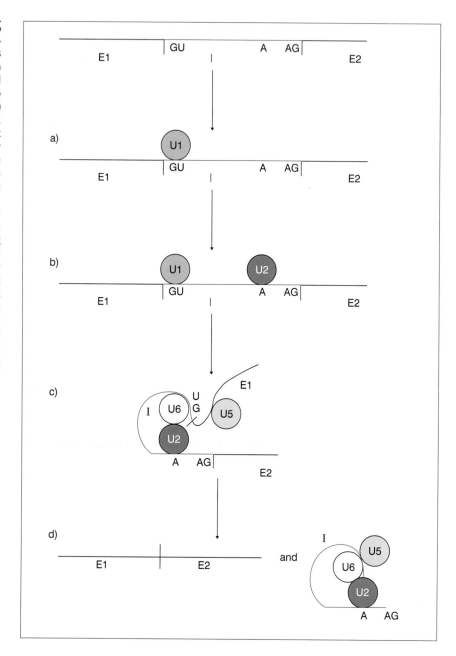

In addition to the U RNAs and the proteins associated with them in the UsnRNPs, the splicesosome also contains other proteins which are not directly associated with the U RNAs. The most intensively studied of these are a series of proteins which are rich in serine and arginine residues and which are therefore known as SR proteins (S = serine, R = arginine in the one letter code for amino acids) (for reviews see Manley and Tacke, 1996; Tollervey and Caceres, 2000).

These proteins are believed to serve both in recruiting the snRNPs to the RNA to be spliced and in determining which splice sites are joined to one another. As the vast majority of RNAs contain multiple exons and introns (rather than the simple two-exon one-intron model we have been considering) this is obviously highly critical in order to ensure that the correct exons are joined to one another so that exon I is joined to exon II which in turn is joined to exon III, rather than, for example, exon I being joined to exon III, resulting in exon II being missing from the mRNA and therefore producing a protein lacking a particular region. As will be described in Section 5.2 this process can be regulated so that different combinations of exons are joined to one another in different situations in a process known as alternative splicing and it appears that the SR proteins play a critical role in the regulation of this process.

3.3.4 Coupling of transcription and RNA processing within the nucleus

Although we have so far discussed transcription, capping, polyadenylation and splicing as if they were entirely separate processes, it is clear that they are, in fact, tightly coupled together within the nucleus (for reviews see Hirose and Manley, 2000; Proudfoot, 2000). The first evidence that this was the case was the finding that deletion of the C-terminal domain (CTD) of RNA polymerase II (see Section 3.2.3) not only reduces transcription but also interferes with post-transcriptional processes, such as capping, splicing and polyadenylation, indicating that RNA polymerase II is also involved in these events.

As described in Section 3.2.3, phosphorylation of the CTD of RNA polymerase II leads to the polymerase together with TFIIF initiating transcription and moving off down the gene, leaving behind the other basal transcription factors such as TFIIB and TFIID. Interestingly, these basal factors are then progressively replaced by factors which are involved in capping, splicing and polyadenylation and which bind specifically to the phosphorylated form of the RNA polymerase II CTD.

Thus, as soon as transcription begins, capping enzymes bind to the CTD of RNA polymerase II, allowing the 5' end of the RNA to be capped as described in Section 3.3.1 (Fig. 3.20b). Subsequently, the SR proteins and other components of the splicing complex (see Section 3.3.3) bind to the CTD, allowing splicing to occur (Fig. 3.20c). Finally, components of the polyadenylation complex (see Section 3.3.2) interact with the transcribing complex and polyadenylate the RNA (Fig. 3.20d). Hence, the transcription complex, as it moves down the gene, attracts the factors which are needed to post-transcriptionally modify the RNA and this occurs whilst the RNA is being transcribed, paralleling the translation of the RNA whilst it is being transcribed, which occurs in prokaryotes (see Section 3.1). Once the coupled processes of transcription, capping, splicing and polyadenylation have taken place, the RNA is fully mature and ready to be transported to the cytoplasm.

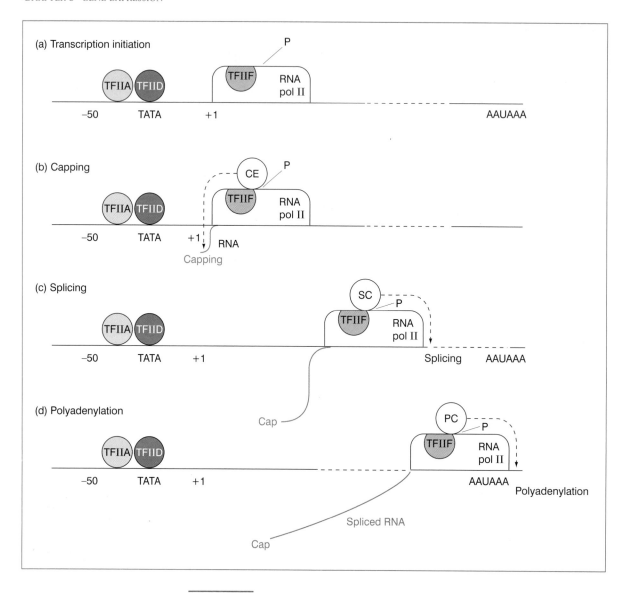

Figure 3.20

As the RNA polymerase II complex transcribes the gene (panel a), it becomes progressively associated with the proteins that mediate capping (panel b, CE = capping enzymes), splicing (panel c, SC = splicing complex) or polyadenylation (panel d, PC = polyadenylation complex). This allows the RNA to be capped, spliced and polyadenylated as it is being transcribed. Note the phosphorylation (P) of the RNA polymerase in the transcribing complex (see also Fig. 3.7) which is essential for the recruitment of the capping, splicing and polyadenylation factors.

3.3.5 RNA transport

Although the processes of capping, polyadenylation and RNA splicing take place in the nucleus, the process of translation, whereby the mature mRNA is converted into protein, takes place in the cytoplasm. Thus the RNA must be transported through the nuclear membrane from the nucleus to the cyto- plasm before it can be translated into protein (for reviews see Lee and Silver, 1997; Nakielny and Dreyfuss, 2000).

A number of proteins which associate with the RNA in the nucleus and accompany it into the cytoplasm have been identified and are believed to play a critical role in mediating the transport of the RNA from the nucleus to the cytoplasm (for reviews see Nigg, 1997; Nakielny and Dreyfuss, 2000; Daneholt, 2001). One of these proteins, hnRNP A1, is a member of the heterogeneous nuclear RNP (hnRNP) proteins which associate with the primary transcripts of protein coding genes even before splicing complexes form on these RNAs (for a review of hnRNP proteins see Dreyfuss *et al.*, 1993). Although some of these proteins, such as hnRNP C, disassociate from the mRNA before it is transported to the cytoplasm hnRNP A1 remains associated with the mRNA as it moves into the cytoplasm (Fig. 3.21). Indeed, in the insect *Chironomus tentans* hnRNP A1 can be directly visualised in the electron microscope associated with mRNA which is passing through the nuclear pore into the cytoplasm (Daneholt, 2001) (Fig. 3.22).

Although hnRNP A1 is thus a candidate for a molecule mediating the transport of the mRNA into the cytoplasm, no direct evidence exists that it is actually essential for this process. However, such direct evidence does

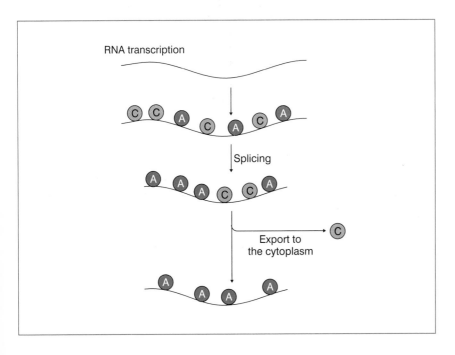

Figure 3.21

Prior to splicing the RNA transcript binds hnRNP proteins such as hnRNP A1 and C. Although hnRNPC is released from the RNA prior to export to the cytoplasm hnRNP A1 moves with the RNA into the cytoplasm.

Figure 3.22

Electron micrographs (top row) and schematic diagrams (bottom row) showing the passage of the mRNA/hnRNP particle from nucleus (nuc) to cytoplasm (cyt) via the nuclear envelope (dotted line) in the salivary glands of *Chironomus tentans*. The mRNA/protein particle is shown within the nucleus (left hand panel), approaching the nuclear envelope (second panel) and partially unfolded particles passing through pores in the nuclear membrane are shown in the central and right hand panels. The scale bar indicates 100 nm.

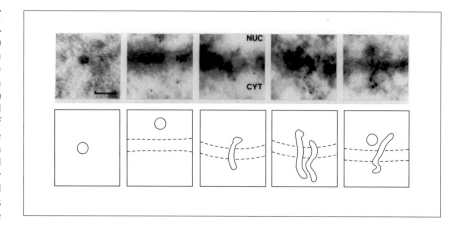

exist in the case of another mRNA binding protein, namely the Gle1 protein found in yeast. Thus this protein was originally identified as being encoded by a gene whose mutation prevents the transport of polyadenylated mRNA from the nucleus to the cytoplasm in yeast, indicating that it is essential for this purpose. Most interestingly, the Gle1 protein contains a short amino acid sequence known as a nuclear export signal (NES). If this sequence is transferred to another unrelated protein, it can direct the transfer of that protein from the nucleus to the cytoplasm. This indicates that, following binding of the Gle1 protein to an mRNA, the NES sequence directs the export of the protein and the associated mRNA from the nucleus to the cytoplasm (Fig. 3.23).

A similar NES sequence has also been identified in the Rev protein of the human immunodeficiency virus (HIV) which is involved in regulating the transport of different HIV mRNAs from the nucleus to the cytoplasm (see Section 5.4). Similarly, the TFIIIA transcription factor (see Section 3.2.2) also contains an NES sequence. As discussed in Section 3.2.2, the TFIIIA molecule binds to an internal region within the transcribed portion of the gene encoding the 5SrRNA. This region is therefore also present in the resulting 5SrRNA transcript and TFIIIA can bind to it at the RNA level also and promote transport of the 5SrRNA from the nucleus to the cytoplasm.

Hence the internal promoter characteristic of the 5SrRNA genes allows TFIIIA to both act as a transcription factor by binding to its recognition sequence at the DNA level and also to act as a transport protein by binding to the corresponding sequence at the RNA level. As discussed in Section 7.6 this binding of TFIIIA to both the 5S DNA and the 5S RNA is also used to regulate the level of transcription of the 5S rRNA genes in oogenesis.

It is likely therefore that Gle1, Rev and TFIIIA act via similar mechanisms to promote the transport to the cytoplasm of different RNA molecules, with their action involving binding to the RNA by an RNA binding region of the protein followed by transport of the RNA/protein complex mediated by the NES signal sequence (Fig. 3.23).

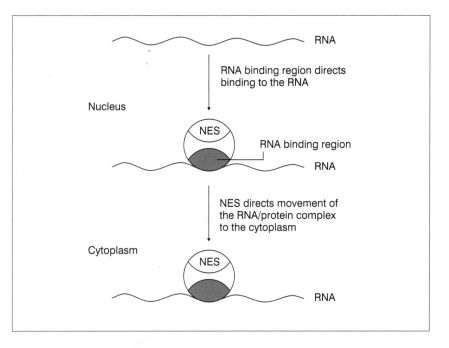

Figure 3.23

Proteins such as Gle I, Rev and TFIIIA bind to the RNA via an RNA binding region and then direct its movement to the cytoplasm via the Nuclear Export Signal (NES).

Interestingly, only RNA that has been properly capped, spliced and polyadenylated is transported to the cytoplasm (for a review see Maquat and Carmichael, 2001). This is evidently of critical importance to prevent the transport of unspliced RNAs containing introns which could not be translated to produce functional protein. This is achieved by a protein known as Aly, which binds to the RNA during the splicing process and remains tightly associated with it (Zhou *et al.*, 2000). As Aly is essential for transport to the cytoplasm, this ensures that only properly spliced RNA leaves the nucleus (for a review see Reed and Magni, 2001).

3.3.6 Translation

The process of translating the mRNA into protein takes place on defined cytoplasmic organelles known as ribosomes which, in eukaryotes, consist of a large subunit of 60S in size and a small subunit of 40S in size (for a review of translation see Pestova *et al.*, 2001). Each of these subunits contains both specific RNAs and proteins. Thus the large subunit consists of 28S, 5.8S and 5S ribosomal RNAs associated with 49 different proteins whilst the small subunit consists of an 18S ribosomal RNA and 33 associated proteins (for reviews see Lake, 1985; Moore, 1988).

Translation is initiated by the binding of the eukaryotic initiation factor (eIF), eIF-4E, to the cap structure at the 5' end of the mRNA (see Section 3.3.1) (Fig. 3.24a) (for reviews see Hentze, 1997; Dever, 1999). Subsequently other eIF factors, such as eIF-4G and eIF-4A/B, bind to the 5' end of the RNA (Fig. 3.24b). Together eIF-4A, eIF-4E and eIF-4G form a complex which is known as eIF-4F. The binding of the various components

Figure 3.24

Mechanism of translational initiation. Initial binding of eIF4E to the cap structure (panel a) is followed by the binding of eIF4G and eIF4A (panel b). The complex of eIF4A, E and G is known as eIF4F. The eIF4F complex is then recognised by a complex consisting of the 40S ribosomal sub unit, the initiator tRNA and eIF2 (panel c). The 40S subunit then migrates along the RNA until it reaches the initiator AUG codon (panel d). This is followed by release of eIF2 and binding of the 60S large ribosomal subunit. Note that, for clarity, eIF4A, E and G are shown as a single eIF4F complex in panel c.

of eIF-4F is followed by the binding of a complex consisting of the small 40S ribosomal subunit, a transfer RNA molecule (tRNA), which carries the amino acid methionine, and the initiation factor eIF-2 (Fig. 3.24c). Subsequently, as described in Section 3.3.1, the 40S subunit migrates down the mRNA until it finds an AUG initiation sequence in the mRNA which is located in the appropriate context for the initiation of translation (Fig. 3.24d). At this point, eIF-2 is released from the complex and the 60S large ribosomal subunit binds to the small subunit to form the complete 80S ribosome which is able to initiate translation (Fig. 3.24e). Two factors, eIF-5 and eIF-5b, play an essential role in the joining of the large and small ribosomal subunits (Pestova *et al.*, 2000).

A key role in the process of translation is played by the transfer RNA (tRNA) molecules. These RNAs are unique in that they can bind an amino acid at their 3′ end (Fig. 3.25) and therefore function to bring the amino acids to the ribosome for incorporation into protein. The tRNAs are RNA molecules from 74–95 bases in length which can fold into a highly defined secondary structure (Fig. 3.25). In this structure one specific loop contains the so-called anticodon. Each tRNA molecule which binds a different amino acid has a distinct anticodon of three bases which can pair with the complementary three bases within the mRNA (for reviews see Normanly and Abelson, 1989; Schimmel, 1987). Thus the initiator tRNA described above contains the anticodon sequence CAU and therefore binds to the initiator AUG codon. This initiator tRNA carries a methionine amino acid linked to it and the chain of the protein molecule thus begins with a methionine residue (Fig. 3.26a).

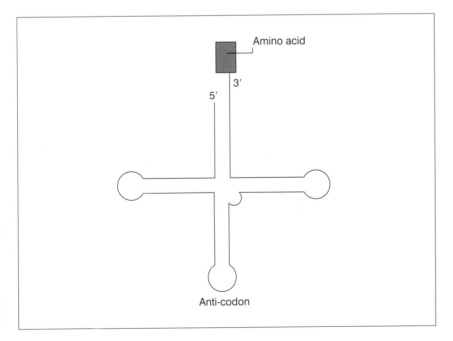

Figure 3.25

Cloverleaf structure of tRNA with stem structures containing paired bases and loops containing unpaired bases. Note the unpaired bases in the anticodon which will bind to the complementary bases in the codon of the mRNA, thereby delivering the appropriate amino acid bound to the 3′ end of the tRNA to the ribosome.

Figure 3.26

Translation elongation. Following the arrival of the ribosome at the initiator AUG codon and binding of the initiator tRNA bearing a methionine amino acid to the AUG codon (panel a, see also Fig. 3.24), a second tRNA is recruited in association with the elongation factor eEF1 and binds to the next codon via its corresponding anticodon (panel b). Note that this codon may have any sequence, depending on the next amino acid to be incorporated into the protein, a CUG codon which recruits tRNA with the anticodon sequence CAG and bearing a leucine amino acid is shown as an example. Subsequently a peptide bond is formed between the methionine and the next amino acid (leucine in this case) with translocation of the methionine to bond with the leucine residue. The elongation factor eEF1 then initiates translocation of the ribosome three bases along the mRNA and the first tRNA is released (panel c). The cycle then repeats itself with a new tRNA being recruited to the next three bases (panel d) with consequent peptide bond formation producing a three amino acid peptide.

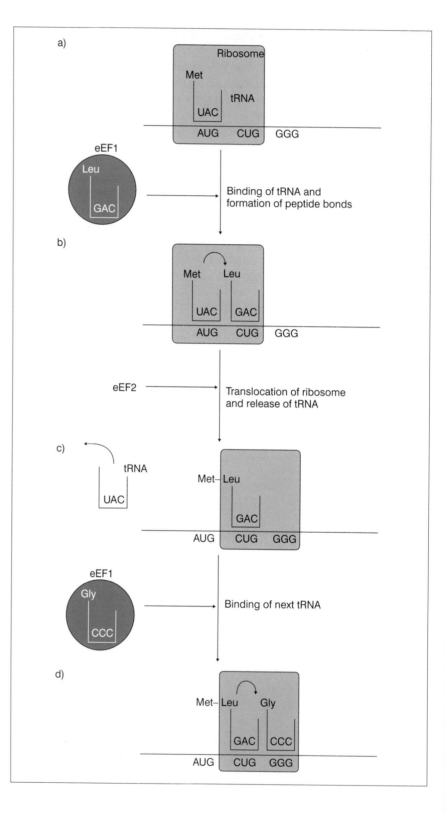

Subsequently, a further tRNA molecule bearing the appropriate anti-codon is recruited to bind to the next three base sequence in the mRNA immediately downstream of the initiating AUG sequence. As each particular tRNA containing a specific anticodon also binds to a specific amino acid, this tRNA will deliver a particular amino acid to the mRNA-bound ribosome, with the nature of the amino acid being specified by the nature of the three base codon immediately following the AUG (Fig. 3.26b).

This second tRNA is recruited to the ribosome by a specific eukaryotic elongation factor (eEF), eEF-1. The structure of the ribosome allows two specific sites within it to be occupied by tRNA molecules which are bound to adjacent three base sequences in the mRNA. Once this has occurred, an enzymatic activity associated with the large ribosomal subunit forms a peptide bond between the two amino acids which are bound to the two adjacent tRNA molecules. In this manner, the growth of the peptide chain of the protein molecule begins.

Subsequently, another elongation factor, eEF-2, catalyses the move of the ribosome three more bases down the mRNA molecule (Fig. 3.26c). The initiator tRNA is then released from the mRNA and a further tRNA molecule is recruited to the ribosome bound mRNA by eEF-1 (Fig. 3.26d). This tRNA contains an anticodon which can bind to the next three bases in the mRNA sequence and has a specific amino acid bound to it. Once again, a peptide bond is formed between the two amino acid chain and this third amino acid, with the resulting three amino acid peptide being transferred to the new tRNA.

In this manner, the amino acid chain of the protein is gradually built up (Fig. 3.26) (for a review see Wilson and Noller, 1998). As each tRNA with a particular anticodon binds a specific amino acid, amino acids are added to the chain in accordance with the anticodon of the tRNA which is recruited. In turn, this is determined by the codon present in the mRNA and so the information in the mRNA is gradually translated into protein in accordance with the genetic code (Table 3.2). Ultimately, the moving ribosome encounters one of the three stop codons which are not recognised by a specific tRNA and a single release factor catalyses the release of the completed polypeptide from the ribosome.

The production of the final protein molecule thus completes the process which began with the transcription of the gene and results in the genetic information encoded in the DNA and subsequently in the RNA being converted into a functional protein molecule.

3.4   CONCLUSIONS

The multi-step process which has been described in this chapter and results in the conversion of the genetic information in the DNA into protein clearly offers numerous opportunities for the regulation of gene expression. The data leading to the conclusion that, in fact, the primary regulation of gene expression occurs at the level of transcription is described in the next

Base 1	Base 2			
	A	**C**	**G**	**U**
A	AAA Lys	ACA Thr	AGA Arg	AUA Ile
	AAC Asn	ACC Thr	AGC Ser	AUC Ile
	AAG Lys	ACG Thr	AGG Arg	AUG Met
	AAU Asn	ACU Thr	AGU Ser	AUU Ile
C	CAA Gln	CCA Pro	CGA Arg	CUA Leu
	CAC His	CCC Pro	CGC Arg	CUC Leu
	CAG Gln	CCG Pro	CGC Arg	CUG Leu
	CAU His	CCU Pro	CGU Arg	CUU Leu
G	GAA Glu	GCA Ala	GGA Gly	GUA Val
	GAC Asp	GCC Ala	GGC Gly	GUC Val
	GAG Glu	GCG Ala	GGG Gly	GUG Val
	GAU Asp	GCU Ala	GGU Gly	GUU Val
U	UAA Stop	UCA Ser	UGA Stop	UUA Leu
	UAC Tyr	UCC Ser	UGC Cys	UUC Phe
	UAG Stop	UCG Ser	UGG Trp	UUG Leu
	UAU Tyr	UCU Ser	UGU Cys	UUU Phe

Table 3.2 The genetic code

chapter. The cases of regulation at other levels, such as RNA splicing, RNA transport, RNA stability and translation into protein, are described in Chapter 5.

REFERENCES

Adams, M.D., Rudner, D.Z. and Rio, D.C. (1996). Biochemistry and regulation of pre-mRNA splicing. *Current Opinion in Cell Biology* 8, 331–339.
Andel, F., Ladurner, A.G., Inouye, C., Tjian, R. and Nogales, E. (1999). Three-dimensional structure of the human TFIID-IIa-IIB complex. *Science* 286, 2153–2156.
Banerjee, A.K. (1980). 5′ terminal cap structure in eukaryotic messenger ribonucleic acids. *Microbiological Reviews* 44, 175–205.
Barabino, S.M.L. and Keller, W. (1999). Last but not least: regulated Poly(A) tail formation. *Cell* 99, 9–11.
Beelman, C.A. and Parker, R. (1995). Degradation of mRNA in eukaryotes. *Cell* 81, 179–183.
Brand, M., Leurent, C., Mallouh, V., Tora, L. and Schultz, P. (1999). Three-dimensional structures of the TAF$_{II}$-containing complexes TFIID and TFTC. *Science* 286, 2151–2153.
Breathnach, R. and Chambon, P. (1981). Organisation and expression of eukaryotic split genes coding for proteins. *Annual Review of Biochemistry* 50, 349–383.
Burley, S.K. and Roeder, R.G. (1996). Biochemistry and structural biology of transcription factor IID (TFIID). *Annual Review of Biochemistry* 65, 769–799.
Chang, W.-H. and Kornberg, R. D. (2000). Electron crystal structure of the transcription factor and DNA repair complex, core TFIIH. *Cell* 102, 609–613.
Chua, K. and Reed, R. (1999). The RNA splicing factor hSlu7 is required for correct 3′ splice-site choice. *Nature* 402, 207–210.

Daneholt, B. (2001). Assembly and transport of a premessenger RNP particle. *Proceedings of the National Academy of Sciences USA* **98**, 7012–7017.

Davidson, E.H. and Britten, R.J. (1979). Regulation of gene expression: possible role of repetitive sequences. *Science* **204**, 1052–1059.

Dever, T.E. (1999). Translation initiation: adept at adapting. *Trends in Biochemical Sciences* **24**, 398–403.

Dreyfuss, G., Matunis, M.J., Pinol-Roma, S. and Burd, C.J. (1993). hnRNP proteins and the biogenesis of mRNA. *Annual Review of Biochemistry* **62**, 89–321.

Gabrielsen, S.O. and Sentenac, A. (1991). RNA polymerase III (C) and its transcription factors. *Trends in Biochemical Sciences* **16**, 412–416.

Geiduschek, E.P. and Kassavetis, G.A. (1995). Comparing transcriptional initiation by RNA polymerases one and three. *Current Opinion in Cell Biology* **7**, 344–351.

Hentze, M.W. (1997). eIF4G: a multipurpose ribosome adaptor. *Science* **275**, 500–501.

Hernandez, N. (1993). TBP: a universal transcription factor. *Genes and Development* **7**, 1291–1308.

Hernandez, N. (2001). Small nuclear RNA genes: a model system to study fundamental mechanisms of transcription. *Journal of Biological Chemistry* **276**, 26733–26736.

Hirose, Y. and Manley, J. L. (2000). RNA polymerase II and the integration of nuclear events. *Genes and Development* **14**, 1415–1429.

Hoeijmakers, J.H.J., Egly, J.-M. and Vermeulen, W. (1996). TFIIH: a key component in multiple DNA transactions. *Current Opinion in Genetics and Development* **6**, 26–33.

Jacob, F. and Monod, J. (1961). Genetic regulatory mechanism in the synthesis of proteins. *Journal of Molecular Biology* **3**, 318–356.

Jacob, S.T. (1995). Regulation of ribosomal gene transcription. *Biochemical Journal* **306**, 617–626.

Jeffreys, A.J. and Flavell, R.A. (1977). The rabbit beta globin gene contains a large insert in the coding sequence. *Cell* **12**, 1097–1108.

Keller, W. (1995). No end yet to messenger RNA 3′ processing. *Cell* **81**, 829–832.

Klug, A., 2001. A marvellous machine for making messages. *Science* **292**, 1844–1846.

Koleskea, J. and Young, R.A., 1995. The RNA polymerase II holoenzyme and its implications for gene regulation. *Trends in Biochemical Sciences* **20**, 113–116.

Kozak, M. (1986). Point mutations define a sequence flanking the AUG initiator codons that modulate translation by eukaryotic ribosomes. *Cell* **44**, 283–292.

Lake, J.A. (1985). Evolving ribosome structure: domains in Archaebacteria, Eubacteria, Eocytes and Eukaryotes. *Annual Review of Biochemistry* **54**, 507–530.

Lamond, A. I. (1999). Running rings around RNA. *Nature* **397**, 655–656.

Landick, R. (2001). RNA polymerase clamps down. *Cell* **105**, 567–570.

Lee, M.S. and Silver, P.A. (1997). RNA movement between the nucleus and the cytoplasm. *Current Opinion in Genetics and Development* **7**, 212–219.

Manley, J.L. and Tacke, R. (1996). SR proteins and splicing control. *Genes and Development* **10**, 1569–1579.

Maquat, L. E. and Carmichael, G. G. (2001). Quality control of mRNA function. *Cell* **104**, 173–176.

Moore, P.B. (1988). The ribosome returns. *Nature* **331**, 223–227.

Nakielny, S. and Dreyfuss, G. (2000). Transport of proteins and RNAs in and out of the nucleus. *Cell* **99**, 677–690.

Neilsen, T.W. (1994). RNA-RNA interactions in the splicesosome: unravelling the ties that bind. *Cell* **78**, 1–4.

Nigg, E.A. (1997). Nucleocytoplasmic transport: signals, mechanisms and regulation. *Nature* **386**, 779–787.

Nikolov, D.B. and Burley, S.K. (1997). RNA polymerase II transcription initiation: a structural view. *Proceedings of the National Academy of Sciences USA* **94**, 15–22.

Normanly, J. and Abelson, J. (1989). tRNA identity. *Annual Review of Biochemistry* **58**, 1029–1049.

Orphanides, G., Lagrange, T. and Reinberg, D. (1996). The general transcription factors of RNA polymerase II. *Genes and Development* **10**, 2657–2683.

Pestova, T.V., Kolupaeva, V.G., Lomakin, I.B., Pilipenko, E.V., Shatsky, I.N., Agol, V.I. and Hellen, C.U.T. (2001). Molecular mechanisms of translation initiation in eukaryotes. *Proceedings of the National Academy of Sciences USA* **98**, 7029–7036.

Pestova, T.V., Lomakin, I.B., Lee, J.H., Choi, S.K., Dever, T.E. and Hellen, C.U.T. (2000). The joining of ribosomal subunits in eukaryotes requires eIF5B. *Nature* **403**, 332–335.

Pieler, T. and Theunissen, O. (1993). TFIIIA: Nine fingers – three hands. *Trends in Biochemical Sciences* **18**, 226–230.

Proudfoot, N. (1996). Ending the message is not so simple. *Cell* **87**, 779–781.

Proudfoot, N. (2000). Connecting transcription to messenger RNA processing. *Trends in Biochemical Sciences* **25**, 290–293.

Pugh, B.F. (1996). Mechanisms of transcription complex assembly. *Current Opinion in Cell Biology* **8**, 303–311.

Reed, R. and Magni, K. (2001). A new view of mRNA export: separating the wheat from the chaff. *Nature Cell Biology* **3**, E201–E204.

Rigby, P.W.J. (1993). Three in one and one in three: it all depends on TBP. *Cell* **72**, 7–10.

Roeder, R.G. (1996). The role of general initiation factors in transcription by RNA polymerase II. *Trends in Biochemical Sciences* **21**, 327–334.

Sakonju, S., Bogenhagen, D.F. and Brown, D. D. (1980). A control region in the centre of the 5S RNA gene directs specific initiation of transcription. *Cell* **19**, 13–25.

Schimmel, P. (1987). Aminoacyl tRNA synthetases: general scheme of structure function relationships in the polypeptides and recognition of transfer RNAs. *Annual Review of Biochemistry* **56**, 125–158.

Schultz, P., Fribourg, S., Poterszman, A., Mallouh, V., Moras, D. and Egly, J.M. (2000). Molecular structure of human TFIIH. *Cell* **102**, 599–607.

Sentenac, A. (1985). Eukaryotic RNA polymerases. *CRC Critical Reviews in Biochemistry* **1**, 31–90.

Sharp, P. (1992). TATA-binding protein is a classless factor. *Cell* **68**, 819–821.

Sharp, P.A. (1994). Split genes and RNA splicing. *Cell* **77**, 805–815.

Staley, J.P. and Guthrie, C. (1998). Mechanical devices of the spliceosome: motors, clocks, springs and things. *Cell* **92**, 315–326.

Tansey, W.P. and Herr, W. (1997). TAFs: guilt by association? *Cell* **88**, 729–732.

Tilghman, S.M., Curtis, P.J., Tiemeier, D.C., Leder, P. and Weissman, C. (1978). The intervening sequence of a mouse beta globin gene is transcribed within the 15S beta globin mRNA precursor. *Proceedings of the National Academy of Sciences USA* **75**, 1309–1313.

Tollervey, D. and Caceres, J. F. (2000). RNA processing marches on. *Cell* **103**, 703–709.

Tuite, M.F. (1996). Death by decapitation for mRNA. *Nature* **382**, 577–579.

Weis, L. and Reinberg, D. (1992). Transcription by RNA polymerase II initiator directed formation of transcription-competent complexes. *FASEB Journal* **6**, 3300–3309.

White, R.J. (1994). *RNA Polymerase III Transcription*. R.G. Landes Co., Austin, TX. p 147.

White, R.J. and Jackson, S.P. (1992). The TATA binding protein: a central role in transcription by RNA polymerases I and III. *Trends in Genetics* **8**, 284–288.

Wilson, K.S. and Noller, H.F. (1998). Molecular movement inside the translational engine. *Cell* **92**, 337–349.

Zawel, L. and Reinberg, D. (1995). Common themes in assembly and function of eukaryotic transcription complexes. *Annual Review of Biochemistry* **64**, 533–561.

Zhou, Z., Luo, M.-J., Straesser, K., Katahira, J., Hurt, E. and Reed, R. (2000). The protein Aly links pre-messenger-RNA splicing to nuclear export in metazoans. *Nature* **407**, 401–405.

4 REGULATION AT TRANSCRIPTION

4.1 INTRODUCTION

The existence of the many different potential regulatory stages discussed in the previous chapter has led to many studies to determine which of these are used in any particular situation. In general, such studies have shown that, in higher eukaryotes, as in bacteria, the primary control of gene expression is at the level of transcription, and the evidence showing that this is the case is discussed in this chapter. A number of cases of post-transcriptional regulation do exist, however, and these are discussed in Chapter 5.

4.2 EVIDENCE FOR TRANSCRIPTIONAL REGULATION

The evidence for the regulation of gene transcription comes from several types of study, which will be considered in turn.

4.2.1 Evidence from studies of nuclear RNA

If regulation of gene expression takes place at the level of transcription, the differences in the cytoplasmic levels of particular mRNAs which occur between different tissues should be paralleled by similar differences in the levels of these RNAs within the nuclei of different tissues. In contrast, regulatory processes in which a gene was transcribed in all tissues and the resulting transcript either spliced or transported to the cytoplasm in a minority of tissues would result in cases where differences in mRNA content occurred without any corresponding difference in the nuclear RNA (Fig. 4.1). Hence a study of the level of particular RNA species in the nuclear RNA of individual tissues or cell types serves as an initial test to distinguish transcriptional and post-transcriptional regulation.

The earliest studies in this area focused on the highly abundant RNA species produced during terminal differentiation, which could be readily studied simply because of their abundance. Thus, even before the discovery of intervening sequences revealed the possibility of regulation at the level of RNA splicing, Gilmour *et al.* (1974) studied the processes regulating the accumulation of globin which occur when Friend erythroleukaemia cells are treated with an inducer of globin production, dimethyl sulfoxide. In this system, the accumulation of cytoplasmic globin RNA, which produces the increase in synthesis of globin protein, was paralleled by increasing accumulation of globin RNA within the nucleus, suggesting that the primary effect of the inducer was at the level of globin gene transcription. Similarly, in the experiments of Groudine *et al.* (1974) globin-specific RNA was readily detectable in the nuclei of globin-synthesising erythroblasts but not in the nuclear RNA of fibroblasts or muscle cells, hence indicating the involvement of transcriptional control in regulating the abundant production of globin RNA only in the red blood cell lineage and not in other cell types.

Figure 4.1

Consequences of transcriptional or post-transcriptional regulation on the level of a specific nuclear RNA in a tissue that expresses the corresponding cytoplasmic mRNA (tissue 1) and one that does not (tissue 2).

Subsequently, the use of Northern blotting (see Section 1.3.1) allowed the separation of different species within nuclear RNA by size, and their visualisation by hybridisation to an appropriate probe. In the case of genes with many intervening sequences, such as that encoding the egg protein ovalbumin, many different RNA species could be observed in the nucleus, including not only the primary transcript and the fully spliced RNA prior to transport to the cytoplasm but also a series of intermediate-sized RNAs from which some of the intervening sequences had still to be removed (Roop *et al.*, 1978; Fig. 4.2).

The identification of such potential precursors of the mature mRNA allowed a study of their expression in tissues either producing or not producing ovalbumin mRNA and protein. Thus cytoplasmic ovalbumin mRNA and protein are present only in the oviduct following stimulation with oestrogen, and disappear when oestrogen is withdrawn. Similarly, the mRNA and protein are absent in other tissues, such as the liver, and cannot be induced by treatment with the hormone in these tissues. Studies of the distribution of both the fully spliced nuclear RNA and the larger precursors (Roop *et al.*, 1978) showed that these species could only be detected in the nuclear RNA of the oviduct following oestrogen stimulation and were absent in the liver nuclear RNA, or in oviduct nuclear RNA following oestrogen withdrawal (Fig. 4.3). Hence the distribution of these precursors in the nucleus exactly parallels that of the cytoplasmic mRNA, a finding entirely consistent with the transcriptional induction of the ovalbumin gene in the oviduct in response to oestrogen. These observations are difficult to reconcile, however, with a model in which the hormone acts to relieve a block in RNA splicing or transport, which exists in untreated oviduct and other tissues, since such models would predict an accumulation of unspliced or untransported RNA for ovalbumin within the nucleus.

These early studies have now been abundantly supplemented by many others measuring the nuclear RNA levels of other specific genes whose expression changes in particular situations, such as those encoding a number of highly abundant mRNAs present only in the soya bean embryo (Goldberg *et al.*, 1981) or that encoding the developmentally regulated mammalian liver protein, α-foetoprotein (Latchman *et al.*, 1984). In general, such studies have led to the conclusion that in most cases alterations in specific

Figure 4.2

Northern blot showing that unspliced and partially spliced precursors (a–g) to the ovalbumin mRNA (mRNA$_{OV}$) are detectable in the nuclear RNA (nRNA) of oestrogen-stimulated oviduct tissue.

mRNA levels in the cytoplasm are accompanied by parallel changes in the levels of the corresponding nuclear RNA species.

Interestingly, such studies carried out with cloned DNA probes for individual RNA species of relatively high abundance are in contrast to reports using $R_o t$ curve analysis (see Section 1.3.2) to study variations in the total nuclear RNA population in different tissues. Such experiments, which examine mainly low-abundance RNAs, have suggested that in some organisms, such as sea urchins (Wold *et al.*, 1978) and tobacco plants (Kamaly and Goldberg 1980), the nuclear RNA is a highly complex mixture of different RNAs, some of which can be retained in the nucleus in one tissue and transported to the cytoplasm in other tissues. These studies, which are indicative of post-transcriptional control in these organisms, are discussed in Section 5.1. It is noteworthy, however, that in mammals a different situation exists. In these organisms the increased number of different sequences in nuclear, compared to cytoplasmic, RNA can be accounted for by the presence of intervening sequences, which are transcribed but not transported to the cytoplasm, without the need to postulate the existence of whole transcripts which are confined to the nucleus in specific tissues.

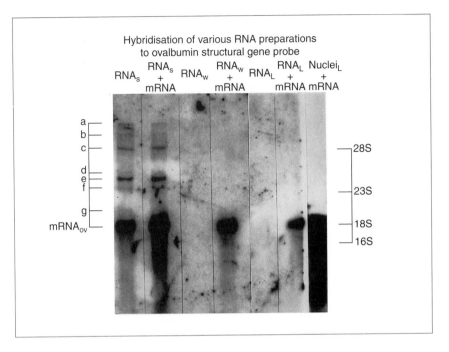

Hybridisation of various RNA preparations
to ovalbumin structural gene probe

Figure 4.3

Northern blot showing that the nuclear precursors to ovalbumin mRNA (seen in Fig. 4.2) are detectable in the nuclear RNA of oestrogen-stimulated oviduct (RNA$_s$) but are absent in the nuclear RNA of oestrogen-withdrawn oviduct (RNA$_w$) and of liver (RNA$_L$). Moreover, ovalbumin mRNA mixed with withdrawn oviduct (RNA$_w$ + mRNA) or liver (RNA$_L$ + mRNA) nuclear RNA is not degraded, showing that the absence of RNA for ovalbumin in these nuclear RNAs is not due to a nuclease specifically degrading the ovalbumin RNA.

Hence it appears from these studies that, at least in mammals and other higher eukaryotes, the regulation of transcription, resulting in parallel changes in nuclear and cytoplasmic RNA levels, is the primary means of regulating gene expression. However, the studies described so far suffer from the defect that they measure only steady-state levels of specific nuclear RNAs. It could be argued that the gene is being transcribed in the non-expressing tissue but that the transcript is degraded within the nucleus at such a rate that it cannot be detected in assays of steady-state RNA levels (Fig. 4.4). This degradation in the non-expressing tissue, but not in the expressing tissue, could be produced directly by specific regulation of the rate of turnover of a particular RNA in the different tissues, producing rapid degradation only in the non-expressing tissue. Alternatively, it might result from a block to splicing or transport of the RNA in the non-expressing tissue, followed by the rapid degradation of this unspliced or untransported RNA, which in the expressing tissue would be processed or transported before it could be degraded. These considerations necessitate the direct measurement of gene transcription itself in the different tissues in order to establish unequivocally the existence of transcriptional regulation. Methods to do this have been devised and these will now be discussed.

4.2.2 Evidence from pulse-labelling studies

The synthesis of RNA from DNA by the enzyme RNA polymerase involves the incorporation of ribonucleotides into an RNA chain. Therefore the synthesis of any particular RNA can be measured by adding a radioactive

Figure 4.4

Model in which lack of expression of genes in particular tissues is caused by rapid degradation of their RNA transcripts.

ribonucleotide (usually uridine labelled with tritium) to the cells and measuring how much radioactivity is incorporated into RNA specific to the gene of interest. Clearly, if the degradation mechanisms discussed in the last section do exist, they will, given time, degrade the radioactive RNA molecule produced in this way and drastically reduce the amount of labelled RNA detected. In order to prevent this, the rate of transcription is measured by exposing cells briefly to the labelled uridine in a process referred to as pulse labelling. The labelled uridine is incorporated into nascent RNA chains that are being made at this time and, even before a complete transcript has had time to form, the cells are lysed and total RNA is isolated from them. This RNA, which contains labelled partial transcripts from genes active in the tissue used, is then hybridised to a piece of DNA derived from the gene of interest, the number of radioactive counts that bind providing a measure of the incorporation of labelled precursor into its corresponding RNA (Fig. 4.5).

This method provides the most direct means of measuring transcription and has been used, for example, to show that the induction of globin production which occurs in Friend erythroleukaemia cells in response to treatment with dimethyl sulfoxide is mediated by increased transcription of

Figure 4.5

Pulse-labelling assay to assess the transcription rate of a specific gene (gene A) by measuring the amount of radioactivity (stars) incorporated into nascent transcripts.

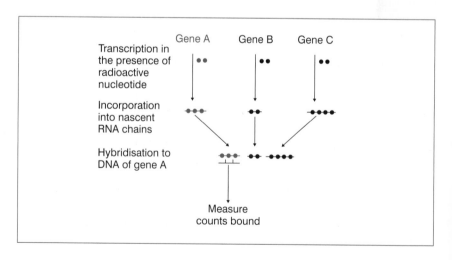

the globin gene (Lowenhaupt *et al.*, 1978). This increased transcription produces the increased levels of globin-specific RNA in the nucleus and cytoplasm of the treated cells, which was discussed earlier in this chapter (see Section 4.2.1).

Although pulse labelling provides a very direct measure of transcription rates, the requirement to use very short labelling times to minimise any effects of RNA degradation limits its applicability. Thus in the experiments of Lowenhaupt *et al.* (1978) it was possible to measure the amount of radioactivity incorporated into globin RNA in the very brief labelling times used (5 or 10 min) only because of the enormous abundance of globin RNA and the very high rate of transcription of the globin gene. With other RNA species, the rates of transcription are insufficient to provide measurable incorporation of label in the short pulse time. More label will, of course, be incorporated if longer pulse times are used, but such pulse times allow the possibility of RNA turnover and are therefore subject to the same objections as the measurement of stable RNA levels.

Hence, although pulse labelling can be used to establish unequivocally that transcriptional control is responsible for the massive synthesis of the highly abundant RNA species present in terminally differentiated cells, it cannot be used to demonstrate the generality of transcriptional control processes and, in particular, their applicability to RNAs which, although regulated in different tissues, never become highly abundant.

This limitation of the pulse-labelling method is especially relevant in view of the existence of the Davidson and Britten model for gene regulation (Davidson and Britten, 1979) which specifically postulates that highly abundant RNA species will be regulated in a different manner to the bulk of the RNA species, which are of moderate or low abundance. Thus, in this model (Fig. 4.6) all genes are postulated to be transcribed in all tissues at a low basal rate and regulation takes place at a post-transcriptional level by deciding which transcripts are processed to functional mRNA and which are degraded. In the case of most genes the level of RNA and protein required in any particular tissue would be met by processing correctly all of the primary transcript produced by this low basal rate of transcription. The level of the abundant RNA species would be such, however, that they could not be produced with this low rate of gene transcription, even by correctly processing all of the primary transcript. Hence, for the genes encoding these RNA species, a special mechanism would operate and their transcription would be dramatically increased in some tissues, as observed by pulse labelling.

This theory postulates that the regulation of gene expression by changes in transcription is confined to the few genes encoding highly abundant RNA species, and that post-transcriptional control processes will regulate the expression of less abundant RNAs whose genes will be transcribed even in tissues where no mRNA is synthesised. In order to test this theory it is necessary to use a method of measuring transcription which, although less direct than pulse labelling, is more sensitive and hence can be applied to a wider variety of cases, including non-highly abundant mRNAs. This method is discussed in the next section.

Figure 4.6

Davidson and Britten model of post-transcriptional regulation, in which all genes are transcribed at a low basal rate and regulation is achieved by controlling whether the resulting transcript is processed to mature mRNA. Transcriptional control is confined to genes (such as gene D) where the level of RNA required cannot be achieved by processing all of the primary transcript produced by the low basal rate of transcription.

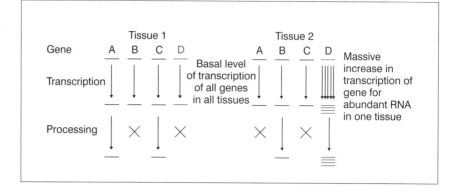

4.2.3 Evidence from nuclear run-on assays

The primary limitation on the sensitivity of pulse labelling is the existence within the cell of a large pool of non-radioactive ribonucleotides, which are normally used by the cell to synthesise RNA. When labelled ribonucleotide is added to the cell, it is considerably diluted in this pool of unlabelled precursor. The amount of label incorporated into RNA in the labelling period is therefore very small, since the majority of ribonucleotides incorporated are unlabelled. The sensitivity of this method is thus severely reduced, resulting in its observed applicability only to genes with very high rates of transcription. Interestingly, however, although transcription takes place in the nucleus, most of the pool of precursor ribonucleotides is present in the cytoplasm. Hence it is possible, by removing the cytoplasm and isolating nuclei, to remove much of the pool of unlabelled ribonucleotide. When this occurs the RNA polymerase ceases transcribing due to lack of ribonucleotide but remains associated with the DNA (Fig. 4.7). Labelled ribonucleotides can then be added directly to the isolated nuclei in a test tube and the RNA polymerase resumes transcribing and runs on to the end of the gene incorporating labelled ribonucleotide into the RNA transcript. Because the labelled ribonucleotides are not diluted in the unlabelled cytoplasmic pool, considerably more label is incorporated into any particular RNA transcript than is observed in pulse-labelling experiments. The label incorporated into any particular transcript is detected by hybridisation to its corresponding DNA, exactly as in pulse-labelling experiments.

This method, which is known as a nuclear run-on assay, is therefore much more widely applicable than pulse labelling, and can be used to quantify the transcription of genes that are never transcribed at levels detectable by pulse labelling. Moreover, very many studies (for a review see Darnell, 1982 and references therein) have now established that the RNA synthesised by isolated nuclei in the test tube is similar to that made by intact whole cells, and that the method is therefore not only sensitive but also provides an accurate measure of transcription, free from artefact.

In initial studies, nuclear run-on assays were used to measure the transcription of highly abundant RNA species. Thus, for example, nuclei

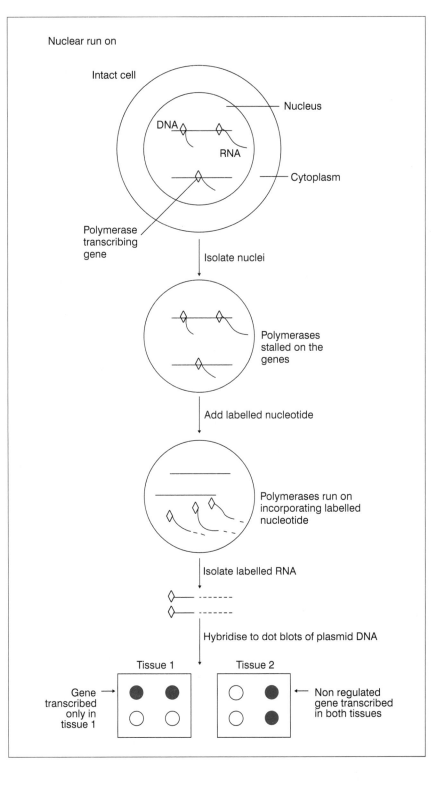

Nuclear run on

Intact cell

Nucleus

DNA

RNA

Cytoplasm

Polymerase
transcribing
gene

Isolate nuclei

Polymerases
stalled on the
genes

Add labelled nucleotide

Polymerases run on
incorporating labelled
nucleotide

Isolate labelled RNA

Hybridise to dot blots of plasmid DNA

Tissue 1

Tissue 2

Gene
transcribed
only in
tissue 1

Non regulated
gene transcribed
in both tissues

Figure 4.7

Nuclear run-on assay. Following
isolation of nuclei, RNA
polymerase ceases transcribing
due to lack of ribonucleotides
but remains associated with the
DNA. When radioactively
labelled ribonucleotide is added,
the polymerase resumes
transcription and runs on to the
end of the gene incorporating
the labelled ribonucleotide into
the RNA transcript. The labelled
RNA made in this way in
different tissues can then be used
to probe filters containing DNA
from genes whose transcription
is to be measured in the different
tissues.

isolated from the erythrocytes of adult chickens (which, unlike the equivalent cells in mammals, do not lose their nuclei) were shown to transcribe the gene encoding the adult β-globin protein, whereas nuclei prepared from similar cells isolated from embryonic chickens failed to transcribe this gene and instead transcribed the gene encoding the form of β-globin made in the embryo (Groudine *et al.*, 1981). Such a finding parallels the observation discussed earlier that the RNA specific for the adult form of β-globin is present only in adult and not in embryronic erythroid cell nuclei (see Section 4.2.1) and indicates that the developmentally regulated production of different forms of globin protein is under transcriptional control.

A similar parallel between the results of nuclear RNA studies and nuclear run-on assays of transcription is also found in the case of the ovalbumin gene. Thus the presence of ovalbumin-specific RNA only in the nuclear RNA of hormonally stimulated oviduct tissue is paralleled by the ability of nuclei prepared from stimulated oviduct cell nuclei to transcribe the ovalbumin gene at high levels in run-on assays. In contrast, nuclei from other tissues or unstimulated oviduct cells failed to transcribe this gene, paralleling the observed absence of ovalbumin RNA in the nuclei and cytoplasm of these tissues (Swaneck *et al.*, 1979). Hence the observed increase in nuclear and cytoplasmic RNA for ovalbumin in response to oestrogen is indeed caused by increased transcription of the ovalbumin gene. Interestingly, in this study the incorporation of label into ovalbumin RNA in the run-on assay was observed to peak after 15 min and did not decrease at longer labelling times of up to 1 hour. This suggested that, unlike intact cells, isolated nuclei do not degrade or process the RNA that they synthesise, and hence allowed the use of longer labelling times, further increasing the sensitivity of this technique.

This increased sensitivity has allowed the use of nuclear run-on assays to demonstrate that transcriptional control of many genes encoding specific proteins is responsible for the previously observed differences in the levels of these proteins in different situations. Cases where such transcriptional control has been demonstrated in this manner are far too numerous to mention individually but involve a range of different tissues and organisms, such as the synthesis of α-foetoprotein in foetal, but not adult, mammalian liver, production of insulin in the mammalian pancreas, the expression of the *Drosophila melanogaster* yolk protein genes only in ovarian follicle cells, the expression of aggregation stage-specific genes in the slime mould *Dictyostelium discoideum* and the expression of soya bean seed proteins such as glycinin in embryonic, but not adult, tissues.

These studies on individual genes for particular proteins have been supplemented by the more general studies of Darnell and colleagues (Derman *et al.*, 1981; Powell *et al.*, 1984). In these experiments the authors studied 12 different genes whose corresponding mRNAs were present in mouse liver cytoplasm but were absent in brain cytoplasm. These included both genes encoding previously isolated liver-specific proteins, such as albumin or transferrin, and those which had been isolated simply on the basis of the presence of their corresponding cytoplasmic RNA in the liver and not in other tissues and for which the protein product had not been identified. Measurement of the transcription rate of these genes in nuclei isolated from brain or liver showed that such transcription was detectable only in the liver nuclei (Fig. 4.8), indicating that the difference in cytoplasmic mRNA level was produced by a corresponding difference in gene transcription. In these experiments the rate of transcription of the 12 genes was also measured in nuclei prepared from kidney tissue. In this tissue mRNAs corresponding to two of the genes were present at a considerably lower level than that observed in the liver, while RNA corresponding to the other 10 genes was undetectable. As with the brain nuclei, the level of transcription detectable in the kidney nuclei exactly paralleled the level of RNA present. Thus, only the two genes producing cytoplasmic mRNA in the kidney were detectably transcribed in this tissue and the level of transcription of these genes was much lower than that seen in the liver nuclei.

These studies on a relatively large number of different liver RNAs of different abundances, when taken together with the studies of many genes encoding specific proteins, indicate that, for genes expressed in one or a few cell types, increased levels of RNA and protein in a particular cell type are brought about primarily by increases in gene transcription and that, at least in mammals, post-transcriptional control mechanisms, such as that postulated by Davidson and Britten (1979), are not the primary means of regulating gene expression, although they may be more important in other animals, such as the sea urchin (see Section 5.1).

4.2.4 Evidence from polytene chromosomes

Although pulse-labelling and nuclear run-on studies of many genes have conclusively established the existence of transcriptional regulation, it is necessary to discuss another means of demonstrating such regulation, in

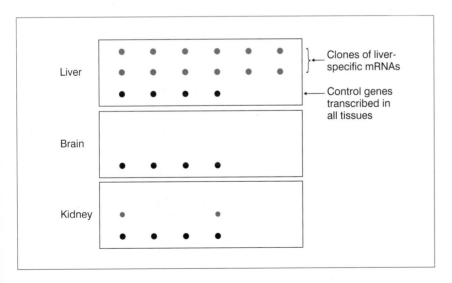

Liver

}← Clones of liver-specific mRNAs

← Control genes transcribed in all tissues

Brain

Kidney

Figure 4.8

Nuclear run-on assay to measure the transcription rates of genes encoding liver-specific mRNAs and of control genes expressed in all tissues. Note that the liver-specific genes are not transcribed at all in brain, while in the kidney the only two of these genes to be transcribed are those that are known to produce a low level of mRNA in the kidney.

which increased transcription can be directly visualised. As described in Section 2.3.2, the chromosomal DNA in the salivary glands of *Drosophila* is amplified many times, resulting in a giant polytene chromosome. Such chromosomes exhibit along their length areas known as puffs in which the DNA has decondensed into a more open state, resulting in the expansion of the chromosome (Fig. 4.9). If cells are allowed to incorporate labelled ribonucleotides into RNA and the resulting RNA is then hybridised back to the polytene chromosomes, it localises primarily to the positions of the puffs. Hence these puffs represent sites of intense transcriptional activity which, because of the large size of the polytene chromosomes, can be directly visualised.

Most interestingly, many procedures which, in *Drosophila*, result in the production of new proteins, such as exposure to elevated temperature (heat shock) or treatment with the steroid hormone ecdysone, also result in the production of new puffs at specific sites on the polytene chromosomes, each treatment producing a different specific pattern of puffs. This suggests that these sites contain the genes encoding the proteins whose synthesis is increased by the treatment, and that this increased synthesis is mediated via increased transcription of these genes, which can be visualised in the puffs.

Figure 4.9

Transcriptionally active puff (arrowed) in a polytene chromosome of *Drosophila melanogaster*. The puff appears in response to treatment with the steroid hormone ecdysone (E) and is not present prior to hormone treatment (O).

In the case of ecdysone treatment this has been confirmed directly by showing that the radioactive RNA synthesised immediately after ecdysone treatment hybridises strongly to the ecdysone-induced puffs but not to a puff which regresses upon hormone treatment. In contrast, RNA prepared from cells prior to ecdysone treatment hybridises only to the hormone-repressed puff and not to the hormone-induced puffs (Fig. 4.10; Bonner and Pardue, 1977). Similarly, RNA labelled after heat shock hybridises intensely to puff 87C, which appears following exposure to elevated temperature and is now known to contain the gene encoding the 70 kDa heat-shock protein (hsp70) which is the major protein made in *Drosophila* following heat shock (Spradling *et al.*, 1975).

Thus the large size of polytene chromosomes allows a direct visualisation of the transcriptional process and indicates that, as in other situations, gene activity in the salivary gland is regulated at the level of transcription.

4.3 REGULATION AT TRANSCRIPTIONAL ELONGATION

4.3.1 Initiation of transcription

In the vast majority of cases where increased transcription of a particular gene has been demonstrated, it is likely that such increased transcription is mediated by an increased rate of initiation of transcription by RNA polymerase which occurs at the region of the gene known as the promoter (see Section 3.2). Hence in a tissue in which a gene is being transcribed actively, a large number of polymerase molecules will be moving along the gene at any particular time, resulting in the production of a large number of transcripts. Such a series of nascent transcripts being produced from a single transcription unit can be visualised in the lampbrush chromosomes of amphibian oocytes, the nascent transcripts associated with each RNA polymerase molecule increasing in length the further the polymerase has proceeded along the gene, resulting in the characteristic nested appearance (Fig. 4.11).

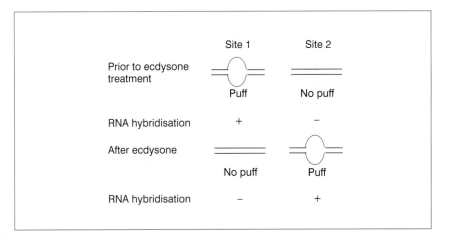

Figure 4.10

The newly synthesised RNA made following ecdysone stimulation can be labelled with ³H uridine and shown to hybridise to the puffs that form following ecdysone treatment. Conversely, a puff that regresses after hormone treatment binds only the labelled RNA synthesised before addition of the hormone.

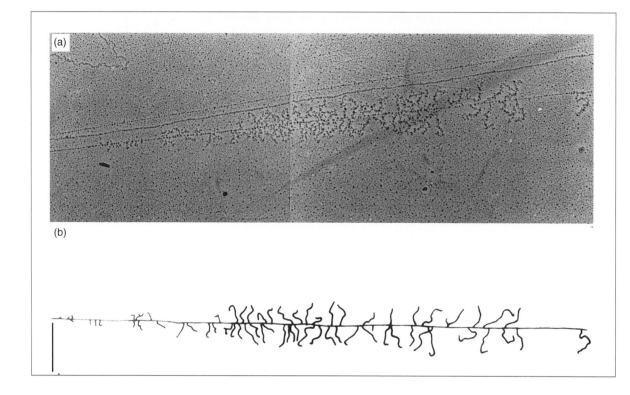

(a)

(b)

Figure 4.11

Electron micrograph (A) and summary diagram (B) of a lampbrush chromosome in amphibian oocytes, showing the characteristic nested appearance produced by the nascent mRNA chains attached to transcribing RNA polymerase molecules. The bar indicates 1 μm.

By contrast, in tissues where a gene is transcribed at very low levels, initiation of transcription will be a rare event and only one or a very few polymerase molecules will be transcribing a gene at any particular time. Similarly, the absence of transcription of a particular gene in some tissues will result from a failure of RNA polymerase to initiate transcription in that tissue (Fig. 4.12). A large number of sequences upstream of the point at which initiation occurs and which are involved in its regulation have now been described, and these will be discussed in Chapter 7.

4.3.2 Transcriptional elongation

Although the majority of cases of transcriptional regulation are likely to occur at the level of initiation, some cases have been described where regulation occurs following the initiation of transcription by RNA polymerase and the production of a truncated RNA of less than full length. In such cases, transcriptional control may operate by releasing a block to elongation of the nascent transcript (for reviews see Uptain *et al.*, 1997; Conaway and Conaway, 1999; Conaway *et al.*, 2000).

This form of regulation is responsible, for example, for the tenfold decline in mRNA levels encoding the cellular oncogene c-*myc* (for discussion of cellular oncogenes see Chapter 9) which occurs when the human pro-myeloid cell line HL-60 is induced to differentiate into a granulocyte type cell. If nuclear run-on assays are carried out using nuclei from undiffer-

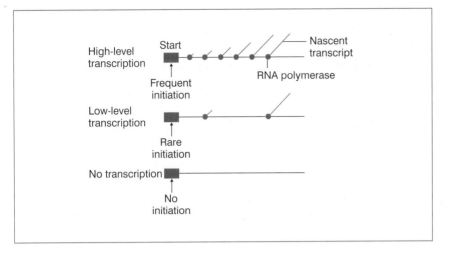

Figure 4.12

Regulation of transcriptional initiation results in differences in the number of RNA polymerase molecules transcribing a gene and therefore in the number of transcripts produced.

entiated or differentiated HL-60 cells, the results obtained vary, depending on the region of the c-*myc* gene whose transcription is being measured (Bentley and Groudine, 1986). Thus the c-*myc* gene consists of three exons, which appear in the messenger RNA and which are separated by intervening sequences that are removed from the primary transcript by RNA splicing. If the labelled products of the nuclear run-on procedure are hybridised to the DNA of the second exon, the levels of transcription observed are approximately tenfold higher in nuclei derived from undifferentiated cells than in nuclei from differentiated cells. Hence the observed differences in c-*myc* RNA levels in these cell types are indeed produced by differences in transcription rates. However, if the same labelled products are hybridised to DNA from the first exon of the c-*myc* gene, virtually no difference in the level of transcription of this region in the differentiated compared to the undifferentiated cells is observed.

Comparison of the rates of transcription of the first and second exons in undifferentiated and differentiated cells indicates that regulation takes place at the level of transcriptional elongation, rather than initiation. Thus, although similar numbers of polymerase molecules initiate transcription of the c-*myc* gene in both cell types, the majority terminate in differentiated cells near the end of exon 1, do not transcribe the remainder of the gene, and hence do not produce a functional RNA. In contrast, in undifferentiated cells most polymerase molecules that initiate transcription, transcribe the whole gene and produce a functional RNA. Hence the fall in c-*myc* RNA in differentiated cells is regulated by means of a block to elongation of nascent transcripts (Fig. 4.13). A 180 bp sequence from the 3′ end of the first exon of c-*myc* has been shown to mediate this effect and to block transcriptional elongation if placed within the transcribed region of another gene (Wright and Bishop, 1989). Interestingly, the rapid inhibition of transcriptional elongation following differentiation is supplemented, several days after differentiation, by an inhibition of c-*myc* transcription at the level of initiation.

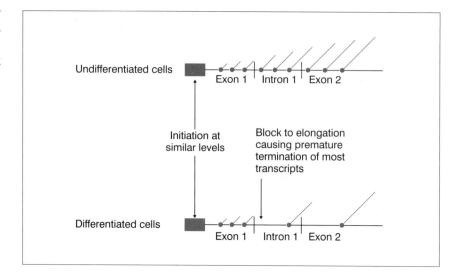

Similar effects on transcriptional elongation have been seen in several other cellular oncogenes, such as c-*myb*, c-*fos* and c-*mos*, indicating that this mechanism is not confined to a single oncogene and may be quite widespread.

Interestingly, a protein affecting the rate of transcriptional elongation has been identified in the case of the human immunodeficiency virus (HIV-1) (for reviews see Cullen, 1991; Greenblatt *et al.*, 1993). Thus early in infection with this virus, a low level of transcription occurs from the HIV-1 promoter and many of the transcripts terminate very close to the promoter producing very short RNAs. Subsequently the viral Tat protein binds to these RNAs at a region known as Tar, which is located at +19 to +42 relative to the start site of transcription at nucleotide +1. This binding produces two effects. Firstly, there is a large increase in transcriptional initiation, leading to many more RNA transcripts being initiated. In addition, however, Tat also overcomes the block to elongation, leading to the production of a greater proportion of long transcripts which can encode the viral proteins.

Hence Tat produces a large increase in the production of HIV transcripts able to encode viral proteins both by stimulating transcriptional initiation and by overcoming a block to transcriptional elongation close to the promoter (Fig. 4.14). As well as the viral Tat protein, cellular proteins which can stimulate transcriptional elongation have also been identified. Indeed such an effect is involved in the increased expression of the gene encoding the heat inducible hsp70 protein that occurs following exposure of cells to elevated temperature and which results from the binding of the heat shock transcription factor (HSF) to the promoter of this gene (see Sections 4.2.4 7.2.2) (for a review see Lis and Wu, 1993) and other transcriptional activators have also been shown to act by stimulating both transcriptional initiation and elongation (Yankulov *et al.*, 1994).

A key target for factors which regulate transcriptional elongation is the phosphorylation of the C-terminal domain (CTD) of RNA polymerase II. Thus, as described in Section 3.2.3, such phosphorylation of the CTD is a key event required for the RNA polymerase II transcription complex to move off down the gene allowing transcriptional elongation to occur. Interestingly, the HIV Tat protein has been shown to be able to interact with a kinase protein that can phosphorylate the CTD. Hence, Tat attracts this kinase to the HIV promoters, ensuring phosphorylation of RNA polymerase II and producing transcriptional elongation (for a review see Conaway *et al.*, 2000).

A similar involvement of CTD phosphorylation is seen in the case of a cellular regulator of transcriptional elongation, the zebrafish protein, Foggy (Guo *et al.*, 2000). Thus, the Foggy protein interacts with the non-phosphorylated form of RNA polymerase II to block transcriptional elongation. However, when RNA polymerase II is phosphorylated it is no longer inhibited by Foggy, allowing transcriptional elongation to proceed (Fig. 4.15). When this activity of Foggy is inactivated by mutation, the resulting zebrafish fail to produce the correct number of dopamine synthesising neurones during development. This indicates that the regulation of transcriptional elongation by Foggy is likely to be critical for the correct expression of genes involved in the production of this cell type and, more generally, demonstrates that regulation at the level of transcriptional elongation can play a key role in controlling gene expression during development.

Thus in different situations, enhanced transcription can arise from increased initiation, or increased elongation. In general, however, it is likely

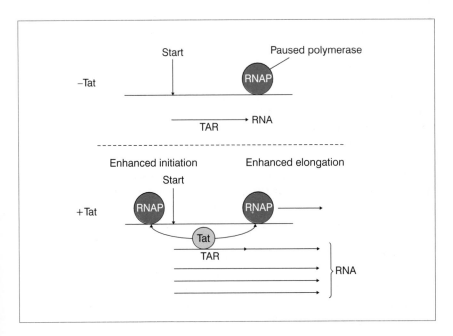

Figure 4.14

The HIV Tat protein acts both by enhancing the rate of transcriptional initiation by RNA polymerase (RNA P) and by enhancing the rate of elongation by overcoming polymerase pausing close to the promoter. Both these effects are achieved by the Tat protein binding to a specific region (TAR) of the nascent RNA.

Figure 4.15

(a) The Foggy protein acts by interacting with the RNA polymerase II transcription complex to prevent transcriptional elongation (E) following transcriptional initiation (I). (b) Phosphorylation of RNA polymerase II prevents this inhibitory effect of Foggy, allowing transcriptional elongation to proceed.

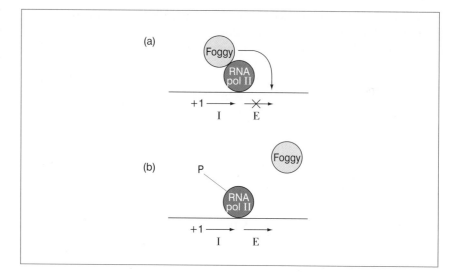

that in most cases transcription is regulated at the level of initiation, with specific regulatory processes acting to enhance the rate at which RNA polymerases initiate transcription of the DNA into RNA. As with DNA amplification or rearrangement (see Chapter 2) control at polymerase elongation is likely to represent a response to the requirements of a particular situation. Thus these processes may, for example, allow a rapid response to a particular situation since the RNA polymerase and associated proteins are already poised on the gene rather than having to bind and initiate transcription *de novo*.

4.4 CONCLUSIONS

The experiments described in this chapter provide conclusive evidence that control of transcriptional initiation is the primary means used to regulate gene expression in eukaryotic organisms. The mechanism by which such transcriptional control is achieved will be discussed in Chapters 6, 7 and 8. Some cases of post-transcriptional control have been described, however, and these will be discussed in the next chapter.

REFERENCES

Bentley, D.L. and Groudine, M. (1986). A block to elongation is largely responsible for decreased transcription of c-*myc* in differentiated HL 60 cells. *Nature* **321**, 702–706.

Bonner, J. J. and Pardue, M.L. (1977). Ecdysone-stimulated RNA synthesis in salivary glands of *Drosophila melanogaster* assay by *in situ* hybridisation. *Cell* **12**, 219–225.

Conaway, J.W. and Conaway, R.C. (1999) Transcription elongation and human disease. *Annual Reviews of Biochemistry* **68**, 301–319.

Conaway, J.W., Shilatifard, A., Dvir, A. and Conaway, R.C. (2000) Control of elongation by RNA polymerase II. *Trends in Biochemical Sciences* **25**, 375–380.

Cullen, B.R. (1991). Regulation of HIV-1 gene expression. *FASEB Journal* 5, 2361–2368.

Darnell, J.E. (1982). Variety in the level of gene control in eukaryotic cells. *Nature* 297, 365–371.

Davidson, E.H. and Britten, R.J. (1979). Regulation of gene expression: possible role of repetitive sequences. *Science* 204, 1052–1059.

Derman, E., Krauter, K., Walling, L., Weinberger, C., Ray, M. and Darnell, J.E. (1981). Transcriptional control in the production of liver specific mRNAs. *Cell* 23, 731–739.

Gilmour, R.S., Harrison, P.R., Windass, J.D., Affara, N.A. and Paul, J. (1974). Globin messenger RNA synthesis and processing during haemoglobin induction in Friend cells. 1. Evidence for transcriptional control in clone M2. *Cell Differentiation* 3, 9–22.

Goldberg, R.B., Hosheck, G., Ditta, G.S. and Breidenbach, R.W. (1981). Developmental regulation of cloned superabundant mRNAs in soybean. *Developmental Biology* 83, 218–231.

Greenblatt, J., Nodwell, J.R. and Mason, S.W. (1993). Transcriptional antitermination. *Nature* 364, 401–406.

Groudine, M., Hoitzer, H., Scherner, K. and Therwath, A. (1974). Lineage dependent transcription of globin genes. *Cell* 3, 243–247.

Groudine, M., Peretz, M. and Weintraub, H. (1981). Transcriptional regulation of haemoglobin switching in chicken embryos. *Molecular and Cellular Biology* 1, 281–288.

Guo, S., Yamaguchi, Y., Schillbach, S., Wada, T., Lee, J., Goddard, A., French, D., Handa, H. and Rosenthal, A. (2000) A regulator of transcriptional elongation controls vertebrate neuronal development. *Nature* 408, 366–369.

Kamaly, J.C. and Goldberg, R.B. (1980). Regulation of structural gene expression in tobacco. *Cell* 19, 935–946.

Latchman, D.S., Brzeski, H., Lovell-Badge, R.H. and Evans, M.J. (1984). Expression of the alpha-foetoprotein gene in pluripotent and committed cells. *Biochimica et Biophysica Acta* 783, 130–136.

Lis, J. and Wu, C. (1993). Protein traffic on the heat shock promoter: parking stalling and trucking along. *Cell* 74, 1–4.

Lowenhaupt, K., Trent, C. and Lingrel, J.B. (1978). Mechanisms for accumulation of globin mRNA during dimethyl sulfoxide induction of mouse erythroleukaemia cells: synthesis of precursors and mature mRNA. *Developmental Biology* 63, 441–454.

Powell, D.J., Freidman, J.M., Oulethe, A.J., Krauter, K.S. and Darnell, J.E. (1984). Transcriptional and post-transcriptional control of specific messenger RNAs in adult and embryonic liver. *Journal of Molecular Biology* 179, 21–35.

Roop, D.R., Nordstrom, J.L., Tsai, S.-Y., Tsai, M.-J. and O'Malley, B.W. (1978). Transcription of structural and intervening sequences in the ovalbumin gene and identification of potential ovalbumin mRNA precursors. *Cell* 15, 671–685.

Spradling, A., Penman, S. and Pardue, M.L. (1975). Analysis of *Drosophila* mRNA by *in situ* hybridization: sequences transcribed in normal and heat shocked cultured cells. *Cell* 4, 395–404.

Swaneck, G.E., Nordstrom, J.L., Kreuzaler, F., Tsai, M.-J. and O'Malley, B.W. (1979). Effect of oestrogen on gene expression in chicken oviduct, evidence for transcriptional control of the ovalbumin gene. *Proceedings of the National Academy of Sciences of the USA* 76, 1049–1053.

Uptain, S.M., Kane, C.M., and Chamberlin, M.J. (1997). Basic mechanisms of transcript elongation and its regulation. *Annual Reviews of Biochemistry* 66, 117–172.

Wold, B.J., Klein, W.H., Hough-Evans, B.R., Britten, R.J. and Davidson, E.H. (1978). Sea urchin embryo mRNA sequences expressed in the nuclear RNA of adult tissues. *Cell* **14**, 941–950.

Wright, S. and Bishop, J.M. (1989). DNA sequences that mediate attenuation of transcription from the mouse proto-oncogene c-*myc*. *Proceedings of the National Academy of Sciences of the USA* **86**, 505–509.

Yankulov, K., Blau, J., Purton, T., Roberts, S. and Bentley, D.L. (1994). Transcriptional elongation by RNA polymerase II is stimulated by transactivators. *Cell* **77**, 749–759.

POST-TRANSCRIPTIONAL REGULATION

5

5.1 REGULATION AFTER TRANSCRIPTION?

Although the evidence discussed in the preceding chapter indicates that, in mammals at least, the primary control of gene expression lies at the level of transcription, a number of cases exist where changes in the rate of synthesis of a particular protein occur without a change in the transcription rate of the corresponding gene. Indeed, in some lower organisms post-transcriptional regulation may constitute the predominant form of regulation of gene expression.

In the sea urchin, for example, the nuclear RNA contains many more different RNA species than are found in the cytoplasmic messenger RNA. Hence a large proportion of the genes transcribed give rise to RNA products that are not transported to the cytoplasm and do not function as a messenger RNA. Interestingly, however, this process is regulated differently in different tissues; an RNA species which is confined to the nucleus in one tissue being transported to the cytoplasm and functioning as a messenger RNA in another tissue. Thus, up to 80% of the cytoplasmic mRNAs found in the embryonic blastula are absent from the cytoplasmic RNA of adult tissues, such as the intestine, but are found in the nuclear RNA of such tissues (Wold *et al.*, 1978).

Although post-transcriptional regulation in mammals does not appear to be as generalised as in the sea urchin, some cases exist where changes in cytoplasmic mRNA levels occur without alterations in the rate of gene transcription. Such post-transcriptional regulation may be more important in controlling variations in the level of mRNA species expressed in all tissues than in the regulation of mRNA species that are expressed in only one or a few tissues. For example, in the experiments of Darnell and colleagues (Powell *et al.*, 1984), which demonstrated the importance of transcriptional control in the regulation of liver-specific mRNAs (see Section 4.2.3), tissue-specific differences in the levels of the mRNAs encoding actin and tubulin (which are expressed in all cell types) were observed in the absence of differences in transcription rates. Clearly, these and other cases where mRNA levels alter in the absence of changes in transcription rates indicate the existence of post-transcriptional control processes and require an understanding of their mechanisms.

In principle, such post-transcriptional regulation could operate at any of the many stages between gene transcription and the translation of the corresponding mRNA in the cytoplasm. Indeed, the available evidence indicates that in different cases regulation can occur at any one of these levels. Each of these will now be discussed in turn.

5.2 REGULATION OF RNA SPLICING

5.2.1 RNA splicing

The finding that the protein-coding regions of eukaryotic genes are split by intervening sequences (introns) which must be removed from the initial transcript by RNA splicing of the protein-coding exons (see Section 3.3.3) led to much speculation that this process might provide a major site of gene regulation. Thus, in theory, an RNA species transcribed in several tissues might be correctly spliced to yield functional RNA in one tissue and remain unspliced in another tissue. An unspliced RNA would either be degraded within the nucleus or, if transported to the cytoplasm, would be unable to produce a functional protein due to the interruption of the protein-coding regions (Fig. 5.1). Several cases of such processing versus discard decisions have now been described in *Drosophila* and a similar pathway in which unspliced RNAs are specifically degraded within the nucleus has recently been characterised in yeast (Bousquet-Antonelli *et al.*, 2000). Hence, in lower organisms, there appear to be regulatory pathways in which a particular transcript is spliced to produce a functional mRNA in one situation whilst remaining unspliced and then being degraded within the nucleus in another situation.

5.2.2 Alternative RNA splicing

Although such processing versus discard decisions are not widespread in mammals, numerous cases of alternative RNA splicing have been described, both in mammals and other organisms. In this process (for reviews see Black,

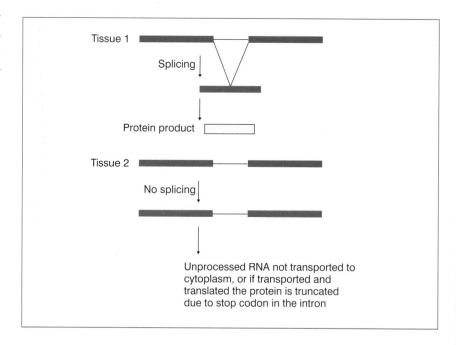

Figure 5.1

The absence of RNA splicing of a transcript in a particular tissue results in a lack of production of the corresponding protein.

2000; Graveley, 2001) a single gene is transcribed in several different tissues, the transcripts from this gene being processed differentially to yield different functional messenger RNAs in the different tissues (Fig. 5.2). In many cases these RNAs are translated to yield different protein products. It is noteworthy that this mechanism of gene regulation involves not only regulation of processing but also regulation of transcription, in that the alternatively processed RNAs are transcribed only in a restricted range of cell types and not in many other cells.

Cases of alternative RNA processing occur in the genes involved in a wide variety of different cellular processes, ranging from genes which regulate embryonic development or sex determination in *Drosophila* to those involved in muscular contraction or neuronal function in mammals. A representative selection of such cases is given in Table 5.1.

For convenience, cases of alternative RNA processing can be divided into three groups (Leff *et al.*, 1986): (a) situations where the 5′ end of the differentially processed transcripts is different, (b) situations where the 3′ end of the differentially processed transcripts is different, and (c) situations where both the 5′ and 3′ ends of the differentially processed transcripts are identical.

Situations where the 5′ end of the transcripts is different

In these cases, two alternative primary transcripts are produced by transcription from different promoter elements and these are then processed differentially. In several situations, such as the mouse α-amylase gene (Fig. 5.3), differential splicing is controlled simply by the presence or absence of a particular exon in the primary transcript. Thus in the salivary gland, where transcription takes place from an upstream promoter, the exon adjacent to this promoter is included in the processed RNA and a downstream exon is omitted. In the liver, where the transcripts are initiated 2.8 kb downstream and do not contain the upstream exon, the processed RNA includes the downstream exon (Young *et al.*, 1981). However, other cases of this type, for example that of the myosin light chain gene (Fig. 5.4), are more complex, with each of the alternative primary transcripts containing both the alternatively spliced exons. In such cases it is assumed that the different primary transcripts fold into different secondary structures which favour the different splicing events. Whether this is the case or not, it is clear that cases

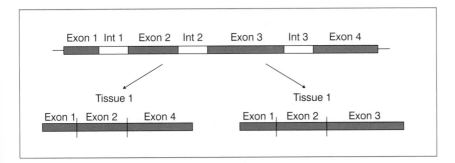

Figure 5.2

Alternative splicing of the same primary transcript in two different ways results in two different mRNA molecules.

Figure 5.3

Alternative splicing at the 5′ end of α-amylase transcripts in the liver and salivary gland. The two alternative start sites for transcription are indicated (TATAA) together with the 5′ region of the mRNAs produced in each tissue.

of alternative splicing arising from differences in the site of transcriptional initiation represent further examples of transcriptional regulation in which the variation in RNA splicing is secondary to the selection of the different promoters in the different tissues. This is not so for the remaining two categories of alternative processing event.

Situations where the 3′ end of the transcripts is different

After the primary transcript has been produced, it is rapidly cleaved and a poly(A) tail is added (see Section 3.3.2). In many genes the process of cleavage and polyadenylation occurs at a different position within the primary transcript in different tissues, and the different transcripts are then differentially spliced.

The best-defined example of this process occurs in the genes encoding the immunoglobulin heavy chain of the antibody molecule, and plays an important role in the regulation of the antibody response to infection. Thus, early in the immune response, the antibody producing B-cell synthesises membrane-bound immunoglobulin molecules whose interaction with antigen triggers proliferation of the B-cell and results in the production of more antibody-synthesising cells. The immunoglobulin produced by these cells is secreted, however, and can interact with antigen in tissue fluids, triggering the activation of other cells in the immune system. The production of membrane-bound and secreted immunoglobin molecules is controlled by the alternative splicing of different RNA molecules differing in their 3′ ends (Fig. 5.5). The longer of these two molecules contains two exons encoding the portion of the protein that anchors it in the membrane. When this molecule is spliced, both these two exons are included, but a region encoding the last 20 amino acids of the secreted form is omitted. In the shorter RNA, the two transmembrane domain-encoding exons are absent and the region specific to the secreted form is included in the final messenger RNA.

If the polyadenylation site used in the production of the shorter immunoglobulin RNA is artificially removed, preventing its use (Danner and

Protein	Species	Nature of transcripts which undergo alternative splicing	Cell types carrying out alternative splicing
(a) Immune system			
Immunoglobulin heavy-chain IgD. IgE, IgG, IgM	Mouse	3′ end differs	B cells
Lyt-2	Mouse	Same transcript	T cells
(b) Enzymes			
Alcohol dehydrogenase	*Drosophila*	5′ end differs	Larva and adult
Aldolase A	Rat	5′ end differs	Muscle and liver
α-amylase	Mouse	5′ end differs	Liver and salivary gland
(2′5′) oligo A synthetase	Human	3′ end differs	B cells and monocytes
(c) Muscle			
Myosin light chain	Rat/mouse/ human/chicken	5′ end differs	Cardiac and smooth muscle
Myosin heavy chain	*Drosophila*	3′ end differs	Larval and adult muscle
Tropomyosin	Mouse/rat/ human/ *Drosophila*	Same transcript	Different muscle cell types
Troponin T	Rat/quail/ chicken	Same transcript	Different muscle cell types
(d) Nerve cells			
Calcitonin/CGRP	Rat/human	3′ end differs	Thyroid C cells or neural tissue
Myelin basic protein	Mouse	Same transcript	Different glial cells
Neural cell adhesion molecule	Chicken	Same transcript	Neural development
Preprotachykinin	Bovine	Same transcript	Different neurones
(e) Others			
Fibronectin	Rat/human	Same transcript	Fibroblasts and hepatocytes
Early retinoic acid- induced gene I	Mouse	Same transcript	Stages of embryonic cell differentiation
Thyroid hormone receptor	Rat	Same transcript	Different tissues

Table 5.1

Cases of alternative splicing which are regulated developmentally or tissue specifically

Leder, 1985), the expected decrease in the production of secreted immunoglobulin is paralleled by a corresponding increase in the synthesis of the membrane-bound form of the protein (Fig. 5.6a). This indicates that the choice of splicing pattern is controlled by which polyadenylation site is used; removal of the upstream site resulting in increased use of the downstream site and increased production of the messenger RNA encoding the membrane-bound form.

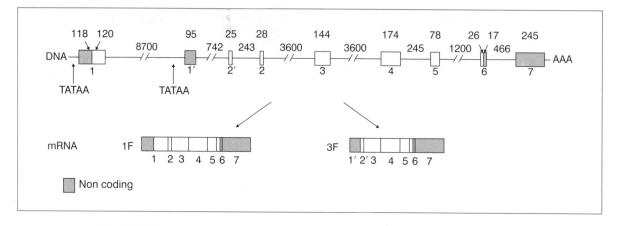

Figure 5.4

Alternative splicing of the myosin light chain transcripts in different muscle cell types produces two mRNAs (1F and 3F) differing at their 5′ ends. The two alternative start sites of transcription used to produce each of the RNAs are indicated (TATAA), together with the intron–exon structure of the gene.

This switch in the polyadenylation site is dependent on an increase in concentration of one of the subunits of the polyadenylation factor CstF (see Section 3.3.2) which occurs during B cell development (Takagaki *et al.*, 1996). Thus CstF binds preferentially to the poly(A) site of the mRNA encoding the membrane form compared to that of the secreted form. Hence, early in B cell development when CstF levels are low, it binds to the poly(A) site for the membrane form and so this mRNA is produced. As the levels of CstF rise binding and cleavage occurs at the poly(A) site for the secreted mRNA (Fig. 5.7).

Figure 5.5

Alternative splicing of the immunoglobulin heavy chain transcript at different stages of B-cell development. The two unspliced RNAs produced by use of the two alternative polyadenylation sites in the gene are shown, together with the spliced mRNAs produced from them.

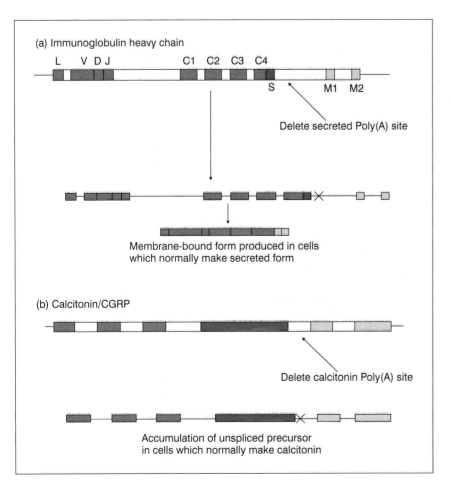

(a) Immunoglobulin heavy chain

Delete secreted Poly(A) site

Membrane-bound form produced in cells
which normally make secreted form

(b) Calcitonin/CGRP

Delete calcitonin Poly(A) site

Accumulation of unspliced precursor
in cells which normally make calcitonin

Figure 5.6

Effect of deleting the more
upstream of the two
polyadenylation sites in the
immunoglobulin heavy chain (a)
and calcitonin/CGRP genes (b)
on the production of the
alternatively spliced RNAs
derived from each of these
genes.

This finding indicates that, in at least some cases of this type, the primary regulatory event is that of determining the site of cleavage and polyadenylation, and that, as with cases of differential promoter usage, alternative RNA splicing is regulated by differences in the structure of the transcript produced in different tissues.

Not all cases of alternative RNA splicing, where the 3′ end of the alternatively spliced RNAs varies, are of this type, however. This conclusion has emerged from the intensive study of the gene encoding the calcium regulatory protein, calcitonin, carried out by Rosenfeld and colleagues (Rosenfeld *et al.*, 1984). When the gene encoding the calcitonin protein (which is a small peptide of 32 amino acids) was isolated it was found that it had the potential to produce an RNA encoding an entirely different peptide of 36 amino acids, which was named calcitonin-gene-related peptide (CGRP). Unlike calcitonin, which is produced in the thyroid gland, CGRP is produced in specific neurons within the brain and peripheral nervous system. These two peptides are produced (Fig. 5.8) by alternative splicing of two distinct transcripts differing in their 3′ ends.

Figure 5.7

Role of the polyadenylation factor CstF in the regulated polyadenylation of the immunoglobulin transcript. At low levels of CstF it binds preferentially to the membrane poly (A) site. Following cleavage at this site, splicing joins exons 4, M1 and M2. At higher levels of CstF it also binds at the weaker secreted poly (A) site leading to loss of exons M1 and M2 by cleavage of the mRNA.

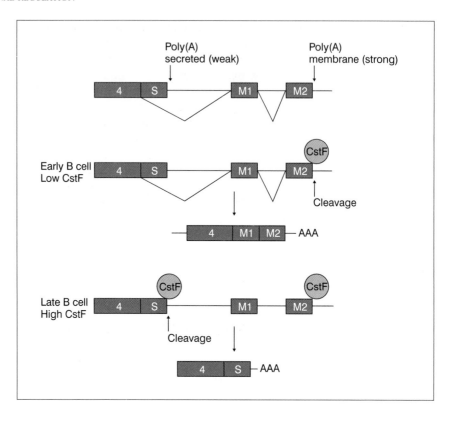

Figure 5.8

Alternative splicing of the calcitonin/CGRP gene in brain and thyroid cells. Alternative splicing followed by proteolytic cleavage of the protein produced in each tissue yields calcitonin in the thyroid and CGRP in the brain.

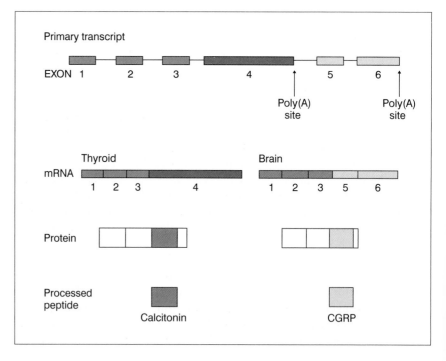

This case differs from that of the immunoglobulin heavy chain, however, in that deletion of the polyadenylation site used in the shorter, calcitonin-encoding RNA does not result in an increase in CGRP expression in cells normally expressing calcitonin. Instead, large unspliced transcripts utilising the downstream (CGRP) polyadenylation site accumulate in these cells (Fig. 5.6b). Although in CGRP-producing cells such transcripts would normally be spliced to yield CGRP messenger RNA, this does not occur in these experiments in cells normally producing calcitonin, and hence these unspliced precursors accumulate. This suggests that in the calcitonin/CGRP gene the use of different polyadenylation sites is secondary to the difference in RNA splicing, suggesting the existence of tissue-specific splicing factors, whose presence or absence in a specific tissue determines the pattern of calcitonin/CGRP RNA splicing.

Situations where both the 5′ and 3′ ends of the differently processed transcripts are identical

The existence of tissue-specific splicing factors which regulate alternative splicing is also indicated by the existence of cases where a transcript with identical 5′ and 3′ ends is spliced differently in different tissues, and which therefore cannot be explained by differential usage of promoters or polyadenylation sites.

Although the initial reports of such cases were confined to the eukaryotic DNA viruses, such as SV40 and adenovirus, a number of cases involving the cellular genes of higher eukaryotes have now been described. In one such case (Fig. 5.9) involving the skeletal muscle troponin T gene (Breitbart *et al.*, 1987) the same RNA can be spliced in up to 64 different ways in different muscle cell types. The existence of tissue-specific splicing factors acting on this gene is indicated by the finding that the artificial introduction and expression of this gene in non-muscle cells or myoblasts results in the complete removal of exons 4–8, whereas in muscle cells (myotubes) the correct pattern of alternative splicing seen with the endogenous gene is reproduced faithfully.

It should be noted, however, that the 64 possible mRNAs produced in this case are certainly not the most dramatic example of multiple mRNAs being produced by alternative splicing. Thus, the vast number of different alternative exons found in the Dscam gene of *Drosophila* could result in over 38 000 different mRNAs being produced which is greater than the number of different genes in this organism (for reviews see Black, 2000; Graveley, 2001)! This illustrates the extraordinary power of alternative splicing to produce multiple mRNAs encoding related but distinct proteins from a single gene.

5.2.3 Mechanism of alternative RNA splicing

RNA sequences involved in alternative splicing

The idea that certain factors are necessary for each particular pattern of splicing in genes, such as calcitonin/CGRP or troponin, clearly begs the question of how such factors act. It is likely that these factors recognise *cis-*

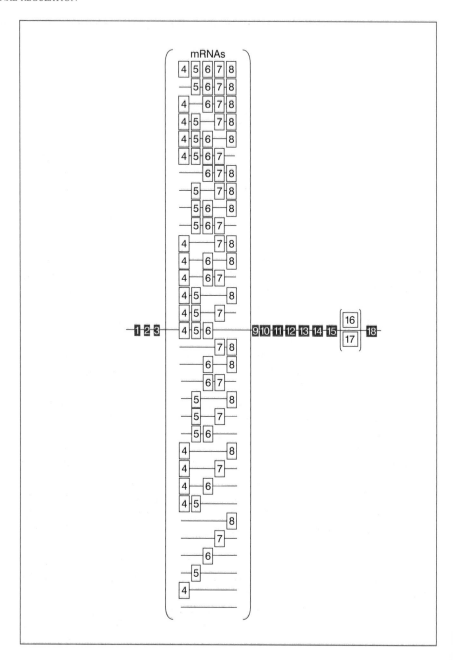

Figure 5.9

Alternative splicing of the four combinatorial exons (4–8) and the two mutually exclusive exons (16 and 17) can result in up to 64 distinct mRNAs from the rat troponin T gene.

acting sequences within the RNA transcript itself. Clearly, the interaction of these factors with such sequences could produce alternative splicing either by promoting splicing at the site of the *cis*-acting sequence at the expense of the alternative splice site (Fig. 5.10a) or by inhibiting splicing at the site of binding and thereby promoting the use of the alternative splice site (Fig. 5.10b).

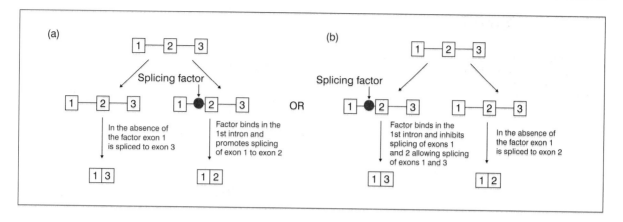

Both of these types of mechanism appear to be used in different cases. Indeed, examples of each of these mechanisms can be seen in the hierarchy of alternatively spliced genes which regulates sex determination in *Drosophila* (for a review see Baker, 1989). In this hierarchy each gene product controls the alternative splicing of the next gene in the pathway, resulting in different protein products in males and females. In turn these products differentially regulate the splicing of the next gene in the hierarchy leading ultimately to the production of a male or female fly (Fig. 5.11).

Thus the *Sxl* gene is differentially spliced in males and females, with the product of the female-specific mRNA not only controlling the splicing of the next gene in the hierarchy, *tra*, but also promoting its own female-specific splicing. Mutations which affect this autocatalytic function of *Sxl* on its own RNA map in the intron between exon 2 and the male-specific exon 3, indicating that the *Sxl* protein acts by inhibiting the male-specific splicing of exons 2 and 3. Similarly *Sxl* prevents the use of the male-specific exon 2 in the *tra* gene by binding within intron 1 and preventing the binding of the constitutive splicing factor U2AF which in turn recruits the U2snRNP particle which is essential for removal of all introns (see Section 3.3.3) (Valcarcel *et al.*, 1993).

In contrast, the action of the *tra* and *tra-2* gene products on the splicing of the *dsx* transcript appears to be mediated by promoting the use of the female-specific splice rather than the constitutive male-specific splicing event. In this case, mutations in *dsx* which affect its splicing map in the intron between exon 3 and the female-specific exon 4, indicating that the alternative splicing factor promotes the splicing of exons 3 and 4 (see below).

As well as responding to the presence or absence of a specific factor, specific RNA sequences can also regulate the pattern of alternative splicing in response to cellular signalling pathways. Thus, when the calcium/calmodulin-dependent protein kinase type IV (CaMK IV) is activated, inclusion of a specific exon (STREX) in transcripts from the BK potassium channel gene is repressed and the exon is spliced out, whereas in the absence of active CaMK IV, this exon is included in the RNA (Fig. 5.12a). This effect is medi-

Figure 5.10

Possible models by which an alternative splicing factor can affect splicing by binding to a *cis*-acting sequence. In (a), the factor acts by promoting the use of the weaker of the two potential splicing sites while in (b) it acts by inhibiting use of the stronger of the two sites so that the other, weaker, site is used.

Figure 5.11

Schematic diagram (not to scale) of the hierarchy of alternatively spliced genes which controls sex determination in *Drosophila*. Female-specific splicing of the *Sxl* transcript produces a protein which promotes the female-specific splicing of its own RNA and that of the *tra* gene. In turn, the protein produced by the *tra* female-specific transcript in conjunction with the *tra-2* gene product promotes the female-specific splicing of the *dsx* transcript. The sites of mutations in the *Sxl* and *dsx* genes which affect their sex-specific splicing are indicated. Modified from Baker (1989).

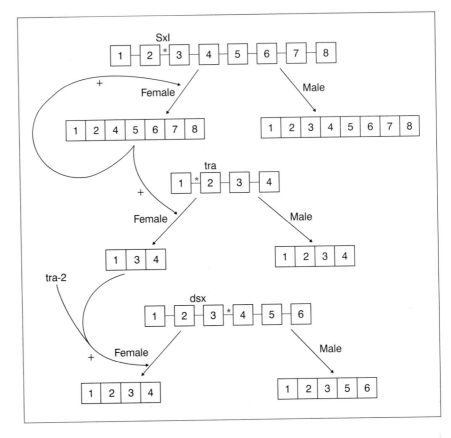

ated by a 54 bp sequence from the 3′ splice site upstream of the regulated STREX exon (Xie and Black, 2001). Thus, mutation of this sequence abolishes regulation of splicing by CaMK IV whilst transferring it to a constitutionally spliced exon renders it sensitive to CaMK IV regulation (Fig. 5.12b). This sequence thus acts as a CaMK IV responsive RNA element (CaRRE) regulating splicing of the STREX exon in response to CaMK IV activation (for a review see O'Donovan and Darnell, 2001).

Nature of alternative splicing factors

The evidence for the existence of alternative splicing factors and the identification of *cis*-acting sequences with which they interact have led to many attempts to identify these factors (for reviews see Wang and Manley, 1997; Smith and Valcárcel, 2000).

In *Drosophila*, which is genetically very well characterised, the main approach to this problem has been a genetic one. Thus, as discussed above, genes such as *Sxl* and *tra*, originally identified by the fact that mutations within them disrupted the process of sex determination, are now known to act by controlling alternative splicing. The products of these genes are alternative splicing factors and their study allows a unique insight into the nature of such factors.

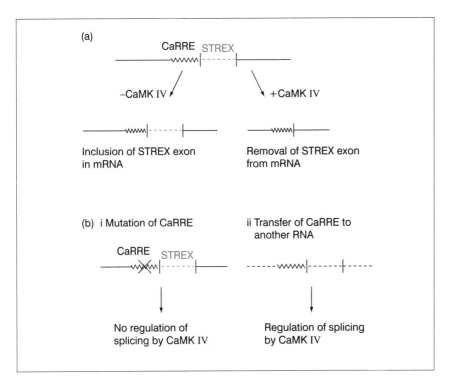

(a)

CaRRE STREX

−CaMK IV

+CaMK IV

Inclusion of STREX exon
in mRNA

Removal of STREX exon
from mRNA

(b) i Mutation of CaRRE

ii Transfer of CaRRE to
another RNA

CaRRE STREX

No regulation of
splicing by CaMK IV

Regulation of splicing
by CaMK IV

Figure 5.12

Regulation of splicing in BK potassium channel gene transcripts by the phosphorylation enzyme CaMK IV. (a) In the absence of active CaMK IV, the STREX exon is spliced out of the BK gene RNA whereas in the presence of active CaMK IV this splicing event is inhibited. (b) The response to CaMK IV is mediated by a 54 bp sequence from the 3' splice site upstream of STREX which is known as the CaMK IV response RNA element (Ca RRE) since inactivation of this sequence prevents regulation of splicing by CaMK IV (i) whereas its inclusion in a completely different RNA (dotted line) causes splicing of that RNA to be regulated by CaMK IV.

The sequencing of the *Sxl* and *tra-2* genes has revealed that the proteins they encode contain one or more copies of a ribonucleoprotein (RNP) consensus sequence that is found in a wide variety of RNA binding proteins, such as those of the mammalian spliceosome, and constitutes an RNA binding domain (for reviews see Bandziulis *et al.*, 1989; Burd and Dreyfuss, 1994). Indeed, recent structural analysis of *Sxl* bound to the *tra* mRNA has shown that the RNA binds to a V-shaped cleft within the *Sxl* protein (Handa *et al.*, 1999) (colour plate 9). Hence Sxl and Tra-2 influence alternative splicing by binding directly to *cis*-acting sequences in the spliced RNAs discussed above.

Hence specific factors have been identified in *Drosophila* which are expressed only in a particular situation and which regulate specific alternative splicing events. Interestingly, however, in vertebrates it appears that many tissue-specific splicing events depend on quantitative variations in the levels of splicing factors which are present in all tissues. Thus the SF2 factor is a member of the SR proteins family discussed in Section 3.3.3 which is present in all cells and is essential for the basic process of splicing itself (for a review of SR proteins see Manley and Tacke, 1996). However, its concentration has been shown to influence which of two competing upstream splice sites is joined to a downstream site. Thus high concentrations of SF2 favour the more proximal of the two sites whilst low concentrations favour the more distal site (Fig. 5.13; for a review see Smith and Valcárcel, 2000). Interestingly the constitutively expressed factor hnRNPA1 which binds to

Figure 5.13

The pattern of splicing can be regulated by the balance in levels between the constitutively expressed SF2 and hnRNPA1 proteins. A high ratio of SF2 to hnRNPA1 favours use of the proximal exon (2) whilst a low ratio favours the distal exon (3).

RNA before it is spliced (see Section 3.3.5) has the opposite effect, favouring the use of the more distal site. Hence, in this situation, the outcome of a specific alternative splicing event could be different in two different tissues, depending on the relative concentration of SF2 and hnRNPA1 in each tissue, even though both factors were present in both tissues (Fig. 5.13).

Such a system, in which the different outcome of alternative splicing events in each tissue is controlled by quantitative differences in the relative levels of constitutively expressed factors, can readily be fitted into the mechanistic models illustrated in Fig. 5.10 simply by suggesting that the two factors have opposite effects on the relative strengths of the two competing splice sites. Indeed it is likely that SR proteins, such as SF2, can influence splice site selection in two ways, both involving the recruitment of the UsnRNP particles which are essential for splicing to occur (see Section 3.3.3). Thus SR proteins can bind to the 5′ splice site and promote the binding of the U1snRNP to this site (Fig. 5.14a). In addition, they can also bind to sequences known as splicing enhancers within the exon downstream of the regulated splice site and stimulate the binding of the U2AF protein to the branch point, thereby recruiting the U2 snRNP particle (see Section 3.3.3) (for reviews see Wang and Manley, 1997; Blencowe, 2000) (Fig. 5.14b).

Thus both tissue-specific factors and variations in constitutively expressed factors can regulate alternative splicing. Indeed, as described above, both Sxl and SR proteins, such as SF2, can act by regulating the recruitment of U2AF. It is not surprising therefore that some cases of alternative splicing are regulated by the interaction of SR proteins with specifically expressed factors. Thus in the case of the *dsx* gene, the splicing of exon 3 to the female-specific exon 4 does not occur in the absence of the female-specific proteins *tra* and *tra-2* (Fig. 5.11). This is because U2AF binds only weakly to the branch point of this intron and the binding site of SR proteins in exon 4 is weak and too far away for SR to enhance U2AF binding (Fig. 5.15). In the presence of *tra* and *tra-2*, however, the interaction of SR with the exon 4 sequence is stabilised, allowing SR in turn to promote binding of U2AF and splicing of exon 3 to exon 4 (Lynch and Maniatis, 1995; Fig. 5.15). Interestingly both the SR

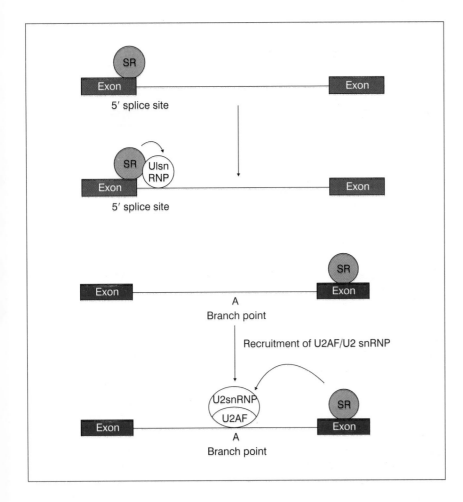

Figure 5.14

SR proteins, such as SF2, can stimulate the use of a particular splice site either by binding to the 5' splice site and promoting binding of the U1snRNP to this site (panel a) or by binding to the downstream exon sequence and promoting recruitment of U2AF and hence of the U2 snRNP to the branch point (panel b).

proteins and *tra/tra-2* contain both an RNA binding domain found in many RNA binding proteins with different functions (for a review see Burd and Dreyfuss, 1994) and a domain rich in serine (S) and arginine (R) residues which gives the SR proteins their name (for a review see Manley and Tacke, 1996) which is so far unique to proteins involved in RNA splicing.

Hence alternative splicing is likely to be controlled by a series of proteins with similar functional domains, some of which are expressed in a highly specific manner whilst others are expressed at different levels in all cell types.

5.2.4 Generality of alternative RNA splicing

The cases discussed above indicate the use of alternative splicing in a wide variety of biological processes. In mammals such splicing has been shown to regulate the immune system's production of antibodies, the production of neuropeptides such as CGRP and the tachykinins, substance P and substance K, as well as the synthesis of the different forms of several of the major

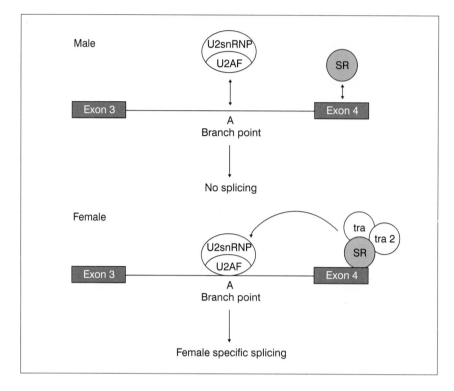

sarcomere muscle proteins. Similarly, in *Drosophila* much of the posterior body plan is determined by developmentally regulated differential splicing of the ultrabithorax gene, while sex determination is also controlled by differential splicing of a hierarchy of genes in males and females (see Section 5.2.3; for a review see Baker, 1989).

The widespread use of alternative splicing in mammals does not refute the initial conclusion that gene regulation occurs primarily at the level of transcription, however. Thus, alternative splicing represents a response to a requirement for the production of related but different forms of a gene product in different tissues. It therefore supplements the regulation of transcription of the gene responsible for producing the different forms. Thus the immunoglobulin heavy chain gene, which produces both membrane-bound and secreted forms of the protein at different stages of B-cell development, is transcribed only in B-cells and not in other cell types, while the transcription of the troponin T gene, which produces multiple different isoforms in different muscle cell types, is confined to differentiated muscle cells. Interestingly alternative splicing is particularly common in non-dividing cells such as muscle cells and nerve cells (for a review see Grabowski, 1998). As discussed in Chapter 6, the reprogramming of cellular commitment and gene transcription often requires a cell division event, allowing alterations in the chromatin structure of the DNA and its associated proteins to occur. Hence alternative splicing may represent a means of conveniently changing the pattern of protein production in cells where such reprogramming cannot be achieved by cell division. Alternative splicing therefore represents a significant supplement to the regulation of transcription, and the flexibility it confers suggests that many more cases will be identified in the future.

5.3 RNA EDITING

The finding that two different protein products can be produced from the same RNA by alternative splicing has been supplemented by the observation that a similar result can be achieved by a post-transcriptional sequence change in the messenger RNA (for reviews see Benne, 1996; Smith and Sowden, 1996; Reenan, 2001). Thus apolipoprotein B, which plays an important role in lipid transport, is known to exist in two closely related forms. A large protein of 512 kDa, known as apo-B100, is synthesised by the liver, while a smaller protein, apo-B48, is produced by the intestine. The smaller protein is identical to the amino-terminal portion of the larger protein. Analysis of the mRNA encoding these proteins revealed a 14.5 kb RNA in both tissues. These two RNAs were identical with the exception of a single base at position 6666, which is a cytosine in the liver transcript and a uracil in the intestinal transcript (Fig. 5.16). This change has the effect of replacing a CAA codon, which directs the insertion of a glutamine residue, with a UAA stop codon, which causes termination of translation of the intestine RNA and hence results in the smaller protein being made.

Only one gene encoding these proteins is present in the genome, and it is not alternatively spliced. In both intestinal and liver DNA this gene has a cytosine residue at position 6666. Hence the uracil in the intestinal transcript must be introduced by a post-transcriptional RNA editing mechanism.

Indeed a cytidine deaminase enzyme has been identified and shown to bind to the apoB mRNA at sequences adjacent to the edited site. This enzyme removes an amine (NH_2) group from the cytosine residue, generating a uracil. This enzyme is expressed in the intestine, where editing of the apoB mRNA occurs, but not in the liver, where editing does not occur. Interestingly, however, it is also present in other tissues, such as the testis, ovary and spleen, where apoB mRNA is not expressed, indicating that this enzyme is likely to edit other transcripts expressed in these types. A similar

Figure 5.16

RNA editing of the apolipoprotein B transcript in the intestine produces an mRNA encoding the truncated protein apo-B48.

C to U editing process has been observed in the mitochondria and chloroplasts of plants, although its mechanistic relationship to that occurring in mammalian cells is unclear (for a discussion see Benne, 1996).

As well as C to U editing by cytidine deaminase, an adenosine deaminase enzyme also exists in mammalian cells which removes an amino residue from adenine to produce an inosine base which is read as a guanine residue by the translation apparatus (for a review see Reenan, 2001). This form of editing was first identified in the gene encoding a receptor for the excitatory amino acid glutamate which is expressed in neuronal cells. The editing of an A residue to a I (G) residue in the transcript of this gene results in it encoding an arginine rather than the glutamine found in other related receptors and alters its properties so that it is permeable to calcium. Interestingly, this editing of the glutamate receptor mRNA is essential for the survival of the animal. Thus, mice lacking the adenosine deaminase enzyme (ADAR2) which carries out this editing are prone to seizures and die young. However, their survival can be restored to normal by inserting a glutamate receptor gene in which the alteration produced by editing has been carried out at the DNA level, removing the need for editing at the RNA level (Higuchi et al., 2000).

Although the glutamate receptor is thus a critical target for RNA editing, editing by adenosine deamination is prevalent, especially in the nervous system, and also occurs in other neuronally expressed receptors such as the kainate receptors and the serotonin receptor (for reviews see Scott, 1997; Reenan, 2001).

As well as affecting the nature of the protein produced from a particular mRNA, A to I editing by adenosine deaminase enzymes can also affect the splicing of the edited RNA. This is seen in the case of the gene encoding the ADAR2 adenosine deaminase enzyme itself. Thus, the ADAR2 transcript contains an AA sequence 47 bases upstream of an AG sequence which is normally used as a 3′ splice site. Editing of the AA sequence to AI (which mimics AG) creates a new 3′ splice site which is used for splicing resulting in the 47 base sequence being retained in the mRNA (Fig. 5.17) (Rueter et al., 1999).

Figure 5.17

In the ADAR2 RNA, splicing normally occurs using an AG sequence as the 3′ splice site (a). RNA editing alters an AA sequence 47 bases upstream to AI. As AI resembles AG, this sequence is used as the 3′ splice site, resulting in an extra 47 bases being retained in the final mRNA (b).

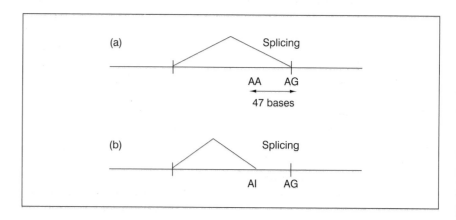

This case is of particular interest in that the editing event is regulated so that it occurs frequently in the brain and lung where ADAR2 transcripts predominantly contain the 47 base extra exon and much less frequently in the heart where this exon is therefore absent from the mRNA. Moreover, it demonstrates that editing by adenosine deaminase can affect RNA splicing as well as altering the encoded protein and that it can affect the RNA encoding the adenosine deaminase itself as well as RNAs encoding other proteins such as the glutamate receptor.

The existence of two distinct editing enzymes (adenosine deaminase and cytidine deaminase) in mammalian cells, each with multiple substrates, indicates that this represents a widely used mechanism which, like alternative splicing, can produce related but distinct proteins from the same transcript. As with alternative splicing, however, this is likely to act as a supplement to transcriptional control, the apoB gene, for example, being transcribed in the liver and the intestine but not in other tissues. Interestingly, as well as these similarities in the role/outcome of RNA editing and alternative splicing, there is evidence that the two processes may occur in parallel. Thus, it has recently been shown that adenosine deaminase enzyme is found in large ribonucleoprotein particles, which also contain Sm and SR splicing factors (Raitskin *et al.*, 2001).

5.4 REGULATION OF RNA TRANSPORT

5.4.1 Transport from nucleus to cytoplasm

The process of RNA splicing takes place within the nucleus, whereas the machinery for translating the spliced RNA is found in the cytoplasm. Hence the spliced mRNA must be transported to the cytoplasm if it is to direct protein synthesis (see Section 3.3.5). The first example of regulation at this level was described in the human immunodeficiency virus (HIV-1).

Thus early in infection of cells with HIV, the virus produces a high level of very short non-functional transcripts and a small amount of full length functional RNA (see Section 4.3.2). The full length RNA is spliced with the removal of two introns so that the predominant transcript which appears in the cytoplasm is a fully spliced mRNA (Fig. 5.18; for a review see Cullen, 1991). This transcript encodes the regulatory proteins Tat and Rev. As discussed in Section 4.3.2, the Tat protein acts to greatly increase the rate of transcription of the viral genome, as well as promoting transcriptional elongation so that predominantly full length RNAs are produced in the nucleus. However, at this second stage of infection most of the mRNAs that appear in the cytoplasm are either unspliced or have had only the first intron removed (Fig. 5.18). As these transcripts encode the viral structural proteins, this allows the high level production of viral particles which is necessary late in infection.

This change in the nature of the HIV RNA in the cytoplasm is dependent on the action of the Rev protein which is made early in infection (for a review see Cullen and Malim, 1991). Most interestingly, Rev does not affect the amounts of the different RNAs in the nucleus. Rather it is an RNA

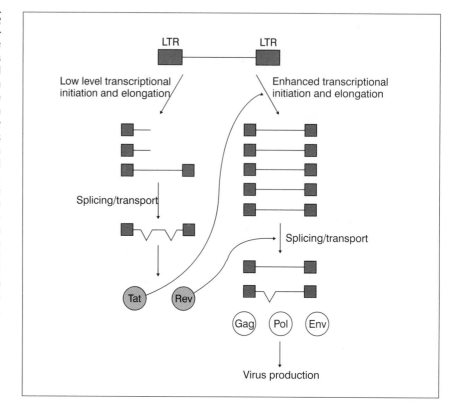

binding protein and binds to a specific site (the Rev response element) in the second intron of the HIV transcripts. This binding promotes the transport of these RNAs to the cytoplasm (Fig. 5.19). As the fully spliced RNA lacks the Rev binding site, its transport is not accelerated and so the proportion of unspliced or singly spliced RNA in the cytoplasm increases. Hence Rev acts at the level of RNA transport, the first regulatory protein to do so.

As noted in Section 3.3.5 the Rev protein resembles several cellular RNA transport proteins in containing a nuclear export signal (NES) which promotes transport of the RNA to which it has bound from the nucleus to the cytoplasm (for a review see Gerace, 1995). Moreover, Rev has been shown to bind to a cellular protein Rip 1 which is localised in the pores of the nuclear membrane and this interaction is likely to play a key role in its ability to mediate export of the HIV RNA (Fig. 5.20). These findings indicate that Rev mediates RNA export via a cellular pathway, involving the NES signal which it shares with cellular RNA transport proteins. It is likely therefore that, as well as the examples of constitutive RNA transport mediated by this system (see Section 3.3.5), there also exist situations in which this system is used to regulate the transport of specific cellular mRNAs in different tissues or in response to specific stimuli. This would, for example, provide a convenient explanation of the existence of RNA species that are confined to the nucleus in specific tissues of the sea urchin (Section 5.1).

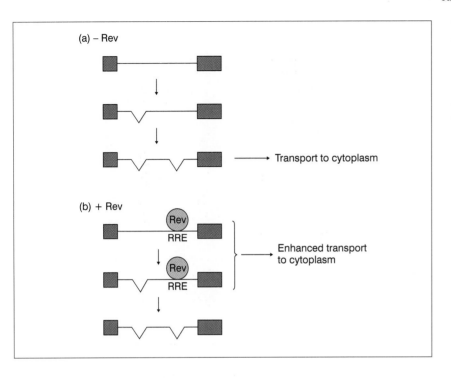

Figure 5.19

Binding of Rev to its response element (RRE) in the HIV RNA promotes the transcript of unspliced or singly spliced RNAs which contain the binding site at the expense of doubly spliced RNA which lacks the binding site and therefore cannot bind Rev.

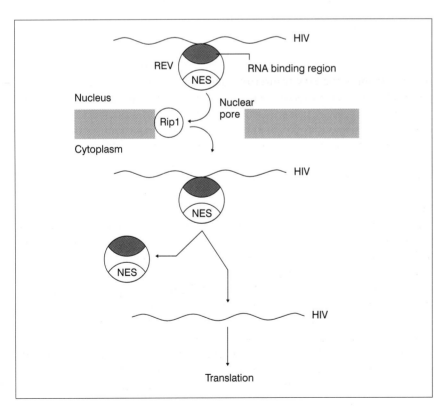

Figure 5.20

The Rev protein binds to the HIV RNA by means of its RNA binding domain and its nuclear export signal (NES) then binds to cellular proteins, such as Rip1, in the nuclear pore and promotes the export of the RNA/protein complex to the cytoplasm.

Figure 5.21

Transport of mRNA within the
cell can involve both its export
from the nucleus to the
cytoplasm (A) as well as in some
cases its preferential localisation
within a particular region of the
cell cytoplasm (B).

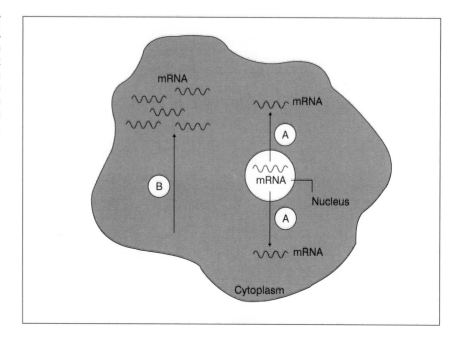

Similarly, a mechanism which prevented the transport of unspliced RNA in genes regulated by processing versus discard decisions (see Section 5.2) would represent a means of preventing wasteful translation of RNA species containing interruptions in the protein-coding sequence.

5.4.2 Transport within the cytoplasm

As well as controlling the transport of mRNA from the nucleus to the cytoplasm (Fig. 5.21a) it is also possible for regulatory processes to control the location of an mRNA species within the cytoplasm of a polarised cell, resulting in the corresponding protein being made only in certain parts of the cell (Fig. 5.21b) (for reviews see Hazelrigg, 1998; Mohr and Richter, 2001). This process is observed, for example, within the oocyte (egg), resulting in different proteins being localised in different regions of the egg. Following fertilisation, this differential distribution of proteins in turn controls embryonic development so that different regions of the embryo develop from different regions of the fertilised egg. Thus, for example, in *Drosophila* the bicoid mRNA is located at the anterior end of the egg whilst the nanos mRNA is localised to the posterior end. The resulting opposite gradients of the Bicoid and Nanos proteins in turn control the polarity of the head, thorax and abdomen of the embryo so that different regions of the body are produced, depending on the relative concentrations of the two proteins (Fig. 5.22).

In some cases, such as the bicoid mRNA, specific sequences have been localised in the 3′ untranslated region of the mRNA which are essential for correct localisation and which can confer the bicoid pattern of localisation on a non-localised mRNA. These sequences in the 3′ untranslated region of the bicoid mRNA are essential for it to bind the staufen protein, to form a

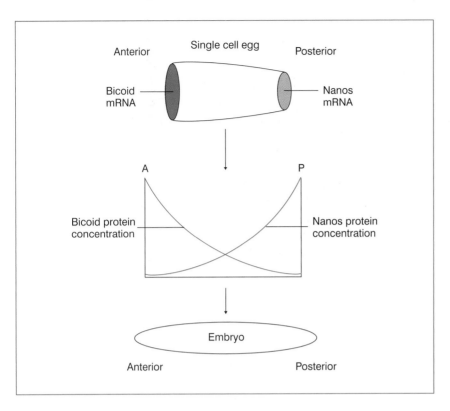

Figure 5.22

Localisation of the bicoid and nanos mRNAs at opposite ends of the *Drosophila* egg sets up opposite gradients in the concentrations of their corresponding proteins which in turn specify the anterior/posterior polarity of the resulting embryo.

ribonucleoprotein complex. In turn, recruitment of the staufen protein is essential for movement of the bicoid mRNA to the anterior end of the egg (Fig. 5.23a). Interestingly, the staufen protein is also expressed in mammalian neurones and appears to play a critical role in directing specific mRNAs to the multiple dendritic processes of the neurone and excluding them from the single axon (Fig. 5.23b) (for a review see Roegiers and Jan, 2000). Hence staufen plays a key role in directing specific mRNAs to particular parts of the cell in very different cell types in widely different organisms.

5.5 REGULATION OF RNA STABILITY

5.5.1 Cases of regulation by alterations in RNA stability

Once the mRNA has entered the cytoplasm, the number of times that it is translated, and hence the amount of protein it produces, will be determined by its stability. The more rapidly degraded an RNA is, the less protein it will produce. Hence an effective means of gene regulation could be achieved by changing the stability of an RNA species in response to some regulating signal. A number of situations where the stability of a specific RNA species is changed in this way have been described (for reviews of this topic see Sachs, 1993; Ross, 1996; Bashirullah *et al.*, 2001). Thus, for example, the mRNA for the milk protein, casein, turns over with a half-life of around 1 h in

Figure 5.23

The Staufen protein plays a key role in localising the bicoid mRNA to the anterior end of the *Drosophila* embryo (a) and in directing specific mRNAs to the dendrites and not the axon of mammalian neurones (b).

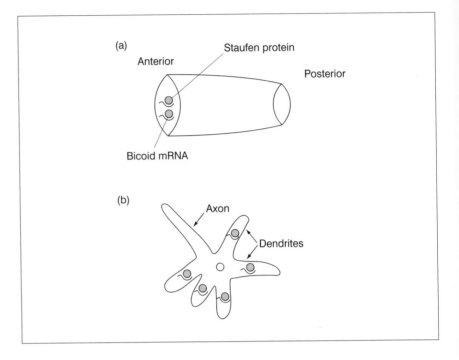

Table 5.2

Regulation of RNA stability

mRNA	Cell type	Regulatory event	Increase or decrease in half-life
Cellular oncogene *c-myc*	Friend erythroleukaemia cells	Differentiation in response to dimethyl sulfoxide	Decrease from 35 to to less than 10 min
c-myc	B cells	Interferon treatment	Decreased
c-myc	Chinese hamster lung fibroblasts	Growth stimulation	Increased
Epidermal growth factor receptor	Epidermal carcinoma cells	Epidermal growth factor	Increased
Casein	Mammary gland	Prolactin	Increased from 1 to 40 h
Vitellogenin	Liver	Oestrogen	Increased 30-fold
Type I pro-collagen	Skin fibroblasts	Cortisol	Decreased
Type I pro-collagen	Skin fibroblasts	Transforming growth factor β	Increased
Histone	Hela	Cessation of DNA synthesis	Decreased from 40 to 80 min
Tubulin	CHO	Accumulation of free tubulin subunits	Decreased 10-fold

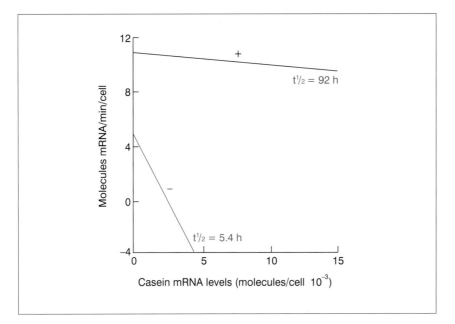

Figure 5.24

Difference in stability of the casein mRNA in the presence (+) or absence (-) of prolactin.

untreated mammary gland cells. Following stimulation with the hormone prolactin, the half-life increases to over 40 h (Guyette *et al.*, 1979), resulting in increased accumulation of casein mRNA and protein production in response to the hormone (Fig. 5.24). Similarly, the increased production of the DNA-associated histone proteins in the S (DNA synthesis) phase of the cell cycle is regulated in part by a fivefold increase in histone mRNA stability that occurs in this phase of the cell cycle (for reviews of the regulation of histone gene expression see Marzluff, 1992; Stein *et al.*, 1992). A representative selection of cases where mRNA stability is altered in a particular situation is given in Table 5.2.

5.5.2 Mechanisms of stability regulation

The first stage in defining the mechanism of changes in RNA stability is to identify the sequences within the RNA that are involved in mediating the observed alterations. This can be achieved by transferring parts of the gene encoding the RNA under study to another gene and observing the effect on the stability of the RNA expressed from the resulting hybrid gene. In a number of cases, short regions have been identified which can confer the pattern of stability regulation of the donor gene upon a recipient gene that is not normally regulated in this manner.

In many cases such regions are located in the 3' untranslated region of the mRNA, downstream of the stop codon that terminates production of the protein (for a review see Wickens *et al.*, 1997). Thus the cell-cycle-dependent regulation of histone H3 mRNA stability is controlled by a 30 nucleotide sequence at the extreme 3' end of the molecule. Similarly, the destabilisation of the mRNA encoding the transferrin receptor in response

Figure 5.25

Similar stem-loop structures in the human ferritin and transferrin receptor mRNAs. Note the boxed conserved sequences in the unpaired loops and the absolute conservation of the boxed C residue, found within the stem, five bases 5′ of the loop.

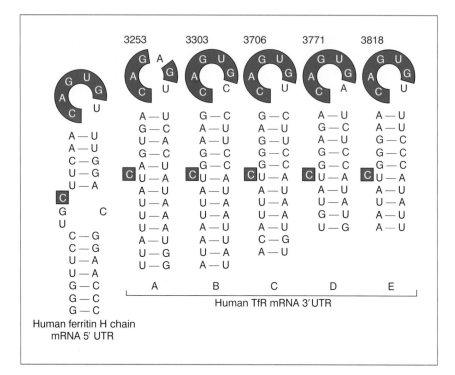

Human ferritin H chain mRNA 5′ UTR

Human TfR mRNA 3′ UTR

to the presence of iron can be abolished by deletion of a 60 nucleotide sequence within the 3′ untranslated region (for reviews see Hentze and Kuhn, 1996; Aisen *et al.*, 2001). Interestingly, both of these sequences have the potential to form stem-loop structures by intra-molecular base pairing (Fig. 5.25), suggesting that changes in stability might be brought about by alterations in the folding of this region of the RNA in response to a specific signal (see Section 3.3.2).

The localisation of sequences involved in the regulated degradation of specific mRNA species to the 3′ untranslated region is in agreement with the important role of this region in determining the differences in stability observed between different RNA species. Such sequences may act either by promoting endonucleolytic cleavage within the RNA transcript or by promoting loss of the poly (A) tail, opening up the RNA to exonucleolytic attack via its free 3′ end (see Section 3.3.2; for reviews see Decker and Parker, 1994; Beelman and Parker, 1995). This suggests that differences in RNA stability, whether between different RNA species or in a single RNA in different situations, may be controlled primarily by this region.

Despite this, cases where other regions of the RNA mediate the observed alterations in stability have also been described. The most extensively studied of such cases concerns the auto-regulation of the mRNA encoding the microtubule protein, β-tubulin, in response to free tubulin monomers (Pachter *et al.*, 1987; Yen *et al.*, 1988). This auto-regulation prevents the wasteful synthesis of tubulin when excess free tubulin, not polymerised into microtubules, is present, and is caused by a destabilisation of the tubulin mRNA. A short sequence, only 13 bases in length, from the 5′ end of the β-tubulin mRNA is responsible for this destabilisation, and can confer the response on an unrelated mRNA. Most interestingly, these bases actually encode the first four amino acids of the tubulin protein, raising the possibility that the trigger for degradation of the tubulin mRNA might be the recognition of these amino acids in the tubulin protein, rather than the corresponding nucleotides in the tubulin RNA. In an elegant series of studies, Cleveland and colleagues (Yen *et al.*, 1988) showed that this was indeed the case. Thus, changing the translational reading frame of this region, such that the identical nucleotide sequence encoded a different amino-acid sequence, abolished the auto-regulatory response (Fig. 5.26a), whereas changing the nucleotide sequence in a manner which did not alter the encoded amino acids (due to the degeneracy of the genetic code) left the response intact (Fig. 5.26b).

This finding provides an explanation for the earlier observation that the tubulin RNA must be translated (at least in part) into protein for degradation to occur. Such a requirement for translation of the RNA to be degraded is also observed, however, in cases where the target sequence for degradation lies at the 3′ end of the molecule, in a region that is not translated into protein. In the case of the histone mRNA, for example, introduction of stop codons resulting in premature termination of translation resulted in the abolition of the selective destabilisation of the RNA following the cessation of DNA synthesis. Such an effect could also be achieved by inserting additional sequences between the normal termination codon and the previously discussed stem-loop structure which is involved in the regulation of degradation (Graves *et al.*, 1987; Fig. 5.27).

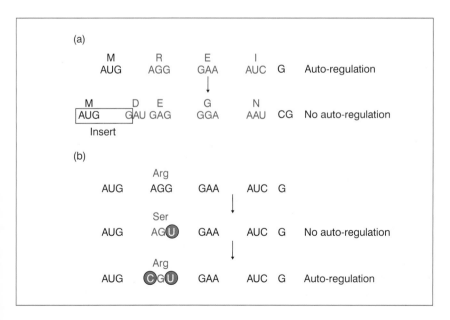

Figure 5.26

Effect of changes in the β-tubulin sequence on auto-regulation of tubulin mRNA stability.

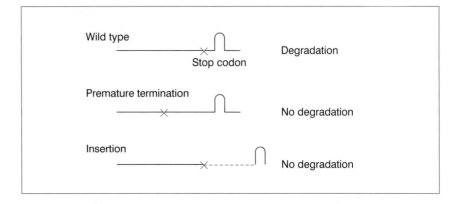

Figure 5.27

Effect of altering the relative position of the stop codon (x) and the stem-loop structure in histone mRNA on its degradation.

These apparently contradictory observations can be combined into a model of RNA stability regulation, if it is assumed that the ribosome translating the RNA species carries with it a nuclease capable of degrading the RNA in response to a specific signal. In the case of tubulin this nuclease is triggered by a recognition event involving the first four amino acids emerging from the ribosome and then it degrades the RNA (Fig. 5.28a). In contrast, in the case of histone mRNA the trigger is supplied by recognition of the 3′ RNA sequence. However, translation of the upstream part of the

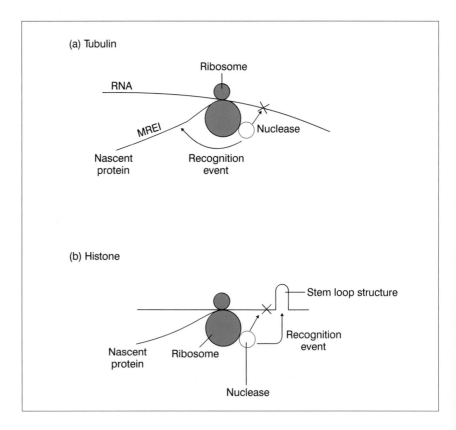

Figure 5.28

Putative mechanism for recognition of degradation signals in nascent tubulin protein (a) or histone mRNA (b) by a ribosome-associated nuclease.

molecule is still required in order to deliver the nuclease-bearing ribosome to the appropriate part of the molecule for recognition and degradation to occur (Fig. 5.28b).

Interestingly, it has also been shown that the processes of RNA degradation and translation can interact in another way. Thus, in the case of the c-*fos* mRNA, a specific instability sequence in the coding region is required for the rapid turnover of the molecule and translation of the mRNA is necessary for such degradation to occur. It has recently been shown that a specific protein complex binds to this instability sequence and also binds to the poly (A) tail of the mRNA (Grosset *et al.*, 2000) (Fig. 5.29a). This stabilises the mRNA by protecting it from deadenylation which would lead to its degradation (see Section 3.3.2). It has been suggested that the passage of the ribosome disrupts this complex, freeing the poly (A) tail and hence resulting in rapid deadenylation and decay of the RNA (Fig. 5.29b).

The finding that translation is required for the degradation of RNA species as diverse as the non-polyadenlylated histone mRNAs and the polyadenylated c-*fos* mRNA suggests that this may represent a generally applicable mechanism controlling RNA degradation (for further discussion of the relationship between mRNA degradation and translation see Jacobson and Peltz, 1996).

5.5.3 Role of stability changes in regulation of gene expression

A consideration of the situations where changes in the stability of a particular RNA occur (see Table 5.2) suggests that the majority have two features in common. First, changes in stability of a particular mRNA are very often accompanied by parallel alterations in the transcription rate of the corresponding gene. Thus prolactin treatment of mammary gland cells results in a two- to fourfold increase in casein gene transcription, and the increased stability of histone mRNA in the S phase of the cell cycle is accompanied by a three- to fivefold increase in transcription of the histone genes. Secondly, cases where RNA stability is regulated are very often those where a rapid and transient change in the synthesis of a particular protein is required. Thus

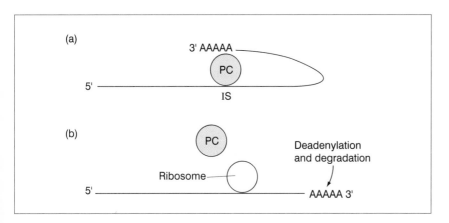

Figure 5.29

In the c-*fos* mRNA a protein complex (PC) binds to both the poly (A) tail and to an instability sequence (IS) which is required for degradation of the mRNA (a). Passage of the ribosome during translation disrupts this binding and leads to exposure of the poly (A) tail, resulting in deadenylation and rapid degradation of the mRNA (b).

synthesis of the histone proteins is necessary only at one particular phase of the cell cycle, when DNA is being synthesised. Following cessation of DNA synthesis a rapid shut-off in the synthesis of unnecessary histone proteins is required. Similarly, following the cessation of hormonal stimulation it would be highly wasteful to continue the synthesis of hormonally dependent proteins such as casein or vitellogenin. In the case of the cellular oncogene c-*myc*, whose RNA stability is transiently increased when cells are stimulated to grow, such continued synthesis would not only be highly wasteful but is also potentially dangerous to the cell. Thus this growth regulatory protein is only required for a short period when cells are entering the growth phase, its continued inappropriate synthesis at other times carrying the risk of disrupting cellular growth regulatory mechanisms, possibly resulting in transition to a cancerous state (see Chapter 9).

These considerations suggest that alterations in RNA stability are used as a significant supplement to transcriptional control in cases where rapid changes in the synthesis of a particular protein are required. Thus, if transcription is shut off in response to withdrawal of a particular signal, inappropriate and metabolically expensive protein synthesis will continue for some time from pre-existing mRNA, unless that RNA is degraded rapidly. Similarly, rapid onset of the expression of a particular gene can be achieved by having a relatively high basal level of transcription with high RNA turnover in the absence of stimulation, allowing rapid onset of translation from pre-existing RNA following stimulation. Hence, as with alternative RNA processing, cases where RNA stability is regulated represent an adaptation to the requirements of a particular situation and do not affect the conclusion that regulation of gene expression occurs primarily at the level of transcription.

5.6 REGULATION OF TRANSLATION

5.6.1 Cases of translational control

The final stage in the expression of a gene is the translation of its messenger RNA into protein (see Section 3.3.6). In theory, therefore, the regulation of gene expression could be achieved by producing all possible mRNA species in every cell and selecting which were translated into protein in each individual cell type. The evidence that different cell types have very different cytoplasmic RNA populations (see Chapter 1) indicates, however, that this extreme model is incorrect. Nonetheless, the regulation of translation, such that a particular mRNA is translated into protein in one situation and not another, does occur in some special cases (for reviews see Curtis *et al.*, 1995; Hentze, 1995; Pain, 1996).

The most prominent of such cases is that of fertilisation. In the unfertilised egg protein synthesis is slow, but upon fertilisation of the egg by a sperm a tremendous increase in the rate of protein synthesis occurs. This increase does not require the production of new mRNAs after fertilisation. Rather, it is mediated by pre-existing maternal RNAs which are present in the unfertilised egg but are only translated after fertilisation. Although in many species such translational control produces only quantitative changes in protein synthesis,

Figure 5.30

Translational control in the clam *Spisula solidissima*. Different proteins are synthesised *in vivo* before fertilisation (track 1) and after fertilisation (track 2). If, however, RNA is isolated either before (track 3) or after fertilisation (track 4) and translated *in vitro* in a cell-free system, identical patterns of proteins are produced. Hence the difference in the proteins produced *in vivo* from identical RNA populations must be due to translational control.

in others it can affect the nature as well as the quantity of the proteins being made before and after fertilisation. Thus in the clam *Spisula solidissima* some new proteins appear after fertilisation, while others which are synthesised in large amounts before fertilisation are repressed thereafter. However, the RNA populations present before and after fertilisation are identical (Fig. 5.30), indicating that translational control processes are operating. Translational control also operates in the egg to ensure that mRNAs which are localised to particular regions of the egg cytoplasm (see Section 5.4.2) are not translated prior to their arrival at the correct region of the cell.

As well as producing parallel changes in the translation of many RNA species, translational control processes may also operate on individual RNAs in a particular cell type. Thus, for example, the rate of translation of the globin RNA in reticulocytes is regulated in response to the availability of the haeme co-factor which is required for the production of haemoglobin. Similarly, the translation of the RNA encoding the iron-binding protein, ferritin, is regulated in response to the availability of iron.

5.6.2 Mechanism of translational control

In principle, translational regulation could operate via modifications in the cellular translational apparatus affecting the efficiency of translation of particular RNAs, or by modifications in the RNA itself which affect the way in which it is translated by the ribosome. Evidence is available indicating that both these types of mechanism are used in different cases.

Thus, in the absence of haeme a cellular protein kinase in the reticulocyte becomes active and phosphorylates the protein initiation factor eIF-2, (see Section 3.3.6) resulting in its inactivation. Since this factor is required for the initiation of protein synthesis, translation of the globin RNA ceases until haeme is available. However, the use of such a mechanism, in which total inactivation of the cellular translational apparatus is used to regulate the translation of a single RNA, is possible only in the reticulocyte, where the globin protein constitutes virtually the only translation product.

In other cell types, where a large number of different RNA species are expressed, such a mechanism is normally used only where large-scale repression of many different RNA species occurs. Thus phosphorylation of eIF-2 (produced by a different kinase to that activated by the absence of haeme) also occurs following exposure of cells to stress, when the translation of most cellular mRNAs is repressed (for a review see Silverman and Williams, 1999).

Interestingly, following exposure to specific stresses, some mRNAs continue to be translated since their protein products are required to mediate response to the stress. Thus, for example, the mRNAs encoding the heat shock proteins are translated even following exposure of cells to elevated temperature and these proteins then produce a protective effect against the damaging effects of stress. This evidently raises the question as to how such translation can occur when general translation initiation factors have been inactivated.

A mechanism by which this is achieved has been elucidated in the case of the cationic amino acid transporter gene (Cat-1) whose mRNA continues to be translated following amino acid starvation in yeast (Fernandez *et al.*, 2001). The Cat-1 mRNA contains an internal ribosome entry site (IRES) which allows the ribosome to bind to the mRNA internally and translate it rather than binding to the 5′ cap as normally occurs (see Section 3.3.1). As such IRES-mediated translation does not require specific initiation factors which are needed for cap-mediated translation, it continues after amino acid starvation, allowing the Cat-1 protein to carry out its function of mediating

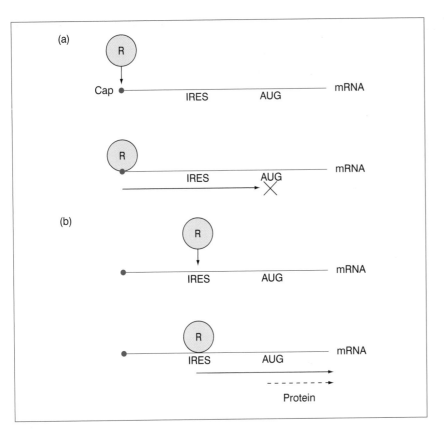

Figure 5.31

Following amino acid starvation, protein synthesis by ribosomes (R) binding to the 5′ cap is decreased due to inactivation of specific initiation factors (a). However, translation initiated by ribosomes binding to an internal ribosome entry site (IRES) continues, so allowing specific proteins to be made even though most protein synthesis is repressed (b).

enhanced amino acid uptake (Fig. 5.31). Since IRES sequences have been found in a variety of viral and cellular mRNAs, it is possible that a similar mechanism may be used to regulate mRNA translation in a number of different situations (for review see Sachs, 2000; Hellen and Sarnow, 2001).

As well as widespread repression of translation, regulation of translation via initiation factors can also produce translational control of specific mRNAs by acting in conjunction with sequences within the mRNA itself which are often located within the 5′ untranslated region upstream of the translational start site (for a review see Jackson and Wickens, 1997). This is seen in the case of specific mRNAs whose translation is activated by treatment of cells with insulin or several growth factors (for a review see Gingras *et al.*, 2001). In this case, in the absence of insulin or growth factor the cap binding translational initiation factor eIF-4E (see Section 3.3.6) is associated with another protein eIF-4E binding protein (eIF-4Ebp). This association prevents the other initiation factors, eIF-4G and eIF-4A, from binding to the RNA. Such binding is particularly necessary for the translation of these insulin-regulated mRNAs since they have a high degree of secondary structure in their 5′ region which must be unwound by the helicase activity of eIF-4A before translation can begin. Following insulin or growth factor treatment, eIF-4Ebp is phosphorylated, releasing it from eIF-4E and

Figure 5.32

In the absence of insulin the
translation initiation factor eIF-4E
is bound by eIF-4Ebp and
translation does not occur due
to the presence of secondary
structure in the mRNA.
Following exposure to insulin,
eIF-4Ebp is phosphorylated
leading to its release from eIF-
4E, eIF-4A and G then bind to
eIF-4E and the helicase activity of
eIF-4A then unwinds the RNA,
allowing translation to occur.

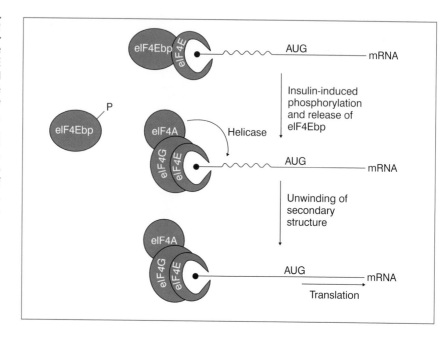

allowing the other factors to bind and unwind the mRNA so that translation can occur (Fig. 5.32).

Interestingly, the mRNA encoding insulin-like growth factor II itself exists in two different forms which are 4.8 kb and 6.0 kb long. The longer mRNA contains sequences upstream of the AUG initiator codon which render it dependent on the helicase activity of eIF-4A and its translation is therefore regulated by this pathway. In contrast, the 4.8 kb mRNA lacks these 5′ sequences and is hence translated in an insulin/growth factor independent manner (Nielsen *et al.*, 1995). Hence in this case, alternative splicing generates two different mRNAs which encode the same protein but are subject to distinct translational regulation.

Several other cases where translational regulation of particular mRNA species is mediated by sequences in their 5′ untranslated region have been defined. Thus, the enhanced translation of the ferritin mRNA in response to iron is mediated by a sequence in this region which can fold into a stem-loop structure (for reviews see Hentze and Kuhn, 1996; Aisen *et al.*, 2001). The structure of this stem-loop is very similar to that found in the 3′ untranslated region of the transferrin receptor mRNA, whose stability is negatively regu-

Table 5.3

Regulation of the transferrin
receptor and ferritin genes

	Effect of iron on protein production	Mechanism	Position of stem-loop
Ferritin	Increased	Increased mRNA translation	5′ untranslated region
Transferrin receptor	Decreased	Decreased mRNA stability	3′ untranslated region

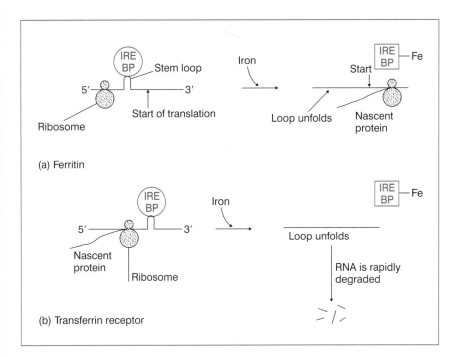

(a) Ferritin

(b) Transferrin receptor

Figure 5.33

Role of iron-induced unfolding of the stem-loop structure in producing increased translation of the ferritin mRNA (panel a) and increased degradation of the transferrin receptor mRNA (panel b). In each case the unfolding of the stem loop is dependent on the dissociation of a binding protein (IRE-BP) whose ability to bind to the stem loop decreases dramatically in the presence of iron.

lated by the presence of iron (see Fig. 5.25 and Table 5.3). This has led to the suggestion that such loops may represent functionally equivalent iron response elements whose opposite effects on gene expression are dependent upon their position (5′ or 3′) within the RNA molecule. This idea was confirmed by transferring the transferrin receptor stem-loop to the 5′ end of an unrelated RNA, resulting in the iron-dependent enhancement of its translation. The identical structure is therefore capable of mediating opposite effects on RNA stability and translation, depending on its position within the RNA molecule. Such an apparent paradox can be explained if it is assumed that the stem-loop structure is an iron response element (IRE) which unfolds in the presence of iron (Fig. 5.33). In the ferritin mRNA, where the element is in the 5′ end, this will allow the binding and movement of the 40S ribosomal subunit along the mRNA until it reaches the initiation codon and translation begins. In contrast, in the transferrin receptor mRNA, where this element is in the 3′ untranslated region, such unfolding renders the RNA susceptible to nuclease degradation at an increased rate. In agreement with this model, an IRE binding protein (IRE-BP) has been identified which binds to the stem-loop element in both the ferritin and transferrin receptor mRNAs. The RNA binding activity of IRE-BP increases dramatically in cells that have been deprived of iron, suggesting that its binding normally stabilises the stem-loop structure.

Not all cases of translational regulation mediated by sequences in the 5′ untranslated region operate via such stem-loop structures, however. Increased expression of the yeast regulatory protein GCN4 in response to amino acid starvation is caused by increased translation of its RNA (for a

Figure 5.34

Presence of short open reading frames capable of producing small peptides in the 5′ untranslated region of the yeast GCN4 RNA. Translation of the RNA to produce these small proteins suppresses translation of the GCN4 protein. The position of the methionine residue beginning each of the small peptides is indicated together with the number of additional amino acids incorporated before a stop codon is reached. When high amino acid levels are present, the small proteins are made and the production of GCN4 is therefore suppressed. When amino acid levels fall, the eIF-2 translational initiation factor is phosphorylated. This prevents the translation of three of the four short peptides and thereby promotes GCN4 production.

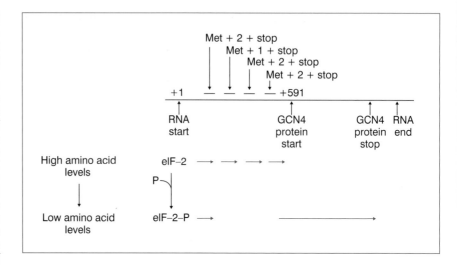

review see Morris and Geballe, 2000). The translational regulation of this molecule is mediated by short sequences within the 5′ untranslated region of the RNA, upstream of the start point for translation of the GCN4 protein. Most interestingly, such sequences are capable of being translated to produce peptides of two or three amino acids (Fig. 5.34). Following translational initiation at the second, third or fourth of these upstream sequences to yield such small peptides, the ribosome fails to reinitiate at the translational start point for GCN4 production and hence this protein is not synthesised.

Following amino acid starvation, the production of these small peptides is suppressed and production of GCN4 correspondingly increased. This switch in initiation also involves the phosphorylation of the eIF-2 factor which is involved in translational regulation in response to haeme or stress (see above). Thus, following amino acid starvation, transfer RNA molecules lacking a bound amino acid accumulate and this activates an enzyme which phosphorylates eIF-2. This decreased activity of eIF-2 caused by its phosphorylation then results in the ribosome not initiating at three of the upstream sites in the 5′ untranslated region, which allows increased initiation at the downstream site leading to enhanced GCN4 production.

Interestingly, alternative initiation codons are also found in the genes encoding two mammalian liver regulatory proteins C/EBPα and C/EBPβ. As discussed in Section 8.4.2, however, in this case the two initiation codons are used to produce two different forms of the proteins. One of these contains an N-terminal sequence which allows it to activate transcription whereas the other form does not and so lacks this ability and can therefore interfere with the stimulatory activity of the activating form. Such alternative translation of the same mRNA thus parallels the use of alternative splicing to produce different protein products from the same gene.

Study of the cases of translational control where the mechanisms have been defined thus makes it clear that, although sequences in the 5′ untranslated region of particular mRNAs are involved in the translational regulation of their expression, the mechanism by which they do so may differ dramatically in different cases.

Although the 5′ untranslated region is an obvious location for sequences involved in mediating translational control, cases where sequences in the 3′ untranslated region play a role in the regulation of translation have been reported. Thus, sequences in this region are involved in modulating the efficiency of translation of specific mRNAs which occurs upon fertilisation of the egg (see Section 5.6.1) as well as in mediating the increased translation of the mRNA encoding lipoxygenase which occurs during erythroid differentiation (Ostareck *et al.*, 2001). In this case, a sequence in the 3′ untranslated region of the lipoxygenase mRNA directly controls the ability of the 60S ribosomal subunit to bind to the mRNA. Thus, early in erythroid differentiation, the 40S ribosomal subunit can bind to the mRNA and, as normally occurs, it moves to the AUG translation initiation codon (see Section 3.3.6). However, a silencing complex, bound within the 3′ untranslated region prevents binding of the 60S ribosomal subunit (Fig. 5.35a). Hence, translation only occurs later in erythroid differentiation when this block is relieved and the 60S subunit can bind and initiate translation (Fig. 5.35b).

A completely different mechanism involving sequences in the 3′ untranslated region regulating translation is seen in the case of *Xenopus* oocytes (for a review see Richter and Theurkauf, 2001). In this case, a number of mRNAs are maintained in a non-translated state in dormant oocytes which are arrested early in meiosis. When the oocyte is stimulated to undergo meiosis,

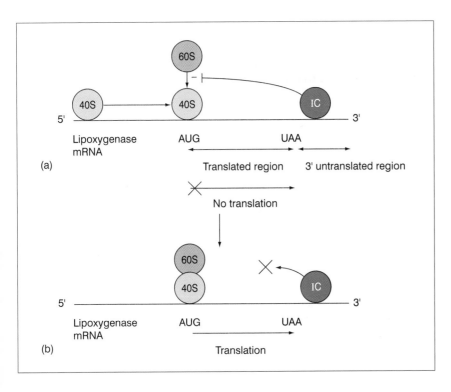

(a)

(b)

Figure 5.35

During erythroid differentiation, translation of the lipoxygenase mRNA is regulated by an inhibitory complex (IC) which binds to the 3′ untranslated region of the mRNA. Early in erythroid differentiation (a), the 40S ribosomal subunit can bind to the mRNA and move to the AUG initiation codon. However, the inhibitory complex prevents the 60S ribosomal subunit from binding and so translation is inhibited. Later in the differentiation process (b), the negative effect of the inhibitory complex is blocked and translation occurs.

the poly (A) tail of these mRNAs is lengthened and the mRNA is translated. Hence, in this case a connection appears to exist between the extension of the poly (A) tail to its full length and the onset of translation.

In the case of one of these mRNAs, encoding cyclin B, this has been shown to occur because in the dormant oocytes the polyadenylation factor CPSF (see Section 3.3.2) is not bound to the AAUAAA sequence which is involved in polyadenylation. However, another factor CPEB is bound to an adjacent region of the RNA together with a further protein known as maskin (Fig. 5.36a). Maskin, in turn, appears to interact with the translation initiation factor eIF-4E and prevent binding of eIF-4G. Hence, maskin acts exactly like the eIF-4E binding protein described above in that it prevents translation by inhibiting binding of eIF-4G.

When the oocyte is stimulated to pass through meiosis, CPEB is phosphorylated, which allows it to recruit CPSF to the AAUAAA sequence and hence polyadenylation occurs. This also results in maskin dissociating from eIF-4E, thereby allowing eIF-4G to bind and translation to begin (Fig. 5.36b).

Hence, mechanisms involving regulation of translation by eIF-4E can involve either a protein (eIF-4bp) which binds only to eIF-4E or one (maskin) which is recruited by an RNA-bound protein (CPEB). Similarly, regulation of translation via sequences located in the 3′ untranslated region can involve different mechanisms just as occurs for processes involving sequences near the 5′ end of the mRNA.

Figure 5.36

In *Xenopus* oocytes arrested early in meiosis, polyadenylation of the cyclin B mRNA is blocked by the lack of CPSF bound to the AAUAAA polyadenylation signal whilst its translation is blocked by the maskin protein which binds to the translation initiation factor eIF-4E and prevents eIF-4G from binding (a) (compare Fig. 5.32). When the oocyte is activated and moves through meiosis, a protein known as CPEB is phosphorylated (b). This results in CPSF being able to bind to CPEB, thereby recruiting it to the AAUAAA sequence and allowing polyadenylation to occur. This also disrupts the association of maskin and eIF-4E, allowing binding of eIF-4G and initiation of translation.

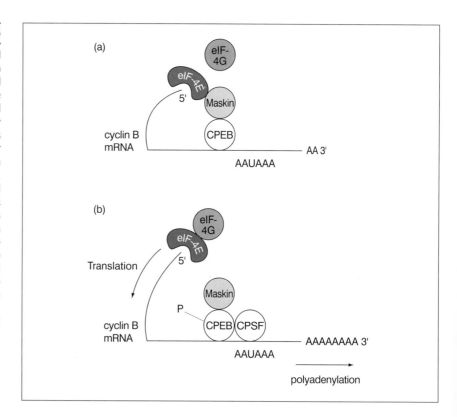

In summary, therefore, it is clear that cases of translational control can be mediated by sequences in various parts of the RNA and can involve secondary structure, use of different translation initiation codons, regulation of ribosome binding or the regulation of polyadenylation.

5.6.3 Significance of translational control

Many of the cases of translational control occur in situations where very rapid responses are required. Thus, following fertilisation a very rapid activation of cellular growth processes is required. Similarly, following heat shock it is necessary to shut down rapidly the synthesis of most enzymes and structural proteins and begin to synthesise the protective heat-shock proteins. Such regulation can be achieved rapidly by translational control, supplemented by increased transcription of the genes encoding the heat-shock proteins. Once again, as with other cases of post-transcriptional regulation, translational control can be viewed as supplementing the regulation of transcription in order to meet the requirements of particular specialised cases. In the case of the heat-shock proteins, this combination of transcriptional and translational control is further supplemented by an increased stability of the heat-shock mRNAs following exposure to elevated temperature, providing a further means of producing a rapid and effective response.

The case of the yeast GCN4 protein provides a different aspect to translational control, however. This protein is a transcriptional regulator, which increases the transcription of several genes encoding the enzymes of amino acid biosynthesis in response to a lack of one or more amino acids. In this case the synthesis of a transcriptional regulatory protein is regulated by translational control. When taken together with the increasing evidence (see Section 8.4.3) that many mammalian transcriptional regulatory molecules may pre-exist in an inactive form and be activated by protein modifications (for example, phosphorylation), this suggests that it may be necessary for the cell to control the expression of some of its transcriptional regulatory molecules at levels other than that of gene transcription. Translational control may be one means by which such regulation is achieved.

5.7 CONCLUSIONS

A wide variety of cases exist in which gene expression can be regulated at levels other than transcription. In some lower organisms such post-transcriptional regulation may constitute the predominant form of gene control. In mammals, however, it appears to represent an adaptation to particular situations, including the need to generate two closely related proteins from one gene by alternative RNA splicing or RNA editing, or to respond rapidly to the withdrawal of hormonal stimulation or stress by the regulation of RNA stability or translation. It is likely that many more cases of post-transcriptional regulation will be described in the future, whilst this form of regulation may be the predominant one for the proteins which themselves regulate the transcription of other genes.

REFERENCES

Aisen, P., Enns, C. and Wessling-Resnick, M. (2001). Chemistry and biology of eukaryotic iron metabolism. *International Journal of Biochemistry and Cell Biology* 33, 940–959.

Baker, B.S. (1989). Sex in flies: the splice of life. *Nature* 340, 521–524.

Bandziulis, R.J., Swanson, M.S. and Dreyfuss, G. (1989). RNA-binding proteins as developmental regulators. *Genes and Development* 3, 431–437.

Bashirullah, A., Cooperstock, R.L. and Lipshitz, H. D. (2001). Spatial and temporal control of RNA stability. *Proceedings of the National Academy of Sciences USA* 98, 7025–7028.

Beelman, C.A. and Parker, R. (1995). Degradation of mRNA in eukaryotes. *Cell* 81, 179–183.

Benne, R. (1996). RNA editing: how a message is changed. *Current Opinion in Genetics and Development* 6, 221–231.

Black, D.L. (2000). Protein diversity from alternative splicing: a challenge for bioinformatics and post-genome biology. *Cell* 103, 367–370.

Blencowe, B.J. (2000). Exonic splicing enhancers: mechanism of action, diversity and role in human genetic diseases. *Trends in Biochemical Sciences* 25, 106–110.

Bousquet-Antonelli, C., Presutti, C. and Tollervey, D. (2000). Identification of a regulated pathway for nuclear pre-mRNA turnover. *Cell* 102, 765–775.

Breitbart, R.E., Andreadis, A. and Nadal-Ginard, B. (1987). Alternative splicing: a ubiquitous mechanism for the generation of multiple protein isoforms from different genes. *Annual Review of Biochemistry* 56, 467–495.

Burd, C.G. and Dreyfuss, G. (1994). Conserved structure and functions of RNA binding proteins. *Science* 265, 615–621.

Cullen, B.R. (1991). Regulation of HIV-1 gene expression. *FASEB Journal* 5, 2361–2368.

Cullen, B.R. and Malim, H. (1991). The HIV-1 Rev protein: prototype of a novel class of eukaryotic transcriptional regulators. *Trends in Biochemical Sciences* 16, 341–348.

Curtis, D., Lehmann, R. and Zamore, P.D. (1995). Translational regulation in development. *Cell* 81, 171–178.

Danner, D. and Leder, P. (1985). Role of an RNA cleavage/poly (A) addition site in the production of membrane bound and secreted IgM mRNA. *Proceedings of the National Academy of Sciences of the USA* 82, 8658–8662.

Decker, C.J. and Parker, R. (1994). Mechanism of mRNA degradation in eukaryotes. *Trends in Biochemical Sciences* 19, 336–340.

Fernandez, J., Yaman, I., Mishra, R., Merrick, W.C., Snider, M.D., Lamers, W.H. and Hatzoglou, M. (2001). Internal ribosome entry site-mediated translation of a mammalian mRNA is regulated by amino acid availability. *Journal of Biological Chemistry* 276, 12285–12291.

Gerace, L. (1995). Nuclear export signals and the fast track to the cytoplasm. *Cell* 82, 341–344.

Gingras, A.-C., Raught, B. and Sonenberg, N. (2001). Regulation of translation initiation by FRAP/mTOR. *Genes and Development* 15, 807–826.

Grabowski, P.J. (1998). Splicing regulation in neurons: tinkering with cell-specific control. *Cell* 92, 709–712.

Graveley, B.R. (2001). Alternative splicing: increasing diversity in the proteomic world. *Trends in Genetics* 17, 100–107.

Graves, R.A., Pandey, N.B., Chodchoy, N. and Marzluff, W.F. (1987). Translation is required for regulation of histone mRNA degradation. *Cell* 48, 615–626.

Grosset, C., Chen, C.-Y.A., Xu, N., Sonenberg, N., Jacquemin-Sablon, H. and Shyu, A.-B. (2000). A mechanism for translationally coupled mRNA turnover: interaction between the Poly (A) tail and a c-*fos* RNA coding determinant via a protein complex. *Cell* **103**, 29–40.

Guyette, W.A., Matusik, R.A. and Rosen, J.M. (1979). Prolactin-mediated transcriptional and post-transcriptional control of casein gene expression. *Cell* **17**, 1013–1023.

Handa, N., Nureki, O., Kurimoto, K., Kim, I., Sakamoto, H., Shimura, Y., Muto, Y. and Yokoyama, S. (1999). Structural basis for recognition of the *tra* mRNA precursor by the Sex-lethal protein. *Nature* **398**, 579–585.

Hazelrigg, T. (1998). The destinies and destinations of RNAs. *Cell* **95**, 451–460.

Hellen, C.U.T. and Sarnow, P. (2001). Internal ribosome entry sites in eukaryotic mRNA molecules. *Genes and Development* **15**, 1593–1612.

Hentze, M.W. (1995). Translational regulation: versatile mechanisms for metabolic and developmental control. *Current Opinion in Cell Biology* **7**, 393–398.

Hentze, M.W. and Kuhn, L.C. (1996). Molecular control of vertebrate iron metabolism : mRNA based regulatory circuits operated by iron, nitric oxide and oxidative stress. *Proceedings of the National Academy of Sciences USA* **93**, 8175–8182.

Higuchi, M., Maas, S., Single, F.N., Hartner, J., Rozov, A., Burnashev, N., Feldmeyer, D., Sprengel, R. and Seeburg, P.H. (2000). Point mutation in an AMPA receptor gene rescues lethality in mice deficient in the RNA-editing enzyme ADAR2. *Nature* **406**, 78–81.

Jackson, R.J. and Wickens, M. (1997). Translational controls impinging on the 5' untranslated region and initiation factor proteins. *Current Opinion in Genetics and Development* **7**, 233–241.

Jacobson, A. and Peltz, S.W. (1996). Interrelationships of the pathways of mRNA decay and translation in eukaryotic cells. *Annual Review of Biochemistry* **65**, 693–739.

Leff, S.E., Rosenfeld, M.G. and Evans, R.M. (1986). Complex transcriptional units: diversity in gene expression by alternative RNA processing. *Annual Review of Biochemistry* **55**, 1091–1117.

Lynch, K.W. and Maniatis, T. (1995). Synergistic interactions between two distinct elements of a regulated splicing enhancer. *Genes and Development* **9**, 284–293.

Manley, J.L. and Tacke, R. (1996). SR proteins and splicing control. *Genes and Development* **10**, 1569–1579.

Marzluff, W.F. (1992). Histone 3' ends: essential and regulatory function. *Gene Expression* **2**, 93–97.

Mohr, E. and Richter, D. (2001). Messenger RNA on the move: implications for cell polarity. *International Journal of Biochemistry and Cell Biology* **33**, 669–679.

Morris, D.R. and Geballe, A.P. (2000). Upstream open reading frames as regulators in mRNA translation. *Molecular and Cellular Biology* **20**, 8635–8642.

Nielsen, F.C., Ostergard, L., Neilsen, J. and Christiansen, J. (1995). Growth-dependent translation of IGFII mRNA by a rapamycin sensitive pathway. *Nature* **377**, 358–362.

O'Donovan, K.J. and Darnell, R.B. (2001). Neuronal signalling through alternative splicing: some CaRRE... Science's STKE http://stke.sciencemag.org/cgi/content/full/OC[rule]sigtrans;2001/94/pe2.

Ostareck, D.H., Ostareck-Lederer, A., Shatsky, I.N. and Hentze, M.W. (2001). Lipoxygenase mRNA silencing in erythroid differentiation: the 3'UTR regulatory complex controls 60S ribosomal subunit joining. *Cell* **104**, 281–290.

Pachter, J.S., Yen, T.J. and Cleveland, D.W. (1987). Auto-regulation of tubulin expression is achieved through degradation of polysomal tubulin mRNAs. *Cell* 51, 283–292.

Pain, V.M. (1996). Initiation of protein synthesis in eukaryotic cells. *European Journal of Biochemistry* 236, 747–771.

Powell, D.J., Freidman, J.M., Oulette, A.J., Krauter, K.S. and Darnell, J.E. (1984). Transcriptional and post-transcriptional control of specific messenger RNAs in adult and embryonic liver. *Journal of Molecular Biology* 179, 21–35.

Raitskin, O., Cho, D.-S.C., Sperling, J., Nishikura, K. and Sperling, R. (2001). RNA editing activity is associated with splicing factors in InRNP particles: the nuclear pre-mRNA processing machinery. *Proceedings of the National Academy of Sciences USA* 98, 6571–6576.

Reenan, R.A. (2001). The RNA world meets behaviour: AfiI pre-mRNA editing in animals. *Trends in Genetics* 17, 53–56.

Richter, J.D. and Theurkauf, W.E. (2001). The message is in the translation. *Science* 293, 60–62.

Roegiers, F. and Jan, Y.N. (2000). Staufen: a common component of mRNA transport in oocytes and neurones? *Trends in Cell Biology* 10, 220–224.

Rosenfeld, M.G., Amara, S.G. and Evans, P.M. (1984). Alternative RNA processing: determining neuronal phenotype. *Science* 225, 1315–1320.

Ross, J. (1996). Control of messenger RNA stability in higher eukaryotes. *Trends in Genetics* 12, 171–175.

Rueter, S.M., Dawson, T.R. and Emeson, R.B. (1999). Regulation of alternative splicing by RNA editing. *Nature* 399, 75–79.

Sachs, A.B. (1993). Messenger RNA degradation in eukaryotes. *Cell* 74, 413–421.

Sachs, A.B. (2000). Cell cycle-dependent translation initiation: IRES elements prevail. *Cell* 101, 243–245.

Scott, J. (1997). Message change for a fat controller. *Nature* 387, 242–243.

Silverman, R.H. and Williams, B.R.G. (1999). Translational control perks up. *Nature* 397, 208–209.

Smith, C.W.J. and Valcárcel, J. (2000). Alternative pre-mRNA splicing: the logic of combinatorial control. *Trends in Biochemical Sciences* 25, 381–388.

Smith, H.C. and Sowden, M.P. (1996). Base-modification mRNA editing through deamination – the good, the bad and the unregulated. *Trends in Genetics* 12, 418–424.

Stein G.S. Stein, J.L., van Wijnen, A.J. and Liam, J.B. (1992). Regulation of histone gene expression. *Current Opinion in Cell Biology* 4, 166–173.

Takagaki, Y., Selpet, R.L., Peterson, M.L. and Manley, J.L. (1996). The polyadenylation factor CstF-64 regulates alternative processing of IgM heavy chain pre mRNA during B cell differentiation. *Cell* 87, 941–952.

Valcárcel, J., Singh, R., Zamore, P.D. and Green, M.R. (1993). The protein sex-lethal antagonises the splicing factor U2AF to regulate splicing of transformer pre-mRNA. *Nature* 362, 171–175.

Wang, J. and Manley, J.L. (1997). Regulation of pre-mRNA splicing in metazoa. *Current Opinion in Genetics and Development* 7, 205–211.

Wickens, M., Anderson, P. and Jackson, R.J. (1997). Life and death in the cytoplasm: messages from the 5' end. *Current Opinion in Genetics and Development* 7, 220–232.

Wold, B. J., Klein, W.H., Hough-Evans, B.R., Britten, R.J. and Davidson, E.H. (1978). Sea urchin embryo mRNA sequences expressed in the nuclear RNA of adult tissues. *Cell* 14, 941–950.

Xie, J. and Black, D.L. (2001). A CaMK IV responsive RNA element mediates depolarisation-induced alternative splicing of ion channels. *Nature* **410**, 936–939.

Yen, T.J., Machlin, P.S. and Cleveland, D.W. (1988). Autoregulated instability of β-tubulin by recognition of the nascent amino-acids of β-tubulin. *Nature* **334**, 580–585.

Young, R.A., Hagenbuchle, O. and Schibler, U. (1981). A single mouse α-amylase gene specifies two different tissue specific mRNAs. *Cell* **23**, 451–458.

6 TRANSCRIPTIONAL CONTROL – CHROMATIN STRUCTURE

6.1 INTRODUCTION

Having established that the primary control of eukaryotic gene expression lies at the level of transcription, it is necessary to investigate the mechanisms responsible for this effect. The fact that regulation at transcription is also responsible for the control of gene expression in bacteria suggests that insights into these procedures obtained in these much simpler organisms (for a review see Travers, 1993) may be applicable to higher organisms. It is possible therefore that regulation of transcription in eukaryotes might occur by means of a protein, present in all tissues, which binds to the promoter region of a particular gene and prevents its expression. In one particular cell type, or in response to a particular signal such as heat shock, this protein would be inactivated either directly (Fig. 6.1a) or by binding of another factor (Fig. 6.1b), and would no longer bind to the gene. Hence transcription would occur only in the one cell type or in response to the signal. This mechanism is based on that regulating the expression of the *lac* operon

Figure 6.1

Model for the activation of gene expression in a particular tissue by inactivation (a) or binding out (b) of a repressor present in all tissues.

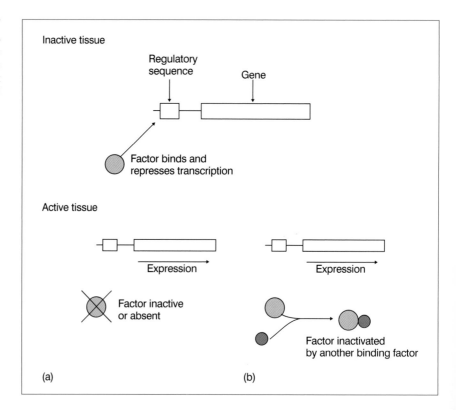

containing the genes encoding proteins required for the metabolism of lactose. This operon is normally repressed by the *lac* repressor protein; binding of lactose to this protein, however, results in its inactivation and allows transcription of the operon (for a review see Muller-Hill, 1996).

Alternatively, the fact that most eukaryotic genes are inactive in most tissues, and become active only in one tissue or in response to a particular signal, suggests that it may be more economical to have a system in which the gene is constitutively inactive in most tissues, without any repressor being required. Activation of the gene would then require a particular factor binding to its promoter. The specific expression pattern of the gene would be controlled by the presence of this factor only in the expressing cell type (Fig. 6.2a) or, alternatively, by the factor's requirement for a co-factor such as a steroid hormone to convert it to an active form (Fig. 6.2b). This mechanism is based on the regulation of the arabinose operon in *Escherichia coli* in which binding of the substrate arabinose to the regulatory araC protein allows it to induce transcription of the genes required for the metabolism of arabinose (for a review see Raibaud and Schwartz, 1984).

Hence, based on the known mechanisms of gene regulation in bacteria, it is possible to produce models of how gene regulation might operate in

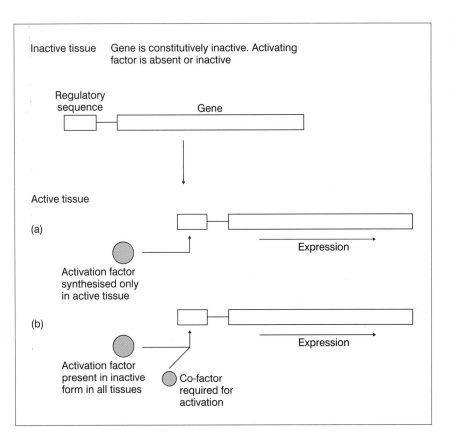

Figure 6.2

Model for the activation of gene expression in a particular tissue by an activator present only in that tissue (a) or activated by a co-factor (b) only in that tissue.

eukaryotes. Indeed, a considerable amount of evidence indicates that many cases of gene regulation do use the activation type mechanisms seen in the arabinose operon. Thus, as will be discussed in Chapters 7 and 8, the effects of glucocorticoid and other steroid hormones on gene expression are mediated by the binding of the steroid to a receptor protein. This activated complex then binds to particular sequences upstream of steroid-responsive genes and activates their transcription.

Even in the case of steroid hormones, however, such mechanisms cannot account entirely for the regulation of gene expression. In the chicken, administration of the steroid hormone oestrogen results in the transcriptional activation in oviduct tissue of the gene encoding ovalbumin, as discussed in Chapter 4. In the liver of the same organism, however, oestrogen treatment has no effect on the ovalbumin gene but instead results in the activation of a completely different gene, encoding the protein vitellogenin. Such tissue-specific differences in the response to a particular treatment are, of course, entirely absent in single-celled bacteria and cannot be explained simply on the basis of the activation of a single DNA-binding protein by the hormone.

Similarly, models of this type cannot explain the data obtained by Becker *et al.* (1987) in their studies of the regulation of the rat tyrosine amino-transferase gene. All the protein factors binding to the regulatory regions of this gene were detectable in both the liver, where the gene is expressed, and in other tissues, where no expression is observed. Moreover, the proteins isolated from these tissues were equally active in binding to the appropriate region of the gene in deproteinised DNA. Only in the liver, however, was actual binding of these factors to the DNA of the gene within its normal chromosomal structure detectable. Once again, these data cannot be explained solely on the basis of models in which proteins stimulate or inhibit gene expression by binding to appropriate DNA sequences. Rather, an understanding of eukaryotic gene regulation will require a knowledge of the ways in which the relatively short-term regulatory processes mediated by the binding of proteins to specific DNA sequences in this manner interact with the much longer-term regulatory processes, which establish and maintain the differences between particular tissues and also control their response to treatment with effectors such as steroids. This baseline regulation of chromatin structure thus renders eukaryotic gene regulation fundamentally different from prokaryotic gene regulation where this does not occur (for a review see Struhl, 1999). The long-term control processes which regulate chromatin structure will be discussed in this chapter, and the DNA sequences and proteins which actually regulate transcription in response to particular signals will be discussed in Chapters 7 and 8.

6.2 COMMITMENT TO THE DIFFERENTIATED STATE AND ITS STABILITY

It is evident that the existence of different tissues and cell types in higher eukaryotes requires mechanisms that establish and maintain such differences, and that these mechanisms must be stable in the long term. Thus, despite some exceptions (see Section 2.2.3), tissues or cells of one type do not, in general, change spontaneously into another cell type. Inspection of mammalian brain tissue does not reveal the presence of cells typical of liver or kidney, and antibody-producing B-cells do not spontaneously change into muscle cells when cultured in the laboratory. Hence, cells must be capable of maintaining their differentiated state, either within a tissue or during prolonged periods in culture.

Indeed, the long-term control processes that achieve this effect must regulate not only the ability to maintain the differentiated state more or less indefinitely but also the observed ability of cells to remember their particular cell type, even under conditions when they cannot express the characteristic features of that cell type. Thus in the experiments of Coon (1966) it was possible to regulate the behaviour of cartilage-producing cells in culture according to the medium in which they were placed. In one medium the cells were capable of expressing their differentiated phenotype and produced cartilage-forming colonies synthesising an extracellular matrix containing chondroitin sulfate. By contrast, in a medium favouring rapid division the cells did not form such colonies and, instead, divided rapidly and lost all the specific characteristics of cartilage cells, becoming indistinguishable from undifferentiated fibroblast-like cells in appearance. Nonetheless, even after 20 generations in this rapid growth medium the cells could resume the appearance of cartilage cells and synthesis of chondroitin sulfate if returned to the appropriate medium. This process did not occur if other cell types or undifferentiated fibroblast cells were placed in an identical medium. Hence the cartilage cells were capable not only of maintaining their differentiated cell type in a particular medium supplying appropriate signals, but also of remembering it in the absence of such signals and returning to the correct differentiated state when placed again in the appropriate medium.

These experiments lead to the idea that cells belong to a particular lineage and that mechanisms exist to maintain the commitment of cells to a particular lineage, even in the absence of the characteristics of the differentiated phenotype. A variety of evidence exists to suggest that cells become committed to a particular differentiated stage or lineage well before they express any features characteristic of that lineage, and that such commitment can be maintained through many cell generations.

Perhaps the most dramatic example of this effect occurs in the fruit fly, *Drosophila melanogaster*. In this organism, the larva contains many discs consisting of undifferentiated cells located at intervals along the length of the body and indistinguishable from each other in appearance. Eventually, these imaginal discs (Hadorn, 1968) will form the structures of the adult, the most anterior pair producing the antennae and others producing the wings, legs, etc. In order to do this, however, the discs must pass through the intermediate pupal state where they receive the appropriate signals to differentiate into the adult structures. If they are removed from the larva and placed directly in the body cavity of the adult, they will remain in an undifferentiated state, since the signals inducing differentiation will not be received (Fig. 6.3).

This process can be continued for many generations. Thus the disc removed from the adult can be split in two, one half being used to propagate the cells by being placed directly in another adult and the other half being tested to see what it will produce when placed in a larva that is allowed to proceed to an adult through the pupal state. When this is done, it is found that a disc which would have given rise to an antenna can still do so when placed in a larva after many generations of passage through adult organisms in an undifferentiated state. This is the case even if the disc is placed at a position within the larva very different from that normally occupied by the imaginal disc producing the antenna. The cells of the imaginal disc within the normal larva must therefore have undergone a commitment event to the eventual production of a particular differentiated state, which they can maintain for many generations, prior to having ever expressed any features characteristic of that differentiated state (Fig. 6.3).

Figure 6.3

The use of larval imaginal discs to demonstrate the stability of the committed state. The disc cells maintain their commitment to produce a specific adult structure even after prolonged growth in an undifferentiated state in successive adults.

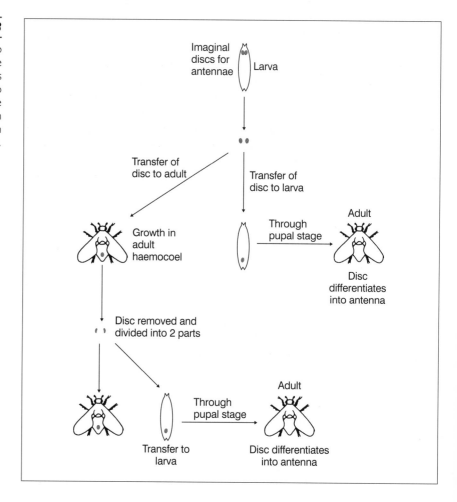

These examples of the stability of the differentiated state and commitment to it imply the existence of long-term regulatory processes capable of maintaining such stability. However, the existence of cases where such stability breaks down, either artificially following nuclear transplantation or more naturally as in lens regeneration (see Section 2.2.3), indicates that such processes, although stable, cannot be irreversible. Indeed, even in the imaginal discs of *Drosophila* the stability of the committed state can be broken down by culture for long periods in adults, the discs giving rise eventually to tissues other than the one intended originally when placed in the larva. This breakdown of the committed state is not a random process but proceeds in a highly characteristic manner. Wing cells are the first abnormal cells produced by a disc that should produce an eye, with other cell types being produced only subsequently. Similarly, a disc that should produce a leg never produces genitalia, although it can produce other cell types. The various transitions that can occur in this system are summarised in Fig. 6.4. Most interestingly, these various transitions are precisely those observed in the homeotic mutations in *Drosophila* (see Section 8.2.2.) which convert one adult body part into another. Hence each of these transitions is likely to be controlled by a specific gene whose activity changes when commitment breaks down.

These examples and the other experiments discussed in Section 2.2 eliminate irreversible mechanisms such as DNA loss as a means of explaining the process of commitment to the differentiated state. It is thought, therefore, that the semi-stability of this process and its propagation through many cell generations is due to the establishment of a particular pattern of association of the DNA with specific proteins. The structure formed by DNA and its associated proteins is known as chromatin. An understanding of how long-term gene regulation is achieved therefore requires a knowledge of the structure of chromatin.

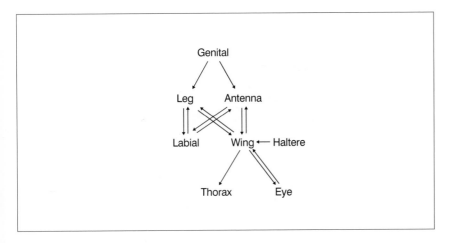

Figure 6.4

Structures that arise from specific imaginal discs of *Drosophila* following the breakdown of commitment.

6.3 CHROMATIN STRUCTURE

If the DNA in a single human individual were to exist as an extended linear molecule, it would have a length of 5×10^{10} km and would extend 100 times the distance from the Earth to the Sun. Clearly this is not the case. Rather, the DNA is compacted by folding in a complex with specific nuclear proteins into the structure known as chromatin (for a review see Wolffe, 1995). Of central importance in this process are the five types of histone proteins (Table 6.1) whose high proportion of positively charged amino acids neutralises the net negative change on the DNA and allows folding to occur. The basic unit of this folded structure is the nucleosome (for a review see Kornberg and Lorch, 1999), in which approximately 200 base pairs of DNA are associated with a histone octamer containing two molecules of each of the four core histones, H2A, H2B, H3, and H4 (Fig. 6.5). In this structure approximately 146 base pairs of DNA make almost two full turns around the histone octamer, and the remainder serves as a linker DNA, joining one nucleosome to another.

The detailed arrangement of the histones and the DNA within the nucleosome has been confirmed by high resolution structural analysis using X-ray crystallography (Luger *et al.*, 1997; for a review see Rhodes, 1997) (colour plate 10). This demonstrates that the histones form H2A-H2B and H3-H4 heterodimers which then associate to form the histone octamer with the DNA wrapped around the surface of the octamer. Interestingly, the amino-terminal tails of the histones extend beyond the surface of the nucleosome particle and are likely to be involved in interactions between nucleosomes. This is of particular interest since these amino-terminal histone tails are subject to specific modifications which are known to alter chromatin structure (see Section 6.6) and these modifications may thus act by affecting the nucleosome–nucleosome interactions which are mediated via these regions of the histone molecule.

Within the nucleosome, the linker DNA is more accessible than the highly protected DNA tightly wrapped around the octamer and is therefore prefer-

Figure 6.5

Structure of DNA and the core histones within a single nucleosome.

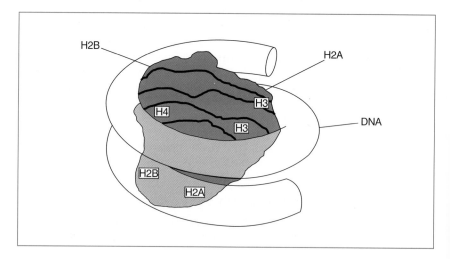

Histone	Type	Molecular weight	Molar ratio
H1	Lysine rich	23000	1
H2A	Slightly lysine rich	13960	2
H2B	Slightly lysine rich	13744	2
H3	Arginine rich	15342	2
H4	Arginine rich	11282	2

Table 6.1

The histones

entially cleaved when chromatin is digested with small amounts of the DNA-digesting enzyme, micrococcal nuclease. Thus, if DNA is isolated following such mild digestion of chromatin, it produces a ladder of DNA fragments of multiples of 200 base pairs on gel electrophoresis, representing the results of cleavage in some but not all linker regions (Fig. 6.6a, track T). This can be correlated with the properties of nucleosomes isolated from the digested chromatin. Thus a partially digested chromatin preparation can be fractionated into individual nucleosomes associated with 200 bases of DNA (Fig. 6.6a, track D), dinucleosomes associated with 400 bases of DNA (Fig. 6.6a, track C), and so on (Finch *et al.*, 1975). The individual mononucleosomes, dinucleosomes or larger complexes in each of these fractions can be observed readily in the electron microscope (Fig. 6.6b). These experiments provide direct evidence for the organisation of DNA into nucleosomes within the cell.

This organisation of DNA into nucleosomes, which can be directly visualised in the electron microscope as the beads on a string structure of chromatin (Fig. 6.7), constitutes the first stage in the packaging of DNA. Subsequently this structure is folded upon itself into a much more compact structure, known as the solenoid (Fig. 6.8). In the formation of this structure histone H1 plays a critical role. As will be seen from Table 6.1, this histone is present at half the level of the other histones and is not part of the core histone octamer. Instead, one molecule of histone H1 seals the two turns which the DNA makes around the core octamer of the other histones. In the solenoid structure, these single histone H1 molecules associate with one another, resulting in tight packing of the individual nucleosomes into a 30 nm fibre (for reviews see Felsenfeld and McGhee, 1986; van Holde and Zlatanova, 1995). This 30 nm fibre is the basic structure of chromatin in cells not undergoing division, although during cell division the DNA is further compacted by extensive looping of the 30 nm fibre to form the readily visible chromosomes. In this structure the fibre loops are linked at their bases via specific DNA sequences known as matrix attachment regions (MAR) to a protein scaffold known as the nuclear matrix (for a review see Saitoh *et al.*, 1995) (Fig. 6.9).

6.4 CHANGES IN CHROMATIN STRUCTURE IN ACTIVE OR POTENTIALLY ACTIVE GENES

6.4.1 Active DNA is organised in a nucleosomal structure

Having established that the bulk of cellular DNA is associated with histone molecules in a nucleosomal structure, an obvious question is whether genes

Figure 6.6a

Separation of mononucleosomes (D), dinucleosomes (C), trinucleosomes (B), and tetranucleosomes (A) by sucrose gradient centrifugation. The upper panel shows the peaks of absorbance produced by the individual fractions of the gradient, the lower panel shows the DNA associated with each fraction separated by gel electrophoresis. Track T shows the DNA ladder produced from a preparation containing all the individual nucleosome fractions.

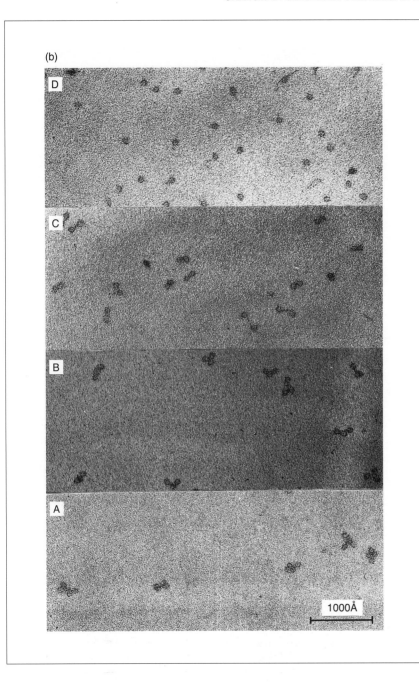

(b)

Figure 6.6b

Electron microscopic analysis of the fractions separated in panel (a). Mononucleosomes are clearly visible in fraction D, dinucleosomes in fraction C, and so on.

Figure 6.7

"Beads on a string" structure of chromatin visualised in the electron microscope in the presence (a) or absence (b) of histone H1. The bar indicates 0.5 µm.

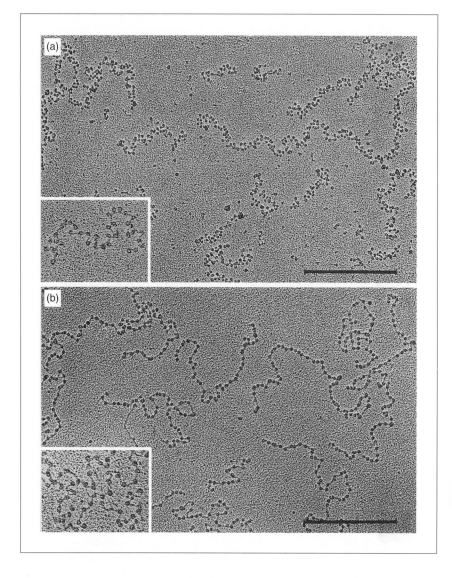

which are either being transcribed or are about to be transcribed in a particular tissue are also organised in this manner or whether they exist as naked, nucleosome-free DNA.

Two main lines of evidence suggest that such genes are organised into nucleosomes. First, if DNA that is being transcribed is examined in the electron microscope, in most cases, the characteristic beads on a string structure (see Section 6.3) is observed, with nucleosomes visible both behind and in front of the RNA polymerase molecules transcribing the gene (Fig. 6.10; McKnight *et al.*, 1978). Thus, although this structure may break down in genes which are being extremely actively transcribed, such as occurs for the genes encoding ribosomal RNA during oogenesis, it is maintained in most

Figure 6.8

The solenoid structure of chromatin.

transcribed genes. Secondly, if DNA organised into nucleosomes is isolated as a ladder of characteristically sized fragments following mild digestion with micrococcal nuclease (see Fig. 6.6), the DNA from active genes is found in these fragments in the same proportion as in total DNA (Lacey and Axel, 1975). Similarly, no enrichment or depletion in the amount of a particular gene found in these fragments is observed when the DNA isolated from a tissue actively transcribing the gene is compared to DNA isolated from a tissue that does not transcribe it. The ovalbumin gene, for example, is found in nucleosome-sized fragments of DNA in these experiments, regardless of whether chromatin from hormonally stimulated oviduct tissue or from liver tissue is used. This is in agreement with the idea that transcribed DNA is still found in a nucleosomal structure rather than as naked DNA, which would be rapidly digested by micrococcal nuclease and hence would not appear in the ladder of nucleosome-sized DNA fragments.

6.4.2 Sensitivity of active chromatin to DNaseI digestion

In order to probe for other differences that may exist between transcribed and non-transcribed DNA, many workers have studied the sensitivity of these different regions to digestion with the pancreatic enzyme deoxyribonuclease I (DNaseI). Although this enzyme will eventually digest all the

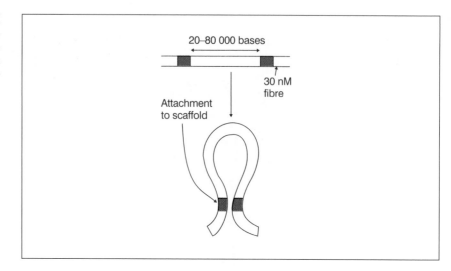

DNA in a cell, if it is applied to chromatin in small amounts for a short period, only a small amount of the DNA will be digested. The proportion of transcribed or non-transcribed genes present in the relatively resistant undigested DNA in a given tissue compared to the proportion in total DNA can therefore be used to detect the presence of differences between active and inactive DNA in their sensitivity to digestion with this enzyme. The fate of an individual gene in this procedure can be followed simply by cutting the DNA surviving digestion with an appropriate restriction enzyme and carrying out the standard Southern blotting procedure (Section 2.2.4) with a labelled probe derived from the gene of interest. The presence or absence of the appropriate band derived from the gene in the digested DNA provides a measure of the resistance of the gene to DNaseI digestion (Fig. 6.11).

This procedure has been used to demonstrate that a very wide range of genes that are active in a particular tissue exhibit a heightened sensitivity to DNaseI, which extends over the whole of the transcribed gene and for some distance upstream and downstream of the transcribed region. For example, when chromatin from chick oviduct tissue is digested, the active ovalbumin gene is rapidly digested and its characteristic band disappears from the Southern blot under digestion conditions which leave the DNA from the inactive globin gene undigested. This difference between the globin and ovalbumin genes is dependent upon their different activity in oviduct tissue rather than any inherent difference in the resistance of the genes themselves to digestion. Thus, if the digestion is carried out with chromatin isolated from red blood cell precursors in which the globin gene is active and that encoding ovalbumin is inactive, the reverse result to that seen in the oviduct is obtained, with the globin DNA exhibiting preferential sensitivity to digestion and the ovalbumin DNA being resistant to such digestion.

It is clear, therefore, that actively transcribed genes, though packaged into nucleosomes, are in a more open chromatin structure than that found for non-transcribed genes, and are hence more accessible to digestion with

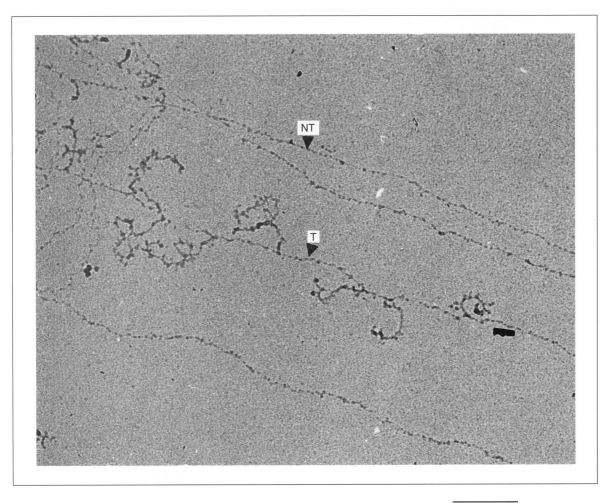

Figure 6.10

Electron micrograph of chromatin from a *Drosophila* embryo. Note the identical "beads on a string" structure of the chromatin that is not being transcribed (NT) and the chromatin that is being transcribed (T) into the readily visible ribonucleoprotein fibrils.

DNaseI. This altered chromatin structure is not confined to the very active genes such as globin and ovalbumin but appears to be a general characteristic of all transcribed genes, whatever their rate of transcription. Thus, if chromatin is digested using conditions that degrade less than 10% of the total DNA, over 90% of transcriptionally active DNA is digested, the DNA of genes encoding rare mRNAs being as sensitive as those encoding abundant mRNAs. Hence, the altered chromatin structure of active genes does not appear to be dependent upon the act of transcription itself, since the genes encoding rare mRNAs will be transcribed only very rarely.

In agreement with this idea, the altered DNaseI sensitivity of previously active genes persists even after transcription has ceased. For example, the ovalbumin gene remains preferentially sensitive to the enzyme when chromatin is isolated from the oviduct after withdrawal of oestrogen when, as we have previously seen, transcription of the gene ceases (see Section 4.2.3). A similar preferential sensitivity to DNaseI is also observed for the genes encoding the foetal forms of globin, which are not transcribed in adult sheep

Figure 6.11

Preferential sensitivity of chromatin containing active genes to digestion with DNaseI, as assayed by Southern blotting with specific probes.

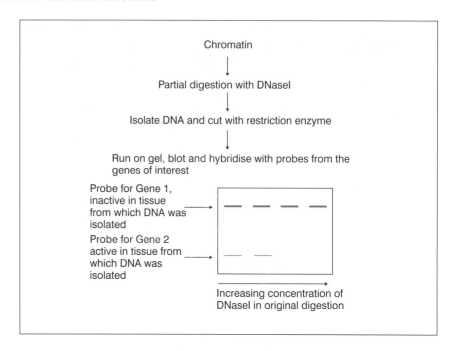

cells, and for the adult globin genes in mature (14-day) chicken erythrocytes, following the switching off of transcription.

As well as being detectable after transcription has ceased permanently, such increased sensitivity can also be detected in genes about to become active prior to the onset of transcription. As discussed previously (see Section 4.2.2), the transcription of the globin gene in Friend erythroleukaemia cells only occurs following treatment of the cells with dimethyl sulfoxide. Increased sensitivity to DNaseI digestion is observed, however, in both the treated cells, which transcribe the gene at high levels, and in the unstimulated cells, which do not.

The altered, more open, chromatin structure detected by increased DNaseI sensitivity does not therefore reflect the act of transcription itself. Rather, it appears to reflect the ability to be transcribed in a particular tissue or cell type. Hence, in cells that have become committed to a particular lineage expressing particular genes, such commitment will be reflected in an altered chromatin structure which will arise prior to the onset of transcription and will persist after transcription has ceased. The ability of imaginal disc cells in *Drosophila* to maintain their commitment to give rise to a particular cell type in the absence of overt differentiation (see Section 6.2) is likely, therefore, to be due to the genes required in that cell type having already assumed an open chromatin structure. Moreover, a breakdown in this process resulting in a change in the chromatin structure of a particular regulatory gene would result in a change in the commitment of the disc cells, as illustrated in Fig. 6.4. Similarly, the altered chromatin structure of the genes required in cartilage cells would be retained in cells cultured in media not supporting expression of these genes, allowing the restoration of the differentiated cartilage phenotype when the cells are transferred to an appropriate medium (see Section 6.2).

As described in Section 6.4.1, active or potentially DNA is still organised in the "beads on a string" structure. However, its enhanced DNaseI sensitivity reflects the fact that it does not form the much more tightly packed solenoid structure (Section 6.3) which is formed by the bulk of DNA which is not about to be or being transcribed. Hence, commitment to a particular pattern of gene expression involves a move from the tightly packed solenoid structure to the more open "beads on a string" structure (Fig. 6.12). As the DNA in this structure is more accessible, it is therefore more readily digested by DNaseI.

Although the greater accessibility to DNaseI of active or potentially active DNA simply represents an experimental tool for dissecting its chromatin structure, the alteration of chromatin to a more open structure in committed cells is likely to be a necessary prerequisite for gene expression and hence of major biological significance. Thus, it will allow the *trans*-acting factors, which actually activate transcription of the gene, access to their appropriate DNA target sequences within it. Hence the different structure of potentially active genes can explain why a particular steroid hormone, such as oestrogen, can induce activity of one particular gene in one tissue and another gene in a different tissue (see Section 6.1). Thus the ovalbumin gene in the oviduct would be in an open configuration, allowing induction to occur, whereas the more closed configuration of the vitellogenin gene would not allow access to the complex of hormone and receptor, and induction would not occur. The reverse situation would apply in liver tissue, allowing induction of the vitellogenin but not the ovalbumin gene.

Therefore, the changes in chromatin structure detected by DNaseI digestion play an important role in establishing the commitment to express the specific genes characteristic of a particular lineage. As well as this change in chromatin structure a variety of biochemical changes have been observed in

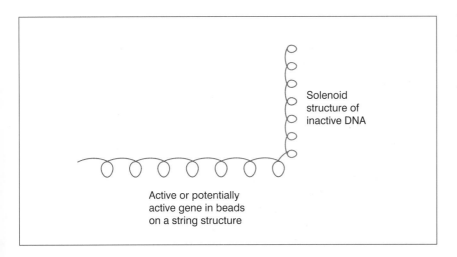

Active or potentially active gene in beads on a string structure

Solenoid structure of inactive DNA

Figure 6.12

Active or potentially active DNA is in the beads on a string chromatin structure whereas transcriptionally inactive DNA is in the more tightly packed solenoid structure.

DNA or its associated proteins in areas which are about to be or are being transcribed. It is likely that these biochemical changes may in turn lead to the alterations in chromatin structure, detected by DNaseI digestion, and they are discussed in the next two sections.

6.5 ALTERATIONS IN DNA METHYLATION IN ACTIVE OR POTENTIALLY ACTIVE GENES

6.5.1 Nature of DNA methylation

Although DNA consists of the four bases adenine, guanine, cytosine, and thymine, it has been known for many years that these bases can exist in modified forms bearing additional methyl groups. The most common of these in eukaryotic DNA is 5-methyl cytosine (Fig. 6.13), between 2–7% of the cytosine in mammalian DNA being modified in this way (for reviews see Bestor and Coxon, 1993; Colot and Rossignol, 1999; Jones and Takai, 2001).

Approximately 90% of this methylated C occurs in the dinucleotide, CG, where the methylated C is followed on its 3′ side by a G residue. Conveniently, this sequence forms part of the recognition sequence (CCGG) for two restriction enzymes, *Msp*I and *Hpa*II, which differ in their ability to cut at this sequence when the central C is methylated. Thus *Msp*I will cut whether or not the C is methylated and *Hpa*II will only do so if the C is unmethylated. This characteristic allows the use of these enzymes to probe the methylation state of the fraction of CG dinucleotides that is within cleavage sites for these enzymes. Hence, if DNA is digested with either *Hpa*II or *Msp*I, both enzymes will give the same pattern of bands only if all the C residues within the recognition sites are unmethylated. In contrast, if any sites are methylated, larger bands will be obtained in the *Hpa*II digest, reflecting the failure of the enzyme to cut at particular sites (Fig. 6.14). If this procedure is used in conjunction with Southern blot hybridisation using a probe derived from a particular gene, the methylation pattern of the *Hpa*II/*Msp*I sites within the gene can be determined.

When this is done it is found that, although some CG sites are always unmethylated and others are always methylated, a number of sites exhibit a

Figure 6.13

Structure of 5-methyl cytosine.

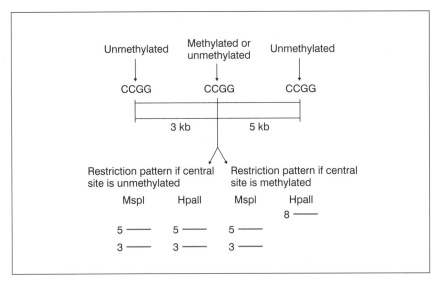

Figure 6.14

Detection of differences in DNA methylation between different tissues, using the restriction enzymes *Msp*I and *Hpa*II.

tissue-specific methylation pattern, being methylated in some tissues but not in others. Such sites within a particular gene are unmethylated in tissues where the gene is active or potentially active and methylated in other tissues (for reviews see Bird 1992; Ng and Bird, 1999). For example, a particular site within the chicken globin gene is methylated in a wide variety of tissues and is therefore not digested with *Hpa*I, but is unmethylated and therefore susceptible to digestion in DNA prepared from erythrocytes (Fig. 6.15). Similarly, the tyrosine amino-transferase gene, which is expressed only in the liver (Section 6.1), is undermethylated in this tissue when compared to other tissues where it is not expressed. The possibility suggested by this finding, that the *trans*-acting factors, which regulate the expression of this gene and are detectable in all tissues, may be inhibited from binding to the gene in tissues where it is methylated, is confirmed by the observation that artificial methylation of the gene prevents the binding of at least one of these factors. Interestingly, clusters of unmethylated CG sites known as CpG islands are also observed in constitutively expressed genes, although such sites remain unmethylated in all tissues consistent with the activity of the gene in all cell types (for a review see Bird, 1987).

In the case of tissue-specific genes which exhibit changes in methylation pattern, under-methylation, like DNaseI sensitivity, is observed prior to the onset of transcription and persists after its cessation. For example, the undermethylation of the chicken globin gene persists in mature erythrocytes, where the gene is not being transcribed but is still sensitive to DNaseI digestion. Most interestingly, the region in which unmethylated C residues are found correlates with that exhibiting heightened DNaseI sensitivity and is also depleted of histone H1. As with DNaseI sensitivity, therefore, under-methylation is a consequence of commitment to a particular pattern of gene expression, and is associated with the change in chromatin structure observed in active or potentially active genes.

Figure 6.15

Tissue-specific methylation of *MspI/HpaII* sites in the chicken globin gene results in the methylation-sensitive enzyme, *HpaII*, producing a band in red blood cell DNA that is identical to that produced by the methylation-insensitive *MspI*, but producing a larger band in brain DNA.

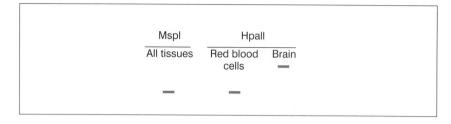

6.5.2 Evidence that DNA methylation plays a role in regulating chromatin structure

Although undermethylation is therefore associated with the more open chromatin structure of active DNA, this association does not prove that it has a key role in creating this structure. Two lines of evidence suggest, however, that this is indeed the case and that alterations in DNA methylation can affect chromatin structure and gene expression.

Introduction of methylated DNA into cells

A number of experiments have shown that, if DNA containing 5-methyl cytosine is introduced into cells, it is not expressed, whereas the same DNA which has been demethylated is expressed. These experiments have been carried out using both eukaryotic viruses and cellular genes, such as those encoding β- and γ-globin, cloned into plasmid vectors. In these experiments the methylated DNA adopts a DNaseI-insensitive structure typical of inactive genes whereas unmethylated DNA adopts the DNaseI-sensitive structure typical of active genes (Fig. 6.16), providing direct evidence for the role of methylation differences in regulating the generation of different forms of chromatin structure.

Effect of artificially induced demethylation

If methylation differences play a crucial role in the regulation of differentiation, it should be possible to change gene expression by demethylating DNA. This has been achieved in a number of cases by treating cells with the cytidine analogue, 5-azacytidine, which is incorporated into DNA but cannot be methylated, having a nitrogen atom instead of a carbon atom at position 5 of the pyrimidine ring. In the most dramatic of these cases, treatment of an undifferentiated fibroblast cell line with this compound results in the activation of a few key regulatory loci and the cells differentiate into multinucleate, twitching, striated muscle cells (Constantinides *et al.*, 1977) (see also Section 8.2.4). In other cases, although not actually producing altered gene expression, demethylation may facilitate it. Thus, if HeLa cells are treated with 5-azacytidine no dramatic effects are observed. If such cells are fused with muscle cells, however, muscle-specific genes are switched on in the treated HeLa cells, a phenomenon which is not observed when untreated HeLa cells are fused with mouse muscle cells (Chiu and Blau, 1985). Hence, treatment of the HeLa cells has altered their muscle-specific genes in such a way as to allow them to respond to *trans*-acting factors

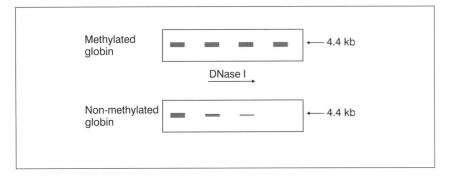

Figure 6.16

Unmethylated DNA introduced into cells adopts an open DNasel-sensitive configuration, whereas the same DNA when methylated and then introduced into cells adopts a more tightly packed DNasel-insensitive form.

present in the mouse muscle cells. This type of regulatory process is exactly what would be predicted if methylation has a role in the alteration of chromatin structure and thereby in facilitating interactions with *trans*-acting regulatory factors. It should be noted, however, that in none of these cases has it been shown directly that 5-azacytidine achieves its effect by inducing demethylation rather than by some other, as yet uncharacterised, action of this compound.

Despite this caveat, the available evidence from these studies, and those discussed above, indicates that DNA methylation plays a central role in the regulation of gene expression, at least in mammals. This conclusion is reinforced by the finding that several different DNA methyltransferase enzymes which methylate DNA are essential for normal embryonic development with the inactivation of any one of these resulting in mice which are non-viable (see, for example, Okano *et al.*, 1999). Hence in mammals DNA methylation is essential for normal embryonic development. Moreover, although originally thought to be absent in *Drosophila*, 5-methyl cytosine has recently also been detected in this organism and shown to be regulated during embryonic development (Lyko *et al.*, 2000), further supporting a key role for this process.

6.5.3 Mechanism by which DNA methylation affects chromatin structure

The evidence suggesting an important role for DNA methylation in modulating the difference in chromatin structure between inactive and active genes raises the question of how this is achieved. It is clearly possible for the methylation differences between inactive or potentially active genes to be recognised by proteins since the proteinaceous enzyme *Hpa*II digests only unmethylated DNA (see Section 6.5.1). Evidently, undermethylation could promote the binding of proteins which produce a more open chromatin structure (Fig. 6.17a). Alternatively, an inhibitory protein could bind to methylated DNA and promote a closed chromatin structure (Fig. 6.17b).

Although both these mechanisms may be used, current evidence supports the second mechanism. Thus, a specific protein MeCP2 has been shown to bind directly to methylated CG but not to unmethylated CG (Nan *et al.*, 1997). Moreover, it has been demonstrated that the binding of MeCP2

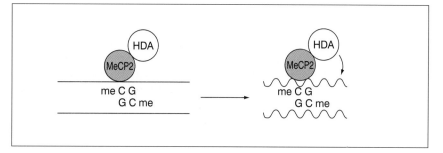

results in the production of a closed chromatin structure and transcriptional repression (for a review see Bird and Wolffe, 1999).

Hence, MeCP2 is able to recognise methyl CG dinucleotides and induce the formation of a closed chromatin structure. The importance of this process is demonstrated by the finding that humans with mutant forms of MeCP2, which are unable to recognise meCG, exhibit Rett syndrome which is a severe developmental disorder leading to mental retardation (for a review see Tucker, 2001). Interestingly, binding of MeCP2 to meCG results in the recruitment of a multi-protein complex which includes a histone deacetylase (for a review see Bird and Wolffe, 1999) that can induce the removal of acetyl groups from histones (Fig. 6.18). As deacetylation of histones is known to be associated with transcriptionally inactive DNA, this illustrates the close link between modification of DNA and modification of histones in determining the structure of chromatin. Therefore, histone modifications associated with changes in chromatin structure are discussed in the next section.

6.6 MODIFICATION OF HISTONES IN THE CHROMATIN OF ACTIVE OR POTENTIALLY ACTIVE GENES

Given the essential role of histones in chromatin structure, it is possible that potentially active chromatin might be marked in some way by modification of the histones within it. In fact, a number of such modifications of these proteins (involving, for example, methylation, phosphorylation, ubiquitination or acetylation) have been reported (for reviews see Strahl and Allis, 2000; Wu and Grunstein, 2000; Jenuwein and Allis, 2001). The most extensive studies to date have involved histone acetylation, which will therefore be discussed in the next section. Subsequent sections will deal with other histone modifications which, although less studied, also appear to have important roles in determining chromatin structure.

6.6.1 Acetylation

In the process of acetylation, the free amino group on specific lysine residues at the N-terminus of the histone molecule are modified by one of the hydrogen atoms in the amino group being substituted by an acetyl group ($COCH_3$). This modification reduces the net positive charge on the histone molecule and occurs primarily for histones H3 and H4. Hyperacetylated forms of these histones, containing several such acetyl groups, have been shown to be localised preferentially in active genes exhibiting DNaseI sensitivity whilst under-acetylation of the histones is characteristic of transcriptionally inactive regions. Furthermore, treatment of cells with sodium butyrate, which inhibits a cellular deacetylase activity and hence increases histone acetylation, has been shown to result in DNaseI sensitivity of some regions of chromatin and to activate the expression of some previously silent cellular genes (Reeves and Cserjesi, 1979). Hence, as with DNA methylation, there is direct evidence that hyperacetylation of histones may play some role in the generation of DNaseI sensitivity and the marking of active or potentially active genes.

More recent evidence in favour of a role for histone acetylation in the regulation of gene expression has been obtained by the identification of several histone acetyl transferase (HAT) enzymes capable of adding acetyl groups to histone molecules (for reviews see Pennisi, 1997; Brown *et al.*, 2000). Most importantly, it has been shown that the CBP transcriptional co-activator which plays a key role in the activation of genes in response to cyclic AMP and other stimuli (see Section 8.3.3) has histone acetyl transferase activity directly linking this enzymatic activity with the ability to stimulate transcription. Similarly the $TAF_{II}250$ subunit of the TFIID transcription factor which is critical for basal transcription of a wide range of genes (see Section 3.2.3) has been shown to have histone acetyl transferase activity (Pennisi, 1997) as has the ATF-2 transcription factor (Kawasaki *et al.*, 2000).

The opposite effect has been observed in the case of the nuclear receptor co-repressor which mediates the inhibitory effect of the thyroid hormone receptor on transcription when thyroid hormone is absent (see Section 8.3.4). Thus this factor has been shown to associate with the Sin3/RPD3 protein complex which has the ability to deacetylate histones (for reviews see Ayer, 1999; Ng and Bird, 2000). The inhibitory effect of the nuclear receptor co-repressor is therefore likely to involve the deacetylation of histones, thereby producing a more tightly packed chromatin structure incompatible with transcription. Hence activating factors can direct the acetylation of histones, thereby opening up the chromatin (Fig. 6.19a) whilst inhibitory factors can direct histone deacetylation, thereby directing a more tightly packed chromatin structure (Fig. 6.19b).

These findings therefore link the study of transcription factors (see Chapter 8) with studies on chromatin structure and suggest that the regulation of histone acetylation/deacetylation, and thereby of chromatin structure, by specific factors plays a key role in the regulation of gene expression (for reviews see Ahringer, 2000; Strahl and Allis, 2000). Similarly, as noted above (Section 6.5.3), a protein which binds specifically to methyl CG dinucleotides can recruit a histone deacetylase activity, thereby linking histone deacetylation to the repressive effect of DNA methylation.

Interestingly, as well as activators recruiting acetylases and repressors recruiting deacetylases, there is evidence that the acetylase/deacetylase enzymes can themselves be regulated (for a review see Stewart and Crabtree, 2000). This is seen in muscle differentiation where in myoblasts the transcriptional activator MEF2 is associated with histone deacetylases. When differentiation from myoblasts to mature myotubes occurs, the histone deacetylase is induced to move to the cytoplasm, thereby freeing MEF2 which can now activate transcription (Fig. 6.20).

As discussed in section 6.3, the amino terminal ends of the histones proteins, which are modified by acetylation, project beyond the nucleosome core and could therefore potentially interact with the amino-terminal ends

Figure 6.19

(a) An activating molecule (A) can direct the acetylation of histones (H) thereby resulting in a change in chromatin structure from a closed (wavy line) to an open (solid line) configuration. (b) An inhibitory molecule can direct the deacetylation of histones, thereby having the opposite effect on chromatin structure.

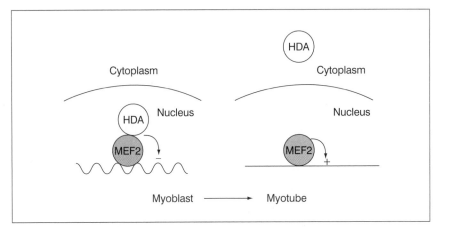

Figure 6.20

In myoblasts, the MEF2 transcriptional activator is associated with a histone deacetylase (HDA) and transcription of myotube-specific genes is repressed. When differentiation is induced, the histone deacetylase moves into the cytoplasm, allowing MEF2 to activate the transcription of myotube-specific genes.

of histones in adjacent nucleosomes or with other non-histone proteins. The effects of histone acetylation on chromatin structure may therefore operate in one of two possible ways (Fig. 6.21). Firstly, such changes might act by affecting the association of the histones with each other. In turn this would result in improved access to the DNA for factors which can stimulate transcription (Fig. 6.21a). Some evidence for this model has been provided in the case of histone acetylation where the binding to DNA of the *Xenopus* transcription factor TFIIIA was greatly enhanced when the nucleosomes contained hyperacetylated histone H3 and H4 compared to when these proteins were not acetylated.

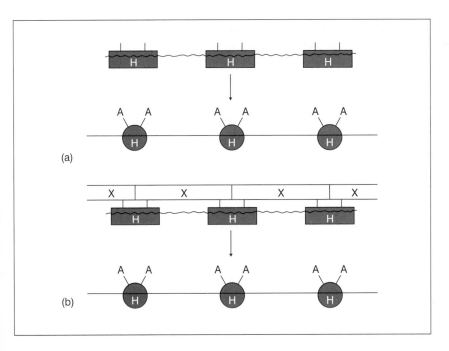

Figure 6.21

Acetylation (A) of histones (H) may activate either by directly altering the conformation of the histones themselves producing a more open chromatin structure (panel a) or indirectly by disrupting their association with an inhibitory molecule (X) (panel b).

An alternative possibility is that these modifications affect the protein–protein interaction of the histones with other regulatory molecules. This would parallel the role proposed for methylation differences in affecting the binding of positively or negatively acting factors to the DNA (see Fig. 6.17; Section 6.5.3). Thus, for example, acetylated histones might be recognised by a positively acting molecule, leading to the destabilisation of the solenoid structure and transcriptional activation. Similarly, acetylation could disrupt an association with an inhibitory molecule involved in maintaining the closed chromatin structure (Fig. 6.21b).

6.6.2 Other histone modifications

Although histone acetylation is the modification which has been most studied in terms of its effect on chromatin structure, other histone modifications have also been associated with altered chromatin structure and these will be discussed in turn.

Ubiquitination

Ubiquitin is a small protein of only 76 amino acids, which forms a conjugate with histone H2A in which the C-terminal carboxyl group of ubiquitin is joined to the free amino group on an internal lysine residue in the histone to form a branched molecule (Fig. 6.22; for a review see Busch and Goldknopf, 1981). This modification reduces the net positive change on the histone molecule, both by neutralising the charged amino group on lysine and by introducing a number of negatively charged amino acids present in the ubiquitin molecule itself. Only a small minority of the H2A in a cell (about 5–10%) exists in this form but, in *Drosophila* at least, this modified form of H2A is localised preferentially in nucleosomes containing active genes. Similarly, $TAF_{II}250$, as well as having histone acetyltransferase activity (see Section 6.6.1), also has the ability to add ubiquitin to histone H1 and this activity has been shown to be essential for the transcriptional activation of specific genes in *Drosophila* (for a review see Wu and Grunstein, 2000).

Hence, there is evidence to suggest that histone ubiquitination may alter chromatin structure and hence gene expression. However, at present such evidence is indirect, in contrast to DNA methylation or histone acetylation where direct evidence for their effect on chromatin structure is available.

Figure 6.22

Linkage of ubiquitin to histone H2A. The carboxyl terminal amino acid (76) of ubiquitin links to the lysine at position 119 of histone H2A. AA indicates the amino-acid backbone of the molecules.

```
                                        119
H2A     NH2 — AA — AA — AA — Lys — AA — AA — COOH
                                   |
                            76   Gly
                                   |
              Ubiquitin    AA
                                   |
                            AA
                                   |
                            NH₂
```

Phosphorylation

Much stronger evidence for a key role in gene regulation is available for histone phosphorylation which, unlike acetylation or ubiquitination, targets serine amino acids in the histone molecule (for reviews see Cheung *et al.*, 2000; Hagmann, 1999). Thus, it has been shown that, when cells are exposed to heat shock, phosphorylated histone H3 is concentrated at the heat shock gene loci which are being actively transcribed and is depleted from other loci which are silenced following heat shock (Nowak and Corces, 2000).

Similarly, when cells are stimulated with growth factors, histone H3 is phosphorylated on the serine residue at position 10 in the amino terminus of the protein. These phosphorylated histones are localised to genes which become transcriptionally active following growth factor stimulation such as the c-*fos* and c-*myc* genes (see Sections 9.4.1 and 9.4.3; Fig. 6.23). Such growth factor-induced phosphorylation of histone H3 does not occur in human patients with Coffin–Lowry syndrome who lack the phosphorylation enzyme, known as Rsk-2, which is responsible for phosphorylating histone H3 following growth factor stimulation (for a review see Hagmann, 1999). These patients suffer from a number of defects including mental retardation as well as facial and other abnormalities. Moreover, they show a lack of gene activation in response to growth factor stimulation, indicating the importance of Rsk-2-mediated phosphorylation of histone H3 in this process.

Interestingly, it appears that phosphorylation of serine at position 10 in histone H3 enhances the ability of histone acetyl transferases to acetylate the

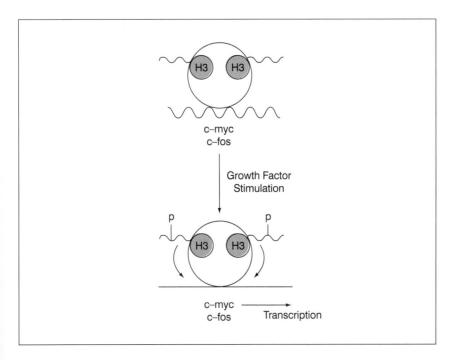

Figure 6.23

Following growth factor stimulation, histone H3 becomes phosphorylated within the nucleosomes bound to the c-*myc* and c-*fos* genes. This is associated with their moving from an inactive closed chromatin structure (wavy line) to a more open structure (solid line), allowing transcription to occur in response to the growth factor.

lysine at position 14 (for a review see Cheung *et al.*, 2000). This has led to the idea of a "histone code" in which a particular pattern of histone modifications is characteristic of active or potentially active genes (for reviews see Paro, 2000; Strahl and Allis, 2000; Jenuwein and Allis, 2001; Fig. 6.24).

Methylation

Histones may also be modified by addition of a methyl (CH_3) group to specific lysine residues. This modification occurs particularly for histones H3 and H4 and, like the other modifications, targets amino acids at the N-terminus of the protein such as the lysine at position 9 of histone H3. Unlike the other modifications, however, methylation of the histones at particular positions is associated with transcriptionally inactive DNA, whereas acetylation and phosphorylation are associated with active or potentially active DNA (for reviews see Berger, 2001; Jenuwein, 2001; Zhang and Reinberg, 2001).

Indeed, there is evidence for an antagonistic interaction between methylation on position 9 and phosphorylation/acetylation at positions 10 and 14, all of which are found in a small region at the amino-terminus of the protein. Thus, for example, it appears that dephosphorylation and deacetylation of positions 10 and 14 promote methylation of the histone molecule on position 9 whilst methylation of the histone on position 9 inhibits acetylation and phosphorylation (Rea *et al.*, 2000). Hence, the histone code described above can be extended to include methylation with the histones associated with active or potentially active DNA being acetylated on lysine 14 and phosphorylated on serine 10 whilst those associated with inactive DNA are methylated on lysine 9 (Fig. 6.25).

It should be noted, however, that not all histone methylation is associated with inactive DNA. Thus, in contrast to methylation on position 9, the methylation of histone H4 on the arginine at position 4 promotes a more open chromatin structure and facilitates transcriptional activation by

Figure 6.24

Phosphorylation of histone H3 on serine residue 10 promotes acetylation on the adjacent lysine 14 resulting in a pattern of phosphorylation/acetylation in active chromatin which is distinct from that of inactive chromatin.

Plate 1

Analysis of differences between two protein samples by two-dimensional gel electrophoresis to separate the proteins (step 1), computer comparison of the two gels to identify proteins which differ between the two samples (step 2) followed by excision of a protein spot which is altered and its identification by mass spectral analysis of the peptides produced by trypsin digestion (step 3).

a)

b)

Plate 2
Gene chip analysis of gene expression patterns using the "Affymetrix gene chip system". Panel (a) shows a single gene chip hybridised to labelled RNA probe with a close-up of a specific region. Panel (b) shows the changes in expression of different genes during the activation of lung fibroblasts. Red and green show up-regulation and down-regulation respectively. Letters represent groups of genes showing similar patterns of expression over time.

Plate 3
"Dolly", the first cloned sheep.

Plate 4
"Cumulina" the first cloned mouse (black) with her foster mother (white) (panel a) and her offspring (panel b).

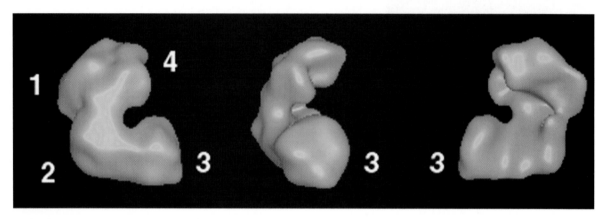

Plate 5
Three-dimensional structure of TFIID. Note the globular domains arranged around a groove into which the DNA fits.

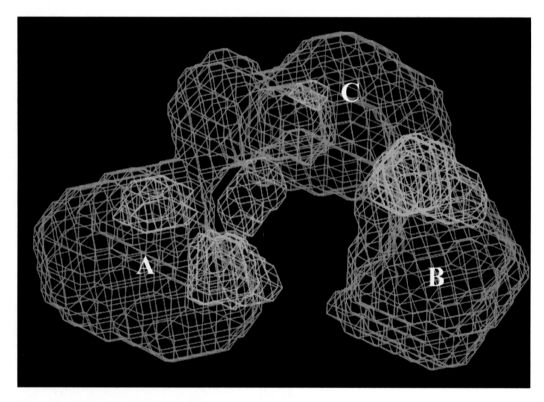

Plate 6
Position of TFIIB (green) and TFIIA (red) complexed with TFIID (blue).

Plate 7
Association of the complex of TFIIB (red), the TBP component of TFIID (green) and TFIIA (yellow and purple) with the DNA (light and dark blue).

Plate 8
The enzyme RNA polymerase II in the act of transcribing DNA. The transcribed strand of the DNA is shown in blue and the non-transcribed strand in green whilst the RNA transcript is shown in red. Note that the DNA enters from the right, unwinds and makes a right-angled turn as it encounters a wall of protein within the polymerase molecule. Compare with the schematic diagram in Fig. 3.8.

Plate 9
Two views of the structure of the Sxl splicing protein bound to an RNA molecule (yellow).

Plate 11
Structure of the Cys$_2$His$_2$ zinc finger of the Xfin factor. The cysteine residues are shown in yellow, the histidine side chains in blue and the zinc atom in white.

Plate 10
Two views of the structure of the nucleosome (panel a) and a schematic diagram indicating the positions of the different histones (panel b).

Plate 13

Two views of the oestrogen receptor binding to DNA. The two zinc fingers are shown in green and blue with the DNA in purple.

Plate 12

Model of the two Cys$_4$ zinc fingers of the glucocorticoid receptor. The two fingers are shown in red and green respectively with the two zinc atoms in white.

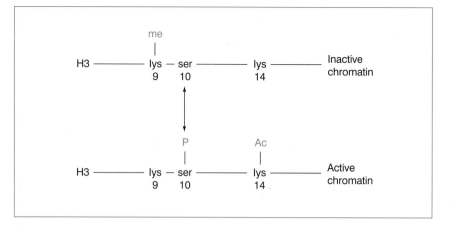

Figure 6.25

The histone H3 in inactive chromatin is modified by methylation of the lysine residue at position 9 in the amino terminus of the protein. The transition to active chromatin is associated with demethylation of this amino acid which is accompanied by phosphorylation of the adjacent serine at position 10 and acetylation of the lysine at position 14.

nuclear hormone receptors (Wang *et al.*, 2001). Moreover, in the region of DNA containing the mating type locus in yeast, it has been shown that transcriptionally active DNA contains histone H3 which is methylated on lysine 4 (the equivalent amino acid to arginine 4 of histone H4), whereas adjacent transcriptionally inactive regions have histone H3 which is methylated on lysine 9 (Noma *et al.*, 2001) and a similar effect has also been demonstrated in higher organisms for the chicken β-globin locus (Litt *et al.*, 2001). Hence, the "histone code" is highly complex with both positive and negative effects of methylation at different positions as well as positive effects of acetylation and phosphorylation.

Evidently, this complex pattern of modifications can be readily fitted into the model shown for acetylation in Fig. 6.21. Thus, the total pattern of methylation, phosphorylation and acetylation is likely to modify chromatin structure either directly, by affecting the interaction of the histones with each other to form the chromatin structure, or indirectly, by affecting their

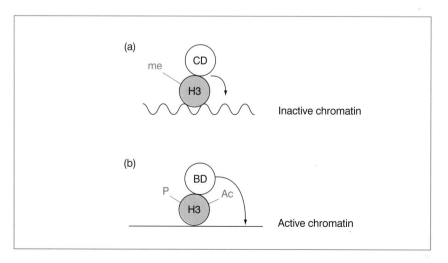

Figure 6.26

(a) Methylation of histone H3 promotes binding of chromodomain-containing proteins (CD) which can induce the closed chromatin structure characteristic of inactive DNA (wavy line).
(b) In contrast, acetylation and phosphorylation promote binding of bromodomain-containing proteins (BD) which induce the more open chromatin structure characteristic of active or potentially active DNA.

interaction with other regulatory proteins. In agreement with the latter possibility, it has recently been shown that methylation of histone H3 on lysine 9 allows it to be recognised by a specific protein HP1 which cannot recognise the unmethylated histone (Lachner *et al.*, 2001; Bannister *et al.*, 2001). HP1 is known to be able to organise the chromatin into a closed inactive structure and contains a region known as a chromodomain which is found in many proteins which inhibit transcription. Hence, recognition of histones methylated on lysine 9 by chromodomain-containing proteins may be a key event in organising the tightly packed structure of inactive chromatin (Fig. 6.26).

Conversely, phosphorylation/acetylation promotes recognition of the histones by a group of activator proteins, containing a region known as a bromodomain. Thus, chromodomain-containing proteins may recognise histones methylated on specific residues and promote the tight packing of chromatin whilst recognition of phosphorylated/acetylated histones by bromodomain-containing proteins promotes the more open chromatin structure characteristic of active or potentially active DNA (Fig. 6.26).

It is likely therefore that histone modifications can affect chromatin structure either directly, by interactions with one another, or indirectly via interactions with other proteins. Regardless of the precise mechanism by which they act, however, it is clear that such modifications do play an important role in the changes in chromatin structure which are necessary for gene activation to occur.

6.7 CHANGES IN CHROMATIN STRUCTURE IN THE REGULATORY REGIONS OF ACTIVE OR POTENTIALLY ACTIVE GENES

6.7.1 DNaseI hypersensitive sites

So far in this chapter we have seen that the region of chromatin containing an active or potentially active gene has a number of distinguishing features, including undermethylation of C residues, histone modifications and increased sensitivity to digestion with DNaseI. Such changes extend over the entire region of the gene and some flanking sequences and, in the case of DNaseI sensitivity, result in an approximately tenfold increase in the rate at which active or potentially active genes are digested.

Following the discovery of such increased DNaseI sensitivity, many investigators studied whether, within the region of increased sensitivity, there might be particular sites which were even more sensitive to cutting with the enzyme and which would therefore be cut even before the bulk of active DNA was digested. The technique used to look for such sites is based on that used to look at the overall DNaseI sensitivity of a particular region of DNA (see Section 6.4.2). Chromatin is digested with DNaseI and a restriction enzyme and then subjected to a Southern blotting procedure using a probe derived from the gene of interest. As we have seen previously, the overall sensitivity of the gene can be monitored by observing how rapidly the specific restriction enzyme fragment derived from the gene disappears with increasing amounts of the enzyme (see Fig. 6.11). To search for hypersensitive sites, however, much lower concentrations of the enzyme are used and the appearance of discrete digested fragments derived from the gene is monitored (Fig. 6.27). Such specific fragments have, at one end, the cutting site for the restriction enzyme used and, at the other, a site at which DNaseI has cut, producing a defined fragment. Since the position at which the restriction enzyme cuts in the gene is known, the position of the hypersensitive site can be mapped simply by determining the size of the fragment produced.

Using this procedure a very wide variety of genes have been shown to contain such hypersensitive sites exhibiting a sensitivity to DNaseI digestion tenfold above that of the remainder of an active gene and therefore about one hundredfold above that seen in inactive DNA (for a review see Gross and Garrard, 1988). A representative list of cases in which such sites have

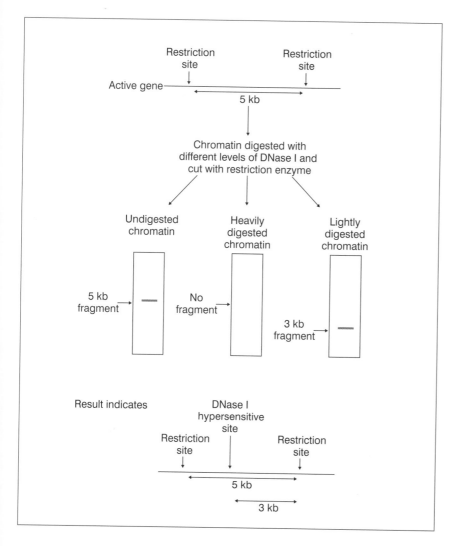

Figure 6.27

Detection of DNaseI hypersensitive sites in active genes by mild digestion of chromatin to produce a digestion product with a restriction site at one end and a DNaseI hypersensitive site at the other (right-hand panel). More extensive digestion will result in the disappearance of the band (central panel) as in the experiment illustrated in Fig. 6.11.

Table 6.2

Examples of genes containing DNaseI-hypersensitive sites

(a)	**Tissue-specific genes**	
	Immune system	Immunoglobulin complement C4
	Red blood cells	α–, β– and ε–globin
	Liver	α–foetoprotein, serum albumin
	Nervous system	Acetylcholine receptor
	Pancreas	Preproinsulin, elastase
	Connective tissue	Collagen
	Pituitary gland	Prolactin
	Salivary gland	*Drosophila* glue proteins
	Silk gland	Silk moth fibroin
(b)	**Inducible genes**	
	Steroid hormones	Ovalbumin, vitellogenin, tyrosine aminotransferase
	Stress	Heat-shock proteins
	Viral infection	β–interferon
	Amino acid starvation	Yeast HIS 3 gene
	Carbon source	Yeast GAL genes, yeast ADH II gene
(c)	**Others**	
	Histones, ribosomal RNA, 5S RNA, transfer RNA, cellular oncogenes *c-myc* and *c-ras*, glucose-6-phosphate dehydrogenase, dihydrofolate	

been detected is given in Table 6.2. As with the increased sensitivity of the gene itself, many hypersensitive sites appear only in tissues where the gene is active. Thus, the increased sensitivity of globin DNA in erythrocytes to digestion is paralleled by the presence of hypersensitive sites within the gene in erythrocyte chromatin but not in that of other tissues. Similarly, the ovalbumin gene in hormonally treated chick oviduct also exhibits hypersensitive sites that are not found in other tissues, including erythrocytes (Fig. 6.28).

As with undermethylation and the sensitivity of the entire gene to digestion, DNaseI hypersensitive sites appear to be related to the potential for gene expression rather than always being associated with the act of transcription itself. Thus the hypersensitive sites near the *Drosophila* heat-shock genes are present in the chromatin of embryonic cells prior to any heat-induced transcription of these genes and one of the sites in the mouse α-foetoprotein gene persists in the chromatin of adult liver after the transcription of the gene (which is confined to the foetal liver) has ceased.

Hence, as with DNA methylation and generally increased sensitivity to DNaseI, the appearance of hypersensitive sites appears to be involved in gene regulation. This idea is reinforced by the location of the hypersensitive sites, which can be precisely mapped as described above. Many sites are located at the 5′ end of the genes, in positions corresponding to DNA sequences that are known to be important in regulating transcription. For example, a site present at the 5′ end of the steroid-inducible tyrosine aminotransferase gene is localised within the DNA sequence that is responsible for the steroid inducibility of the gene (see Section 7.2.3). Even in cases where

Figure 6.28

Detection of a DNaseI hypersensitive site in the ovalbumin gene in oviduct tissue (O) but not in erythrocytes (E). Track 4 shows the detection of a lower band caused by cleavage at a hypersensitive site when oviduct chromatin is digested with DNaseI. Note the progressive appearance of this band as increasing amounts of DNaseI are used to digest the oviduct chromatin (tracks 5–10). No cleavage is observed when similar amounts of DNaseI are used to cut erythrocyte chromatin (tracks 11 and 12). The hypersensitive site in oviduct chromatin is also cleaved, however, with micrococcal nuclease (tracks 2 and 3). Track 1 shows the pattern produced by micrococcal nuclease cleavage of naked DNA (N).

hypersensitive sites are located far from the site of transcriptional initiation, they appear to correspond to other regulatory sequences, such as enhancers, which can act over large distances (see Section 7.3).

In the case of the *Drosophila* gene encoding the glue protein Sgs4, the fortuitous existence of a mutant strain of fly has indicated the functional importance of hypersensitive sites. In normal flies this gene contains two hypersensitive sites, 405 and 480 bases upstream of the start of transcription. In the mutant, both sites are removed by a small DNA deletion of 100 base pairs. Despite the fact that this gene still has the start site of transcription and 350 bases of upstream sequences, no transcription occurs (Fig. 6.29), indicating the regulatory importance of the region containing the hypersensitive sites.

It is clear, therefore, that hypersensitive sites represent another marker for active or potentially active chromatin and are a feature likely to be of particular importance in gene regulation, being associated with many DNA sequences that regulate gene expression. It is therefore necessary to consider the nature and significance of these sites.

In some cases, these sites are likely to be formed where DNA is entirely

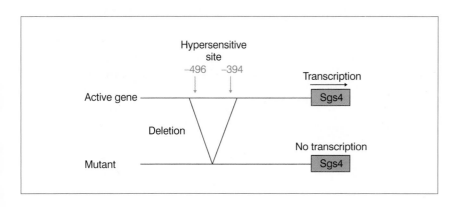

Figure 6.29

Deletion of a region containing the two hypersensitive sites upstream of the *Drosophila sgs4* gene abolishes transcription.

free of nucleosomes and is therefore highly sensitive to DNaseI digestion This is seen, for example, in the case of the DNaseI hypersensitive sites within the enhancer element that regulates transcription of the eukaryotic virus SV40. Thus, when this virus enters cells its DNA, which is circular and only 5000 bases in size, becomes associated with histones in a typical nucleosomal structure, which can be visualised in the electron microscope as a mini-chromosome. When this is done, however, the region containing the hypersensitive sites remains nucleosome-free and is seen as naked DNA (Fig. 6.30). A similar lack of nucleosomes in the region of hypersensitive sites is also found in the chicken β-globin gene, the 5′ hypersensitive site of this gene being excisable as a 115 bp restriction fragment lacking any associated nucleosome.

Although such cases indicate that hypersensitive sites can be produced by the complete loss of nucleosomes in a particular region, other cases exist where such sites are produced by an alteration in the structure of a nucleosome rather than its complete displacement (Fig. 6.31). It is clear, however, that whether caused by nucleosome displacement or structural alterations, the changes occurring in hypersensitive sites facilitate the entry of transcription factors or the RNA polymerase itself and hence allow the onset of transcription (Fig. 6. 31). In view of the critical role played by the changes in nucleosomes which occur at hypersensitive sites, it is necessary therefore to consider the proteins which produce such alterations in nucleosome positioning or structure.

6.7.2 Chromatin remodelling by proteins capable of displacing nucleosomes or altering their structure

In a number of cases protein factors have been identified which can displace nucleosomes or alter their structure, thereby facilitating the subsequent

Figure 6.30

Electron micrograph of the SV40 mini-chromosome consisting of DNA and associated histones. Note the region of the enhancer and hypersensitive sites, which appears as a thin filament of DNA free of associated proteins.

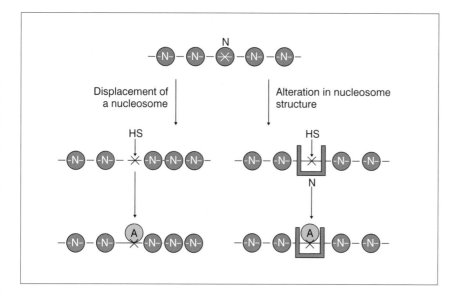

Figure 6.31

A hypersensitive site (HS) can be created either by displacement of a nucleosome (a) or by alteration of its structure (b). In either case such an alteration allows a transcriptional activator (A) to bind to its binding site (X) and activate transcription.

binding of molecules which stimulate transcription (for reviews see Kingston and Narlikar, 1999; Jones and Kadonaga, 2000). Thus in the case of the heat-shock genes whose transcription is induced by elevated temperature (see Section 4.2.4), hypersensitive sites are produced by the binding of the GAGA protein factor to its upstream DNA binding sites in the gene promoter which results in the displacement of a nucleosome (Fig. 6.32) (for a review see Aalfs and Kingston, 2000). This binding of the GAGA factor occurs in cells prior to heat treatment and hence hypersensitive sites are present prior to heat treatment. Following heat treatment, a protein factor,

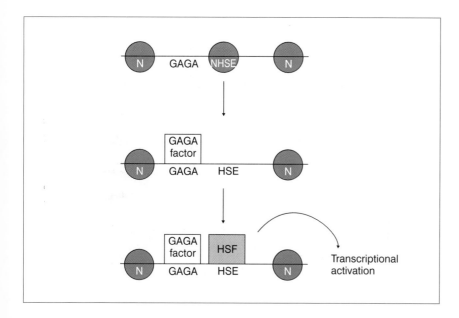

Figure 6.32

The binding of the GAGA factor to its binding site in the heat-shock genes displaces a nucleosome (N) thereby exposing the HSE binding site for the heat-shock factor (HSF). HSF then binds and activates transcription.

known as the heat-shock factor (HSF), binds to this region of DNA and transcription is stimulated (see Section 7.2.2). In this case, HSF is only capable of binding following heat shock and hence stimulation of transcription only occurs following such treatment (Fig. 6.33a).

In other cases, however, where the necessary transcription factors are present in all tissues, transcription may follow immediately the nucleosome-free region is generated, allowing these factors access. Thus, in the case of

Figure 6.33

Two mechanisms for transcriptional activation. (a) The heat-shock transcription factor (HSF) is activated by heat and binds to a pre-existing nucleosome-free region. (b) The receptor–steroid complex displaces a nucleosome, creating a hypersensitive site, and allowing an active transcription factor to bind. X indicates the position of a hypersensitive site. Note that, although the figure shows the displacement of a nucleosome, it is also possible that generation of the hypersensitive site may involve the alteration of nucleosome structure as illustrated in Fig. 6.31.

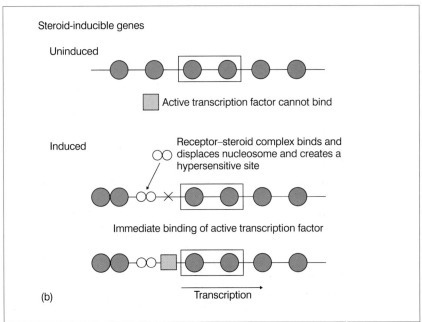

glucocorticoid-responsive genes, the critical regulatory event is the binding of the glucocorticoid receptor/steroid complex to a particular DNA sequence (the GRE or glucocorticoid response element), which displaces a nucleosome or alters its structure and generates a DNaseI hypersensitive site. Ubiquitous factors present in all tissues, such as NFI and the TATA box binding factor TBP (see Section 3.2.4), immediately bind to this region and transcription begins (Fig. 6.33b) (for further details of the glucocorticoid receptor and its mode of action see Sections 7.2.5, 8.2.3 and 8.4.3).

Although these two situations appear different in terms of the time at which the hypersensitive site appears relative to the onset of transcription, they illustrate the basic role of hypersensitive sites, namely the displacement of nucleosomes or alteration of their structure and the generation of a site of access for regulatory proteins.

Interestingly, the glucocorticoid receptor/steroid complex does not directly alter nucleosomal structure. Rather, it acts by recruiting a multi-protein complex known as the SWI/SNF complex which is able to hydrolyse ATP and use this energy to alter nucleosomal structure, so facilitating the subsequent binding of activator molecules (for reviews see Aalfs and Kingston, 2000; Sudarsanam and Winston, 2000).

The role of the SWI/SNF complex is not confined to steroid responsive genes and it appears that it can be recruited to a wide variety of different genes by regulatory proteins which bind to these genes. The SWI/SNF complex then acts to alter the chromatin structure of these genes, thereby facilitating their subsequent activation by other transcription factors (Fig. 6.34). In agreement with this idea, the brahma mutation in *Drosophila* inactivates the SWI2 component of the complex which produces the ATP hydrolysing activity and results in a failure to activate the homeotic genes which play a key role in determining body pattern (see Section 8.2.2) (for a review see Simon, 1995).

Hence the SWI/SNF complex is involved in altering the chromatin structure and thereby facilitating the transcriptional activation of genes as diverse as steroid-responsive genes and homeotic genes. Moreover, once such nucleosome disruption has been produced by SWI/SNF it can dissociate from the gene since the alteration in nucleosome structure and DNAseI hypersensitive site produced by SWI/SNF persists even after it has dissociated (Owen-Hughes *et al.*, 1996).

Interestingly, the GAGA factor described above also plays a general role in chromatin remodelling and does not act solely on the heat-shock genes. Thus inactivation of the gene encoding the GAGA factor in *Drosophila* results in a fly with an altered body pattern known as trithorax which is similar to the brahma mutation in that a wide range of homeotic genes are not activated (Farkas *et al.*, 1994). Moreover, the GAGA factor is also associated with a multi-protein complex known as nucleosome remodelling factor (NURF) which resembles the SWI/SNF factor in its ability to hydrolyse ATP and alter nucleosome structure.

Hence the alteration of chromatin structure at regulatory regions is brought about by multi-protein complexes such as SWI/SNF or NURF

Figure 6.34

A regulatory factor (X) binds to its DNA binding site (X) and recruits the SWI/SNF complex. This complex hydrolyses ATP to ADP and inorganic phosphate (Pi) and uses the energy generated to alter the structure of a nucleosome which was masking the binding site for the transcriptional activator (Y). This allows Y to bind to its site and activate transcription.

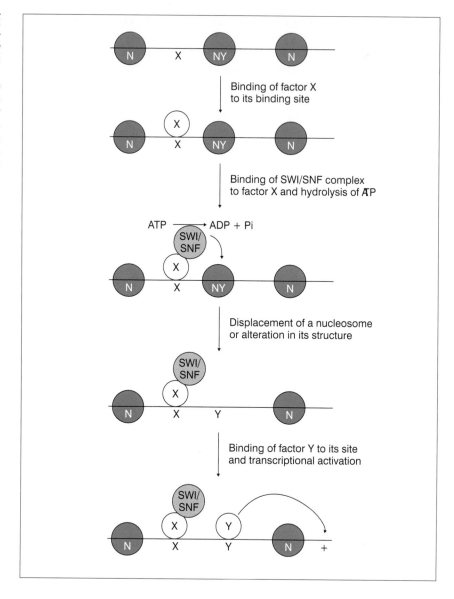

Figure 6.34

A regulatory factor (X) binds to its DNA binding site (X) and recruits the SWI/SNF complex. This complex hydrolyses ATP to ADP and inorganic phosphate (Pi) and uses the energy generated to alter the structure of a nucleosome which was masking the binding site for the transcriptional activator (Y). This allows Y to bind to its site and activate transcription.

which are active on a wide range of genes. These complexes are likely to be recruited to specific gene promoters by interaction with DNA binding proteins such as the glucocorticoid receptor or GAGA which have already bound to the DNA in a gene-specific manner and in turn will facilitate the subsequent binding of other activating molecules by altering nucleosome positioning or structure in an ATP-dependent manner.

In addition, the SWI/SNF complex has recently been shown to be associated with the RNA polymerase II holoenzyme which as discussed in Section 3.2.3 is a complex of RNA polymerase II with several basal transcription factors such as TFIIB, TFIIF and TFIIH. Moreover, it has been shown that the recruitment of the RNA polymerase II complex containing SWI/SNF factors can result in the opening up of chromatin indicating that the SWI/SNF complex can also function when brought to the DNA in this way (Gaudreau *et al.*, 1997).

Multiple mechanisms may therefore act to recruit the SWI/SNF and NURF complexes to a particular gene. Whatever the manner in which they are recruited, however, it is clear that these ATP-dependent chromatin remodelling complexes play a key role in opening up the regulatory regions of specific genes so that trans-acting transcription factors can bind and regulate their expression.

Moreover, although we have discussed the opening of chromatin structures by these complexes and its role in transcriptional activation, chromatin remodelling complexes can also produce a more closed chromatin structure leading to transcriptional repression (for a review see Tyler and Kadonaga, 1998). Hence, these complexes appear to play a key role in the regulation of gene expression via the alteration of chromatin structure.

6.8 OTHER SITUATIONS IN WHICH CHROMATIN STRUCTURE IS REGULATED

In this chapter we have discussed the role of changes in the chromatin structure of individual genes in mediating commitment to a particular differentiated state and thereby allowing the tissue-specific activation of gene expression. However, differences in chromatin structure are also involved in regulating gene expression in two other well characterised processes, namely X-chromosome inactivation and genomic imprinting. Both of these processes involve differences in expression between the two copies of a specific gene which are present on different homologous chromosomes in a diploid cell (for reviews see Lyon, 1993; Lewin, 1998).

6.8.1 X-chromosome inactivation

The fact that females of mammalian species have two X chromosomes whereas males have one X and one Y chromosome creates a problem of how to compensate for the difference in dosage of genes on the X chromosome which occurs because females have two copies whereas males have only one. In mammals this problem is solved by the process of X-chromosome inactivation (for reviews see Migeon, 1994; Lee and Jaenisch, 1997; Panning and Jaenisch, 1998).

Thus, during the process of embryonic development one of the two X chromosomes in each female cell undergoes an inactivation process so that the expression of virtually all the genes on this chromosome is inactivated whilst those on the other chromosome remain active. This results in each female cell having only one active copy of X chromosome genes, paralleling the situation in male cells which have only one X chromosome (Fig. 6.35).

This process occurs apparently randomly with cells in the early female embryo inactivating either the X chromosome inherited from the father (the paternal chromosome) or that inherited from the mother (the maternal chromosome). However, once one or other of the X chromosomes has been inactivated, this inactivation is propagated stably through cell division to all the progeny of that cell (Fig. 6.35). Such stability is dependent upon the fact that the inactive and active X chromosomes have a different chromatin structure.

Thus, within the inactive X chromosome the DNA is tightly packed into a highly condensed structure known as heterochromatin and this high density structure results in the inactive X chromosome being visible as a distinct element within the cell known as a Barr body. In turn such a condensed structure has been shown to involve decreased sensitivity to digestion with DNaseI, reduced histone acetylation and enhanced methylation on C residues which can be observed when genes on the inactive X chromosome are compared to the equivalent gene on the active chromosome. Thus, for example, 60 of the 61 CpG dinucleotides in the CpG island (see Section 6.5.1) located around the promoter of the PGKI gene are methylated on the C residue when the inactive X chromosome is studied, whereas all these sites are unmethylated on the active X. Moreover, treatment with 5-azacytidine, leading to demethylation (see Section 6.5.2), can reactivate previously inactivated X-chromosome genes.

Interestingly, in *Drosophila* compensation for the reduced number of X chromosomes in male cells is achieved by increasing the transcriptional activity of the genes on the single male X chromosome rather than by inactivating one of the female X chromosomes. As in the case of X inactivation, however, this effect appears to involve alterations in chromatin structure since the male X chromosome has a higher level of acetylated histones than either of the female chromosomes and a histone acetyl transferase enzyme (see Section 6.6.1) specifically associates with the male X chromosome (Pennisi, 1997) (for a review of the comparative mechanisms of dosage compensation in mammals and *Drosophila* see Park and Kuroda, 2001).

Chromatin structure thus plays a critical role in differentially regulating the activity of genes on the X chromosomes in females and males and, in

Figure 6.35

X-chromosome inactivation results in one or other of the two X chromosomes becoming inactivated (solid) in each cell whilst the other remains active (open).

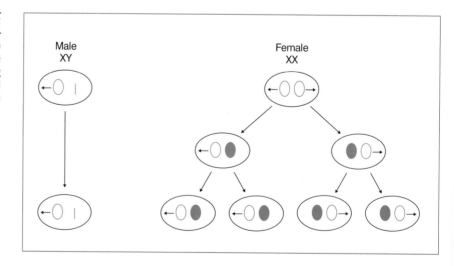

particular, in the maintenance of X-chromosome inactivation through cell division. Interestingly the onset of X-chromosome inactivation in the embryo requires a particular region of the chromosome known as the X-inactivation centre. If this region is deleted then X inactivation does not occur. A gene, known as XIST, has been mapped to the X-inactivation centre and it has been shown that the inactivation of one copy of the XIST gene in mutant mice results in a failure of X-chromosome inactivation on the chromosome which lacks an active XIST gene, with the other X chromosome being preferentially inactivated (Fig. 6.36a). Hence XIST is essential for inactivation of the X chromosome on which it is located (for reviews see Lee and Jaenisch, 1997; Panning and Jaenisch, 1998; Kelley and Kuroda, 2000).

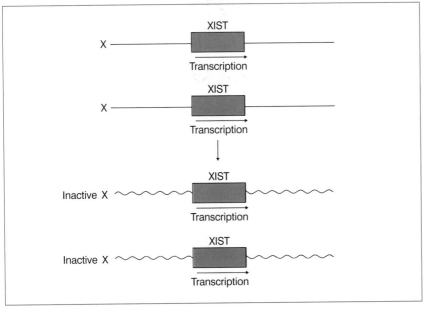

Figure 6.36

Inactivation of one copy of the XIST gene in mouse mutants results in the chromosome containing the intact XIST gene being preferentially inactivated (panel a); whilst mutant mice which transcribe the XIST gene on both X chromosomes show inactivation of both X chromosomes (panel b).

Interestingly, the XIST gene is transcribed only on the inactive X chromosome and not on the active chromosome, the opposite pattern to all other genes. It is likely therefore that transcription of XIST on one of the two X chromosomes early in development in some way results in a change in chromatin structure which is propagated along the rest of the chromosome, switching off all other genes (Fig. 6.37). In agreement with this idea, mutant mice which show expression of XIST on both chromosomes also show inactivation of both X chromosomes, exactly as would be expected if the expression of the XIST gene does indeed inactivate genes on the X chromosome containing it (Fig. 6.36b). Moreover, if an active XIST gene is placed on a non-X chromosome, the transcription of the genes on this chromosome is inactivated and the histones associated with it become deacetylated (for a review see Willard and Salz, 1997). Hence XIST gene transcription is sufficient to inactivate the genes to which it is linked, regardless of the nature of these genes.

The initiation and maintenance of X-chromosome inactivation thus requires the ability to produce and stably propagate an altered chromatin structure which we have previously seen to be critical in tissue-specific gene regulation. Moreover, like the regulation of the chromatin structure of individual genes, it involves processes such as methylation of C residues in the DNA and deacetylation of histones.

Figure 6.37

In normal mice, transcription of the XIST gene on one of the two X chromosomes is associated with its inactivation via a change in chromatin structure.

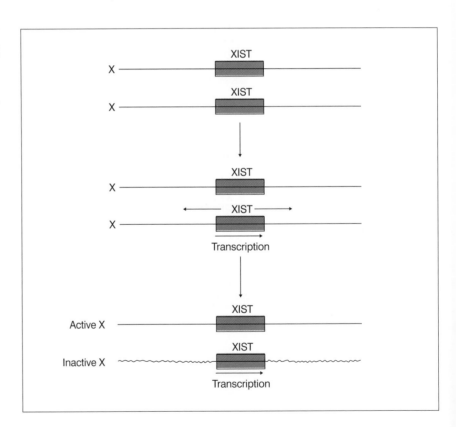

6.8.2. Genomic imprinting

Genomic imprinting (for reviews see Surani, 1998; Ferguson-Smith and Surani, 2001) resembles X-chromosome inactivation in that one of the two copies of specific genes is inactivated whilst the other remains active. This process differs, however, in that only a few genes scattered on different autosomes (non-sex chromosomes) are inactivated and that the same copy is always inactivated in all cells and in all organisms, whether male or female. Thus, it is always the maternally inherited copy of some of these, such as the genes encoding the insulin-like growth factor 2 (IGF2) protein, the SmN splicing protein and the U2AF35 related protein, which are inactivated with the paternally inherited gene remaining active. Conversely the paternally inherited copies of others, such as the Igf2 receptor gene and the H19 gene, are inactivated with the maternally inherited gene remaining active (Fig. 6.38).

As with X inactivation therefore this process results in all cells having only one functional copy of each of these genes, although in the case of genomic imprinting all cells express the same copy. This process cannot be reversed during embryonic development even when genetic crosses are used to produce embryos with two imprinted copies of the gene. Thus embryos which inherit two maternal and no paternal copies of the Igf2 gene or two maternal but no paternal copies of the SmN gene die due to the lack of an

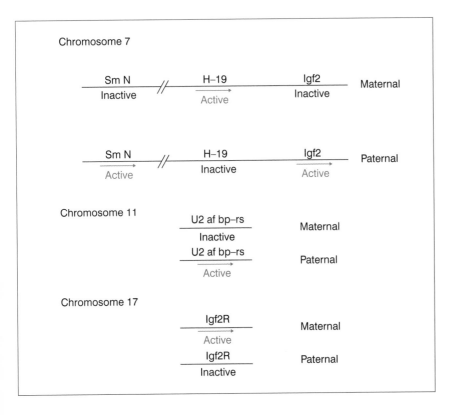

Figure 6.38

Imprinting results in the inactivation of the maternally inherited SmN, U2AFbp-rs and Igf2 genes and of the paternally inherited Igf2R and H19 genes.

active gene. These embryos therefore die due to the absence of a functional protein even though two copies of the gene capable of encoding this protein are present in each cell of the embryo.

Despite the lethal effects which can result when imprinting goes wrong, it is unclear what the normal function of this process is. It has been suggested, for example, that imprinting may represent a means of preventing the parthenogenetic development of the unfertilised egg to produce a haploid embryo with no paternal contribution. Alternatively, it may have evolved because of the conflict between the maternal and paternal genomes in terms of the transfer of nutrients from mother to offspring. Thus, whilst it is in the paternal interest to promote the growth of the individual foetus the maternal interest is to restrict the growth of any individual foetus so that other foetuses fathered by different males either concurrently (in multi-foetal litters) or subsequently can develop fully (for a discussion see Moore and Haig, 1991). Other theories range from the need for the cell to distinguish between the two copies of each chromosome which have been inherited from the mother or the father (for a review see Pardo-Manuel de Villena *et al.*, 2000) to a mechanism for producing differences between male and female offspring (for a review see Pagel, 1999).

This lack of a clear functional role for genomic imprinting is in contrast to the clear role of X-chromosome inactivation in compensating for the extra X chromosome in females compared to males. Indeed it has even been suggested that imprinting may not have a function at all but may represent a vestige of an evolutionarily ancient defence system to inactivate foreign DNA with the imprinted genes having some feature which causes this system to consider them as foreign (see Barlow, 1993 for further discussion of this idea).

Whatever its precise function (if any), it is clear that genomic imprinting resembles X-chromosome inactivation in that specific differences exist in the methylation pattern of CG dinucleotides between the active and inactive copies of the imprinted gene. Moreover, embryos which lack a functional DNA methyltransferase and which therefore cannot methylate their DNA (see Section 6.5.2) fail to carry out genomic imprinting. Hence specific methylation of one of the two copies of the gene is critical for imprinting to occur (for a review see Feil and Khosla, 1999).

Interestingly, in some cases, the role of C-methylation in regulating imprinting can be quite complex. Thus, in the case of the H19 gene, which is silenced on the paternal chromosome, a simple correlation exists with the regulatory region of the gene being methylated on the paternal chromosome but not on the maternal, as would be expected (Fig. 6.39). However, in the case of the adjacent Igf2 gene, which is silenced on the maternal chromosome, no specific methylation is seen on this chromosome compared to the paternal one. Rather, the activity of the Igf2 gene is controlled by an imprinting control region (ICR) which is located between the two genes and which is methylated on the paternal chromosome where the Igf2 gene is active.

This paradox is resolved by the finding that, when the ICR is methylated, a positive regulatory element (known as an enhancer: see Section 7.3)

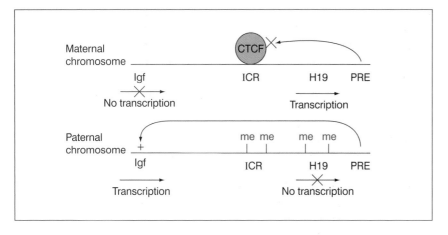

Figure 6.39

Role of C residue methylation (me) in controlling imprinting of the closely linked Igf2 and H19 genes. On the paternal chromosome, the H19 gene is methylated and is therefore only expressed from the maternal chromosome where it is unmethylated. In contrast, demethylation of the imprinting control region (ICR) on the paternal chromosome prevents binding of an insulator protein (CTCF) and allows a distinct positive regulatory element (PRE) to specifically activate Igf2 expression only on the paternal chromosome.

located downstream of the H19 gene acts at a distance to activate Igf2 gene expression (Fig. 6.39). In contrast, when the ICR is unmethylated on the maternal chromosome, a protein known as CTCF binds to it. This protein acts as an insulator (see Section 7.4.2), preventing the positive regulatory element from activating Igf2 and the gene is therefore silent. Hence, in this chromosomal region methylation of C residues on the paternal chromosome results in silencing of the H19 gene and expression of the Igf2 gene on this chromosome (for reviews see Allshire and Bickmore, 2000; Reik and Murrell, 2000).

It is clear therefore that, as with X-chromosome inactivation, imprinting involves similar modifications and changes in chromatin structure to those which are used to regulate cellular commitment and gene transcription in specific cell types.

6.9 CONCLUSIONS

A variety of changes take place in the chromatin of genes during the process of commitment to a particular pathway of differentiation. Such changes involve both modification of the DNA itself by undermethylation, to the histones with which it is associated, and to the general packaging of the DNA in chromatin. In the last few years, the study of these changes has moved from the simple description of changes, such as DNaseI sensitivity and DNA methylation, to a more mechanistic approach. This has identified two key processes regulating chromatin structure, namely histone modifications, particularly acetylation (see Section 6.6) and ATP-dependent remodelling of chromatin structure by complexes such as SWI/SNF and NURF (Section 6.7.2) (for reviews see Kingston and Narlikar, 1999; Aalfs and Kingston, 2000).

The combination of all the various processes discussed in this chapter results in three levels of chromatin structure within the cell (Fig. 6.40). Thus, although the bulk of inactive DNA is organised into a tightly packed solenoid structure, active or potentially active genes are organised into a

Figure 6.40

Levels of chromatin structure in active and inactive DNA.

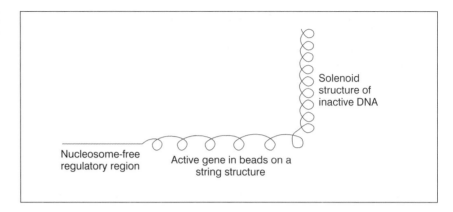

Solenoid structure of inactive DNA

Nucleosome-free regulatory region

Active gene in beads on a string structure

more open "beads on a string" structure, and short regions within the gene are either nucleosome-free or have structurally altered nucleosomes.

The role of these changes in allowing cells to maintain a commitment to a particular differentiated state and to respond differently to inducers of gene expression is well illustrated in the case of the steroid hormones and their effect on gene expression. Thus the difference in the response of different tissues to treatment with oestrogen (see Section 6.1) is likely to be due to the fact that in one tissue certain steroid-responsive genes will be inaccessible within the solenoid structure and will therefore be incapable of binding the receptor–hormone complex that is necessary for activation. In other genes, which are in the "beads on a string" structure and therefore more accessible, such binding of the complex to defined sequences in the gene will occur. Even in this case, however, gene activation will not occur as a direct consequence of this interaction, as might be the case in bacteria. Rather, the binding will result in the displacement of a nucleosome from the region of DNA, generating a hypersensitive site and allowing other regulatory proteins to interact with their specific recognition sequences and cause transcription to occur.

Both the action of the receptor–hormone complex and the subsequent binding of other transcription factors to nucleosome-free DNA clearly involve the interaction of regulatory proteins with specific DNA sequences. The next two chapters will discuss the DNA sequences and proteins involved in these interactions.

REFERENCES

Aalfs, J.D. and Kingston, R.E. (2000). What does "chromatin remodelling" mean? *Trends in Biochemical Sciences* **25**, 548–555.

Ahringer, J. (2000). NuRD and SIN3 histone deacetylase complexes in development. *Trends in Genetics* **16**, 351–356.

Allshire, R. and Bickmore, W. (2000). Pausing for thought on the boundaries of imprinting. *Cell* **102**, 705–708.

Ayer, D.E. (1999). Histone deacetylases: transcriptional repression with SINers and NuRDs. *Trends in Cell Biology* **9**, 193–198.

Bannister, A.J., Zegerman, P., Partridge, J.F., Miska, E.A., Thomas, J.O., Allshire,

R.C. and Kouzarides, T. (2001). Selective recognition of methylated lysine 9 on histone H3 by the HP1 chromo domain. *Nature* **410**, 120–124.

Barlow, D.P. (1993). Methylation and imprinting: from host defence to gene regulation? *Science* **260**, 309–310.

Becker, P.B., Ruppert, S. and Schutz, G. (1987). Genomic fingerprinting reveals cell type-specific binding of ubiquitous factors. *Cell* **51**, 435–443.

Berger, S.L. (2001). The histone modification circus. *Science* **292**, 64–65.

Bestor, T.H. and Coxon, A. (1993). The pros and cons of DNA methylation. *Current Biology* **3**, 384–386.

Bird, A. (1992). The essentials of DNA methylation. *Cell* **70**, 5–8.

Bird, A.C. (1987). CpG islands as gene markers in the vertebrate nucleus. *Trends in Genetics* **3**, 342–347.

Bird, A.P. and Wolffe, A.P. (1999). Methylation-induced repression – belts, braces and chromatin. *Cell* **99**, 451–454.

Brown, C.E., Lechner, T., Howe, L. and Workman, J.L. (2000). The many HATs of transcription coactivators. *Trends in Biochemical Sciences* **25**, 15–19.

Busch, H. and Goldknopf, I.L. (1981). Ubiquitin-protein conjugates. *Molecular and Cellular Biochemistry* **40**, 173–187.

Cheung, P., Allis, C.D. and Sassone-Corsi, P. (2000). Signalling to chromatin through histone modifications. *Cell* **103**, 263–271.

Chiu, C.-P. and Blau, H. (1985). 5-azacytidine permits gene activation in a previously non-inducible cell type. *Cell* **40**, 417–424.

Colot, V. and Rossignol, J.L. (1999). Eukaryotic DNA methylation as an evolutionary device. *BioEssays* **21**, 402–411.

Constantinides, P.G., Jones, P.A. and Gevers, W. (1977). Functional striated muscle cells from non-myoblast precursors following 5-azacytidine treatment. *Nature* **267**, 364–366.

Coon, H.G. (1966). Clonal stability and phenotypic expression of chick cartilage cells *in vitro*. *Proceedings of the National Academy of Sciences of the USA* **55**, 66–73.

Farkas, G. Gausz, J. Gallioni, M., Reuter, G. Yurkovics, H. and Karch, F. (1994). The trithorax-1 gene encodes the *Drosophila* GAGA factor. *Nature* **371**, 806–808.

Feil, R. and Khosla, S. (1999). Genomic imprinting in mammals an interplay between chromatin and DNA methylation? *Trends in Genetics* **15**, 431–435.

Felsenfeld, G. and McGhee, J.D. (1986). Structure of the 30 nm chromatin fibre. *Cell* **44**, 375–377.

Ferguson-Smith, A.C. and Surani, M.A. (2001). Imprinting and the epigenetic asymmetry between parental genomes. *Science* **293**, 1086–1089.

Finch, J. T., Noll, M. and Kornberg, R.D. (1975). Electron microscopy of defined lengths of chromatin. *Proceedings of the National Academy of Sciences of the USA* **72**, 3320–3322.

Gaudreau, L. Schmid, A., Blaschke, D. Ptashne, M. and Horz, W. (1997). RNA polymerase II holoenzyme recruitment is sufficient to remodel chromatin at the yeast PH05 promoter. *Cell* **89**, 55–62.

Gross, D.S. and Garrard, W.T. (1988). Nuclease hypersensitive sites in chromatin. *Annual Review of Biochemistry* **57**, 159–197.

Hadorn, E. (1968). Transdetermination in cells. *Scientific American* **219**, 110–120.

Hagmann, M. (1999). How chromatin changes its shape. *Science* **285**, 1200–1203.

Jenuwein, T. (2001). Re-SET-ting heterochromatin by histone methyltransferases. *Trends in Cell Biology* **11**, 266–273.

Jenuwein, T. and Allis, C.D. (2001). Translating the histone code. *Science* **293**, 1074–1080.

Jones, K.A. and Kadonaga, J.T. (2000). Exploring the transcription-chromatin interface. *Genes and Development* **14**, 1992–1996.

Jones, P.A. and Takai, D. (2001). The role of DNA methylation in mammalian epigenetics. *Science* **293**, 1068–1070.

Kawasaki, H., Schiltz, L., Chiu, R., Itakura, K., Taira, K., Nakatani, Y. and Yokoyama, K.K. (2000). ATF-2 has intrinsic histone acetyltransferase activity which is modulated by phosphorylation. *Nature* **405**, 195–200.

Kelley, R.L. and Kuroda, M.I. (2000). Noncoding RNA genes in dosage compensation and imprinting. *Cell* **103**, 9–12.

Kingston, R.E. and Narlikar, G.J. (1999). ATP-dependent remodelling and acetylation as regulators of chromatin fluidity. *Genes and Development* **13**, 2339–2352.

Kornberg, R.D. and Lorch, Y. (1999). Twenty-five years of the nucleosome, fundamental particle of the eukaryote chromosome. *Cell* **98**, 285–294.

Lacey, E. and Axel, R. (1975). Analysis of DNA of isolated chromatin subunits. *Proceedings of the National Academy of Sciences of the USA* **72**, 3978–3982.

Lachner, M., O'Carroll, D., Rea, S., Mechtler, K. and Jenuwein, T. (2001). Methylation of histone H3 lysine 9 creates a binding site for HP1 proteins. *Nature* **410**, 116–120.

Lee, J.T. and Jaenisch, R. (1997). The (epi) genetic control of mammalian X-chromosome inactivation. *Current Opinion in Genetics and Development* **7**, 274–280.

Lewin, B. (1998). The mystique of epigenetics. *Cell* **93**, 301–303.

Litt, M. D., Simpson, M., Gaszner, M., Allis, C.D. and Felsenfeld, G. (2001). Correlation between histone lysine methylation and developmental changes at the chicken β-globin locus. *Science* **293**, 2453–2455.

Luger, A, M„der, A.W., Richmond, R.K., Sargent, D.F. and Richmond, T.J. (1997). Crystal structure of the nucleosome core particle at 2.8 Å resolution. *Nature* **389**, 251–260.

Lyko, F., Ramsahoye, B.H. and Jaenisch, R. (2000). DNA methylation in *Drosophila melanogaster*. *Nature* **408**, 538–540.

Lyon, M.F. (1993). Epigenetic inheritance in mammals. *Trends in Genetics* **9**, 123–128.

McKnight, S.L., Bustin, M. and Miller, O.L. (1978). Electron microscope analysis of chromosome metabolism in the *Drosophila melanogaster* embryo. *Cold Spring Harbor Symposium on Quantitative Biology* **42**, 741–754.

Migeon, B.R. (1994). X-chromosome inactivation: molecular mechanisms and genetic consequences. *Trends in Genetics* **10**, 230–235.

Moore, T. and Haig, D. (1991). Genomic imprinting in mammalian development. *Trends in Genetics* **7**, 45–49.

Muller-Hill, B.W. (ed.). (1996). *The lac Operon: A Short History of a Genetic Paradigm*. de Grayter, Berlin.

Nan, X., Campoy, F.J. and Bird, A. (1997). MeCP2 is a transcriptional repressor with abundant binding sites in genomic chromatin. *Cell* **88**, 471–481.

Ng, H.-H. and Bird, A. (1999). DNA methylation and chromatin modification. *Current Opinion in Genetics and Development* **9**, 158–163.

Ng, H.-H. and Bird, A. (2000). Histone deacetylases: silencers for hire. *Trends in Biochemical Sciences* **25**, 121–126.

Noma, K.-I., Allis, C.D. and Grewal, S.I.S. (2001). Transitions in distinct histone H3 methylation patterns at the heterochromatin domain boundaries. *Science* **293**, 1150–1155.

Nowak, S.J. and Corces, V.G. (2000). Phosphorylation of histone H3 correlates with transcriptionally active loci. *Genes and Development* **14**, 3003–3013.

Okano, M., Bell, D.W., Haber, D.A. and Li, E. (1999). DNA methyltransferases Dnmt3a and Dnmt3b are essential for de novo methylation and mammalian development. *Cell* **99**, 247–257.

Owen H., Utley, T., Cole, R.T., Peterson, J.C.L. and Workman, J.L. (1996). Persistent site-specific remodelling of a nucleosome array by transient action of the SWI/SNF complex. *Science* **273**, 513–516.

Pagel, M. (1999). Mother and father in surprise genetic agreement. *Nature* **397**, 19–20.

Panning, B. and Jaenisch, R. (1998). RNA and the epigenetic regulation of X chromosome inactivation. *Cell* **93**, 305–308.

Pardo-Manuel de Villena, F., de la Casa-Esperón, E. and Sapienza, C. (2000). Natural selection and the function of genome imprinting: beyond the silenced minority. *Trends in Genetics* **16**, 573–579.

Park, Y. and Kuroda, M.I. (2001). Epigenetic aspects of X-chromosome dosage compensation. *Science* **293**, 1083–1085.

Paro, R. (2000). Formatting genetic text. *Nature* **406**, 579–580.

Pennisi E. (1997). Opening the way to gene activity. *Science* **275**, 155–157.

Raibaud, O. and Schwartz, M. (1984). Positive control of transcription initiation in bacteria. *Annual Review of Genetics* **18**, 173–206.

Rea, S., Eisenhaber, F., O'Carroll, D., Strahl, B.D., Sun, Z.-W., Schmid, M., Opravil, S., Mechtler, K., Ponting, C.P., Allis, C.D. and Jenuwein, T. (2000). Regulation of chromatin structure by site-specific histone H3 methyltransferases. *Nature* **406**, 593–599.

Reeves, R. and Cserjesi, P. (1979). Sodium butyrate induces new gene expression in Friend erythroleukemic cells. *Journal of Biological Chemistry* **254**, 4283–4290.

Reik, W. and Murrell, A. (2000). Silence across the border. *Nature* **405**, 408–409.

Rhodes, D. (1997). The nucleosome core all wrapped up. *Nature* **389**, 231–233.

Saitoh, N. Goldberg, I. and Earnshaw, W.C. (1995). The SMC proteins and the coming of age of the chromosome scaffold hypothesis. *BioEssays* **17**, 759–766.

Simon, J. (1995). Locking in stable states of gene expression: transcriptional control during *Drosophila* development. *Current Opinion in Cell Biology* **7**, 376–385.

Stewart, S. and Crabtree, G.R. (2000). Regulation of the regulators. *Nature* **408**, 46–47.

Strahl, B.D. and Allis, C.D. (2000). The language of covalent histone modifications. *Nature* **403**, 41–45.

Struhl, K. (1996). Chromatin structure and RNA polymerase II connection: implications for transcription. *Cell* **84**, 179–182.

Sudarsanam, P. and Winston, F. (2000). The Swi/Snf family nucleosome-remodelling complexes and transcriptional control. *Trends in Genetics* **16**, 345–351.

Surani, M.A. (1998). Imprinting and the initiation of gene silencing in the germ line. *Cell* **93**, 309–312.

Travers, A. (1993). *DNA-Protein Interactions*. Chapman and Hall, London.

Tucker, K.L. (2001). Methylated cytosine and the brain: a new base for neuroscience. *Neuron* **30**, 649–652.

Tyler, J.K. and Kadonaga, J.T. (1999). The "dark side" of chromatin remodelling: repressive effects on transcription. *Cell* **99**, 443–446.

van Holde, K. and Zlatanova, J. (1995). Chromatin higher order structure: chasing a mirage? *Journal of Biological Chemistry* **270**, 8373–8376.

Wang, H., Huang, Z.-Q., Xia, L., Feng, Q., Erdjument-Bromage, H., Strahl, B.D., Briggs, S.D., Allis, C.D., Wong, J., Tempst, P. and Zhang, Y. (2001). Methylation of histone H4 at arginine 3 facilitating transcriptional activation by nuclear hormone receptor. *Science* **293**, 853–857.

Willard, H.F. and Salz, H.K. (1997). Dosage compensation remodelling chromatin with RNA. *Nature* **386**, 228–229.

Wolffe, A. (1995). *Chromatin: Structure and Function* 2nd edition. Academic Press, London.

Wu, J. and Grunstein, M. (2000). 25 years after the nucleosome model: chromatin modifications. *Trends in Biochemical Sciences* **25**, 619–623.

Zhang, Y. and Reinberg, D. (2001). Transcription regulation by histone methylation: interplay between different covalent modifications of the core histone tails. *Genes and Development* **15**, 2343–2360.

TRANSCRIPTIONAL CONTROL – DNA SEQUENCE ELEMENTS

7

7.1 INTRODUCTION

7.1.1 Relationship of gene regulation in prokaryotes and eukaryotes

As discussed in Chapter 6, various alterations occur in the chromatin structure of a particular gene prior to the onset of transcription. Once such changes have occurred, the actual onset of transcription takes place through the interaction of defined proteins (transcription factors) with specific DNA sequences adjacent to the gene. This final stage of gene regulation is clearly analogous to the activation or repression of gene expression in prokaryotes, which was discussed briefly in Section 6.1. However, before the manner in which transcription factors and DNA sequences act to regulate gene expression in higher eukaryotes can be considered, it is necessary to mention one aspect of eukaryotic systems that does not exist in bacteria.

7.1.2 Co-ordinately regulated genes are not linked in eukaryotes

In prokaryotes, when several genes encoding particular proteins are expressed in response to a particular signal, the genes are found tightly linked together in an operon. In response to the activating signal, all the genes in the operon are transcribed as one single polycistronic (multi-gene) mRNA molecule, and translation of each protein from internal AUG initiation codons (see Section 3.3.1) results in the desired coordinate production of the proteins encoded by the individual genes. A typical example of this is seen in the case of the lactose-inducible genes located in the *lac* operon. The efficient use of lactose requires not only the enzyme β-galactosidase, which cleaves the lactose molecule, but also a permease enzyme, to facilitate uptake of lactose by the cell, and a modifying *trans*-acetylase enzyme. The genes encoding these three molecules must be activated in response to lactose and this is achieved by linking them together in an operon whose expression is dependent upon the presence of lactose (Fig. 7.1).

In higher eukaryotic organisms this does not occur. Thus operon-type structures transcribed into polycistronic RNAs do not exist in higher organisms. Rather, individual genes are transcribed by RNA polymerase II into individual monocistronic RNAs encoding single proteins (see Section 3.3.1). Hence, genes whose protein products are required in parallel are present in the genome as individual genes whose expression must be coordinated. Moreover, such coordinately expressed genes are not even closely associated in the genome in a manner that might be thought to facilitate their regulation, but are very often present on different chromosomes within the eukaryotic nucleus. Thus, the production of a functional antibody molecule by the mammalian B-cell requires the synthesis of both the immunoglobulin heavy- and light-chain proteins, which together make up the functional anti-

Figure 7.1

Structure of the *lac* operon of *E. coli*, in which the three genes are transcribed into one single RNA and translated into individual proteins following binding of the ribosome at three sites in the RNA molecule.

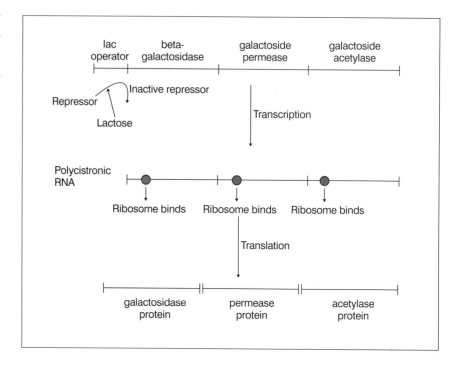

body molecule. The genes encoding the heavy- and light-chain proteins are, however, found on separate chromosomes; the heavy-chain locus being on chromosome 14 in humans whereas light-chain genes are found on both chromosomes 2 and 22. Similarly, the production of a functional globin molecule requires the association of an α-globin-type protein and a β-globin-type protein. Yet the genes encoding the various members of the α-globin family are on chromosome 16 in humans whereas the genes encoding the β–globins are found on chromosome 11.

This difference between prokaryotes and eukaryotes is likely to reflect a need for greater flexibility in regulating gene expression in eukaryotes. Thus the bacterial arrangement, although providing a simple mechanism for coordinating the expression of different proteins, also means that, in general, expression of one protein will also necessitate expression of the other coordinately expressed proteins. In contrast, the eukaryotic system allows α-globin to be produced in parallel with β-globin in the adult organism but to be associated with another β-globin-like protein, namely γ-globin, in the foetus.

7.1.3 The Britten and Davidson model for the coordinate regulation of unlinked genes

This greater flexibility does, however, necessitate some means of coordinately regulating gene expression, allowing the production of α- and β-globin in the adult reticulocyte, γ and β-globin in the embryonic reticulocyte, and the heavy and light chains of immunoglobulin in the antibody-producing B-cell. A model of the mechanism of such coordinate regulation was put forward by Britten and Davidson (1969). They proposed that genes regulated in parallel with one another in response to a particular signal would contain a common regulatory element which would cause the activation of the gene in response to that signal (Fig. 7.2). Individual genes could contain more than one regulatory element, some of which would be shared with other genes, which in turn could possess elements not present in the first gene. Specific signals causing gene activation would act by stimulating a specific integrator gene whose product would activate all the genes containing one particular sequence element. This mechanism would allow the observed activation of distinct but overlapping sets of genes in response to specific signals via the activation of particular integrator genes.

Although this model was proposed over 30 years ago, when our understanding of eukaryotic gene regulation was considerably more limited than at present, it continues to serve as a useful framework for considering gene regulation. In modern terms, the integrator gene would be considered as encoding a transcription factor (see Chapter 8) which binds to the regulatory sequences and activates expression of the corresponding genes. As discussed in Section 8.4, the activation of such a factor by a particular signal has now been shown to operate both by *de novo* synthesis of the factor (in an individual tissue or in response to a particular signal), as envisaged by Britten and Davidson, and by direct activation of a pre-existing protein, for example, by phosphorylation in response to the activating signal. The concept that such factors act by binding to DNA sequences held in common between coordinately regulated genes is now well established, and it is therefore necessary to consider the nature of these sequences. In the original model of Britten and Davidson (1969) it was considered that this role would be played by a class of repeated DNA sequences about 200–300 base pairs in length which had been shown to be present in many copies in the genome. It is now clear, however, that the target DNA binding sites for transcription

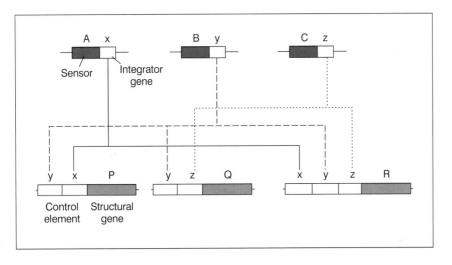

Figure 7.2

Britten and Davidson model for the coordinate expression of unlinked genes in eukaryotes. Sensor elements (A, B, C) detect changes requiring alterations in gene expression and switch on appropriate integrator genes (x, y, z) whose products activate the structural genes (P, Q, R) containing the appropriate control elements. Note the flexibility of the system whereby a particular structural gene can be activated with or without another structural gene by selecting which integrator gene is activated.

factors are much shorter sequences, approximately 10–30 base pairs in length, and these short sequence elements will now be discussed (for a review see Maniatis *et al.*, 1987).

7.2 Short sequence elements located within or adjacent to the gene promoter

7.2.1 Short regulatory elements

In prokaryotes the sequence elements that play an essential role in transcriptional regulation are located immediately upstream of the point at which transcription begins, and form part of the promoter which directs transcription of the gene. Such sequences can be divided into two classes, namely those found in all genes, which play an essential role in the process of transcription itself, and those found in one or a few genes, which mediate their response to a particular signal (Schmitz and Galas, 1979). It would be expected by analogy, therefore, that the sequences that play a role in the regulation of eukaryotic transcription would be located similarly, upstream of the transcription start site within or adjacent to the promoter. Hence, a comparison of such sequences in different genes should reveal both basic promoter elements necessary for all transcription (which would be present in all genes) and those necessary for a particular pattern of regulation (which would be present only in genes exhibiting a specific pattern of regulation). The role of such sequences can be confirmed either by destroying them by deletion or mutation or by transferring them to other genes in an attempt to confer the specific pattern of regulation of the donor gene upon the recipient.

7.2.2 The 70 kDa heat-shock protein gene

To illustrate this method of analysis, we will focus upon the gene encoding the 70 kDa molecular weight heat-shock protein (*hsp70*) and compare it with other genes showing different patterns of regulation. As discussed in Section 4.2.4, exposure of a very wide variety of cells to elevated temperature results in the increased synthesis of a few heat-shock proteins, of which *hsp70* is the most abundant. Such increased synthesis is mediated in part by increased transcription of the corresponding gene, which can be visualised as a puff within the polytene chromosome of *Drosophila* (for a review see Ashburner and Bonner, 1979). Hence, examination of the sequences located upstream of the start site for transcription in this gene should identify potential sequences involved in its induction by temperature elevation, as well as those involved in the general mechanism of transcription. The sequences present in this region of the *hsp70* gene which are also found in other genes are listed in Table 7.1, and their arrangement is illustrated in Fig. 7.3 (Williams *et al.*, 1989; for further information on these sequences see Davidson *et al.*, 1983, Jones *et al.*, 1988 and references therein).

A comparison of this type reveals a number of sequence motifs shared by the *hsp70* gene and other non-heat-inducible genes as well as one which is

Figure 7.3

Transcriptional control elements in the human *hsp70* gene promoter. The protein binding to a particular site is indicated above the line and the corresponding DNA element below the line. These elements are described more fully in Table 7.1.

unique to heat inducible genes. These include the TATA box which was discussed in Section 3.2.3 and several others which will be considered in turn.

The CCAAT and Spl boxes

The basal transcriptional complex containing TBP, RNA polymerase II and other associated factors can produce only a low rate of transcription. This rate is enhanced by the binding of other transcription factors to sites upstream of the TATA box which enhances either the stability or the activity of the basal complex. The binding sites for several of these factors are present in a wide variety of different genes with different patterns of activity. These sites act as targets for the binding of specific factors which are active in all cell types and their binding therefore results in increased transcription in all tissues (for a review see Latchman, 1998).

An example of this is the Spl box, two copies of which are present in the *hsp70* gene promoter (Fig. 7.3). This GC-rich DNA sequence binds a transcription factor known as Spl which is present in all cell types. Similarly, the CCAAT box, which is located upstream of the start site of transcription of a

Name	Consensus	Other genes containing sequences
TATA box	TATA A/T A A/T	Very many genes
CCAAT box	TGTGGCTNNNAGCCAA	α- and β-globin, herpes simplex virus thymidine kinase, cellular oncogenes c-*ras*, c-*myc*, albumin, etc
Spl box	GGGCGG	Metallothionein IIA. type II procollagen. dihydrofolate reductase, etc.
CRE	T/G T/A CGTCA	Somatostatin, fibronectin, α-gonadotrophin, c-*fos*, etc.
AP2 box	CCCCAGGC	Collagenase, class I antigen H-2k[b], metallothionein II A
Heat-shock consensus	CTNGAATNTTCTAGA	Heat-inducible genes *hsp83*, *hsp27*, etc.

Table 7.1

Sequences present in the upstream region of the *hsp* 70 gene which are also found in other genes wide variety of genes (including the *hsp70* gene) which are regulated in different ways, is also believed to play an important role in allowing transcription of the genes containing it by binding constitutively expressed transcription factors.

The heat-shock element

In contrast to these very widespread sequence motifs, another sequence element in the *hsp70* gene is shared only with other genes whose transcription is increased in response to elevated temperature. This sequence is found 62 bases upstream of the start site for transcription of the *Drosophila hsp70* gene and at a similar position in other heat-inducible genes (Davidson *et al.*, 1983). This heat-shock element (HSE) is therefore believed to play a critical role in mediating the observed heat inducibility of transcription of these genes. In order to confirm that this is the case, it is necessary to transfer this sequence from the *hsp70* gene to another gene, which is not normally heat inducible, and show that the recipient gene now becomes inducible. This was achieved by Pelham (1982) who transferred the HSE onto the non-heat-inducible thymidine kinase (*tk*) gene taken from the eukaryotic virus, herpes simplex. When the hybrid gene was introduced into cells and the temperature subsequently raised, increased thymidine kinase production was detected, showing that the HSE had rendered the *tk* gene inducible by elevated temperature (Fig. 7.4).

This experiment therefore proves that the common sequence element found in the heat-inducible genes is responsible directly for their heat inducibility. The manner in which these experiments were carried out also permits a further conclusion with regard to the way in which this sequence acts. Thus, the HSE used by Pelham was taken from the *Drosophila hsp70* gene and, in this cold-blooded organism, would be activated normally by the thermally stressful temperature of 37°C. The cells into which the hybrid gene was introduced, however, were mammalian cells which grow normally at 37°C and only express the heat-shock genes at the higher temperature of 42°C. In these experiments the hybrid gene was induced only at 42°C, the heat-shock temperature characteristic of the cell into which it was introduced, and not at 37°C, the temperature characteristic of the species from which the DNA sequence came. This means that the HSE does not possess some form of inherent temperature sensor or thermostat which is set to go off at a particular temperature, since in this case the *Drosophila* sequence would activate transcription at 37°C, even in mammalian cells. Rather, it must act by being recognised by a cellular protein which is activated in response to elevated temperature and, by binding to the heat-shock element, produces increased transcription. Evidently, although the elements of this response are conserved sufficiently to allow the mammalian protein to recognise the *Drosophila* sequence, the mammalian protein will, of course, only be activated at the mammalian heat-shock temperature and hence induction will only occur at 42°C.

Hence these experiments not only provide evidence for the importance of the HSE in causing heat-inducible transcription, but also indicate that it acts

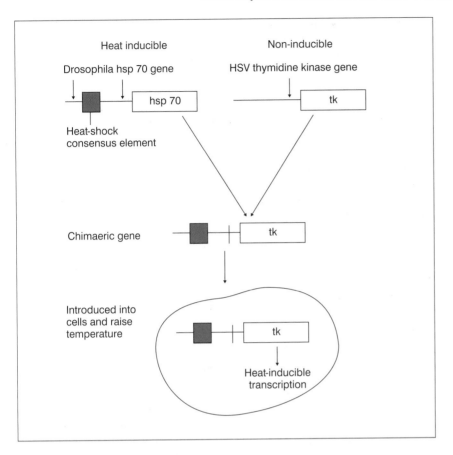

Figure 7.4

Demonstration that the heat-shock consensus element mediates heat inducibility. Transfer of this sequence to a gene (thymidine kinase) which is not normally inducible renders this gene heat inducible.

by binding a protein. This indirect evidence that the HSE acts by binding a protein can be confirmed directly by using a variety of techniques which allow the proteins binding to a specific DNA sequence to be analysed (see Section 7.2.4 for a description of some of these methods). When an analysis of this type is carried out on the upstream regions of the heat-shock genes, it can be shown that in non-heat-shocked cells, TBP is bound to the TATA box and the GAGA factor is bound to upstream sequences (Fig. 7.5a). The binding of the GAGA factor is believed to displace a nucleosome and create the DNaseI hypersensitive sites observed in these genes in non-heat-shocked cells (see Section 6.7.2). By contrast, in heat-shocked cells, where high level transcription of the gene is occurring, an additional protein, which is bound to the HSE, is detectable on the upstream region (Fig. 7.5b). Hence the induction of the heat-shock genes is indeed accompanied by the binding of a protein, known as the heat-shock factor (HSF), to the HSE, as suggested by the experiments of Pelham (1982). The binding of this factor to a gene whose chromatin structure has already been altered to render it potentially activatable, results in stimulation of transcription exactly as discussed in Chapter 6 and illustrated in Fig. 6.33. In agreement with this, the purified heat-shock factor can bind to the HSE and stimulate the transcription of the

Figure 7.5

Proteins binding to the promoter of the *hsp70* gene before (a) and after (b) heat shock.

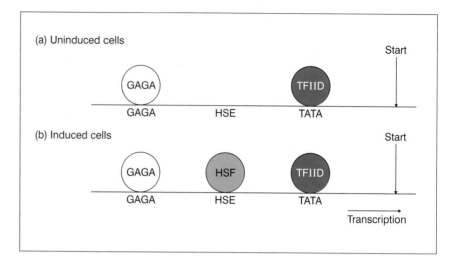

Figure 7.5

Proteins binding to the promoter of the *hsp70* gene before (a) and after (b) heat shock.

hsp70 gene in a cell-free nuclear extract while having no effect on the transcription of the non-heat-inducible actin gene (Parker and Topol, 1984).

The activation of these heat-inducible genes can therefore be fitted very readily into the Britten and Davidson scheme for the regulation of gene expression. The HSE represents the common sequence present in the similarly regulated genes, whereas the heat-shock factor would represent the product of the integrator gene which regulates their expression. Unlike the original model, however, it is clear that the heat-shock factor is not synthesized *de novo* in response to thermal stress. Rather it is present in unstressed cells (Parker and Topol, 1984) and can activate the heat-shock genes following exposure to elevated temperature even in the presence of protein synthesis inhibitors preventing its *de novo* synthesis (Zimarino and Wu, 1987). It is clear, therefore, that upon heat shock a previously inactive factor is activated to a DNA-binding form by a post-translational modification involving an alteration of the protein itself. This activation has been produced in an isolated cell-free nuclear extract by elevated temperature, and appears to involve a temperature-dependent change in the protein from a monomer to a trimer which allows it to bind to the heat-shock element and activate transcription (for a review see Morimoto, 1998). For further discussion of the mechanism of HSF activation see Section 8.4.3.

7.2.3 Other response elements

Although some modification of the Britten and Davidson model, with regard to the activation of the transcription factor itself, is therefore necessary, the heat-shock system does provide a clear example of the role of short common sequences within the promoters of particular genes in mediating their common response to a specific stimulus. A number of similar elements, which are found in the promoters of genes activated by other signals, have now been identified and have been shown to be capable of transferring the specific response to another marker gene (reviewed by Davidson *et al.*, 1983). A selection of such sequences is listed in Table 7.2.

As indicated in Table 7.2, these sequences act by binding specific proteins which are synthesised or activated in response to the inducing signal. Such transcription factors are discussed further in Chapter 8. It is noteworthy, however, that many of the sequences in Table 7.2 exhibit dyad symmetry, a similar sequence being found in the 5′ to 3′ direction on each strand. The oestrogen response element for example, has the sequence:

5′ AGGTCANNNTGACCT 3′
3′ TCCAGTNNNACTGGA 5′

the two halves of the ten-base palindrome being separated by three random bases. Such symmetry in the binding sites for these transcription factors indicates that they bind to the site in a dimeric form consisting of two protein molecules.

Various sequences that confer responses to several different signals have therefore been identified. Exactly as suggested by Britten and Davidson, one gene can possess more than one such element, allowing multiple patterns of regulation. Thus comparison of the sequences listed in Table 7.2 with those contained in the *hsp70* gene, listed in Table 7.1, reveals that, in addition to the HSE, this gene also contains the cyclic AMP response element (CRE)

Consensus sequences	Response to	Protein factor	Gene containing sequences
CTNGAATNTT CTAGA	Heat	Heat-shock transcription factor	*hsp70, hsp83, hsp27,* etc.
T/G T/A CGTCA	Cyclic AMP	CREB/ATF	Somatostatin, fibronectin α- gonadotrophin *c-fos*, *hsp70*
TGAGTCAG	Phorbol esters	AP1	Metallothionein IIA, α-antitrypsin, collagenase
CC(A/T)₆GG	Growth factor in serum	Serum response factor	*c-fos, Xenopus* γ-*actin*
RGRACANNN TGTYCY	Glucocorticoid	Glucocorticoid receptors	Metallothionein IIA, tryptophan oxygenase, uteroglobin, lysozyme
RGGTCANNN TGACCY	Oestrogen	Oestrogen receptor	Ovalbumin, conalbumin vitellogenin
RGGTCAT GACCY	Thyroid hormone	Thyroid hormone receptor	Growth hormone, myosin heavy chain
TGCGCCCGCC AGTTTCNN TTTCNC/T	Heavy metals Interferon-α	Mep-1 Stat-1 Stat-2	Metallothionein genes Oligo A synthetase, guanylate-binding protein
TTNCNNNAA	Interferon-γ	Stat-1	Guanylate-binding protein, Fc γ receptor

N, any base; R, purine; Y, pyrimidine

Table 7.2

Sequences that confer response to a particular stimulus

which mediates the induction of a number of genes, such as that encoding somatostatin, in response to treatment with cyclic AMP. Similarly, while genes may share particular elements, flexibility is provided by the presence of other elements in one gene and not another, allowing the induction of a particular gene in response to a given stimulus which has no effect on another gene. For example, although the *hsp70* gene and the metallothionein IIA gene share a binding site for the transcription factor AP2, only the metallothionein gene has a binding site for the glucocorticoid receptor, which confers responsivity to glucocorticoid hormone induction, and hence only this gene is inducible by hormone treatment.

In some cases, the sequence elements that confer responses to a particular stimulus can be shown to be related to one another. Thus the sequence mediating response to glucocorticoid treatment is similar to that which mediates response to another steroid, namely oestrogen. Similarly, both the oestrogen- and thyroid-hormone-responsive elements contain identical sequences showing dyad symmetry forming a palindromic repeat of the sequence GGTCA. In the oestrogen-responsive element, however, the two halves of this dyad symmetry are separated by three bases which vary between different genes, whereas in the thyroid-hormone-responsive element the two halves are contiguous (Table 7.3a; for review see Gronemeyer and Laudet, 1995).

In addition, as well as being arranged as palindromic repeats, the GGTCA core sequence can also be arranged as two direct repeats (Table 7.3b). In this arrangement, the spacing between the two repeats can again regulate which hormone produces a response with a spacing of four bases forming an alternative thyroid hormone response element whilst a spacing of one base confers responsivity to 9-cis retinoic acid, two or five bases to all trans-retinoic acid and a spacing of four bases to vitamin D (Table 7.3b; see Gronemeyer and Moras, 1995; Mangelsdorf and Evans, 1995).

Table 7.3

Relationship of various hormone response elements

(a) Palindromic repeats	
Glucocorticoid	RGRACANNNTGTYCY
Oestrogen	RGGTCANNNTGACCY
Thyroid	RRGTCA - - - TGACCY

(b) Direct repeats	
9-*cis* retinoic acid	AGGTCAN$_1$AGGTCA
All-*trans* retinoic acid	AGGTCAN$_2$AGGTCA
	AGGTCAN$_5$AGGTCA
Vitamin D3	AGGTCAN$_3$AGGTCA
Thyroid hormone	AGGTCAN$_4$AGGTCA

N indicates that any base can be present at that position
R indicates a purine, i.e. A or G
Y indicates a pyrimidine, i.e. C or T
A dash indicates that no base is present, the gap having been introduced to align the sequence with the other sequences.

Such similarities in these different hormone response elements are paralleled by a similarity in the individual cytoplasmic receptor proteins which form a complex with each of these hormones and then bind to the corresponding DNA sequence. All of these receptors can be shown to be members of a large family of related DNA-binding proteins whose hormone and DNA binding specificities differ from one another. Hence a particular DNA sequence confers a response to a particular hormone because it binds the appropriate receptor which also binds that hormone. The exchange of particular regions of these receptor proteins with those of other family members has provided considerable information on the manner in which sequence-specific binding to DNA occurs, and this is discussed in Section 8.2.3.

Although the sequence elements shown in Table 7.2 are all involved in the response to particular inducers of gene expression, it is clear that other short sequence elements or combinations of elements are involved in controlling the tissue-specific patterns of expression exhibited by eukaryotic genes. Thus the octamer motif (ATGCAAAT), which is found in both the immunoglobulin heavy- and light-chain promoters, can confer B-cell-specific expression when linked to a non-regulated promoter (Wirth *et al.*, 1987). Similarly, short DNA sequences that bind liver-specific transcription factors have been identified in the region of the rat albumin promoter, known to be involved in mediating the liver-specific expression of this gene.

7.2.4 DNA binding by short sequence elements

In discussing short sequence elements which confer a particular pattern of gene regulation we have assumed that they act by binding specific regulatory proteins and indeed the work of Pelham (1982) (see Section 7.2.2) provided indirect evidence that this is the case for the HSE. However, it is necessary to prove directly that this is so for the HSE and other short DNA sequences. A variety of methods have therefore been devised for demonstrating that regulatory protein(s) bind to a particular DNA sequence and for characterising these factors. Two of the most important of these will be described (for further discussion of these and other methods, see Latchman, 1998 and for practical details of how to carry them out see Latchman, 1999).

The DNA mobility shift assay

When a particular DNA sequence is first identified as being involved in a specific pattern of gene regulation, the next step is usually to carry out a DNA mobility shift assay. In this technique a short oligonucleotide containing the DNA sequence is radioactively labelled and mixed with a whole cell or purified nuclear extract. After incubating the two together, the mixture is run on a non-denaturing polyacrylamide gel and autoradiographed to detect the position of the radioactivity. If a protein has bound to the labelled DNA sequence, it will move more slowly in the gel than if no protein has bound. Hence, protein binding is detected by the appearance of a retarded band, giving the technique its alternative names of band shift or gel retardation assay (Fig. 7.6, track A).

Figure 7.6

DNA mobility shift assay. Panel (a) shows schematically the factors binding to a radioactively labelled oligonucleotide (*) whilst panel (b) shows what will be seen when the gel is visualised by autoradiography. In track A, a protein (X) in the extract used has bound to the DNA resulting in a retarded radioactive complex of lower mobility. In track B, the cell extract used does not have the DNA binding protein so no retarded complex forms. In track C, a large excess of unlabelled competitor DNA has been added. It binds protein X but no retarded band is seen on autoradiography since the competitor DNA is not radioactive. In track D, an antibody to protein X has been added so that a supershifted complex of even lower mobility is formed.

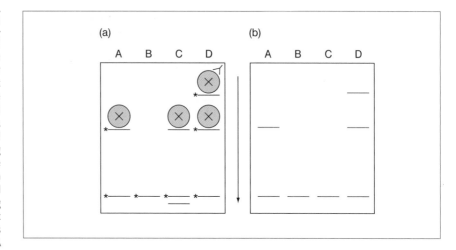

Once such a band has been detected, one can, for example, carry out the assay using extracts of different cell types or of cells treated in different ways to see if the presence or absence of the DNA binding activity correlates with the pattern of gene activity conferred by the DNA sequence (Fig. 7.6, tracks A and B). Similarly, it is possible to characterise the DNA binding specificity of the binding factor by adding a large excess of unlabelled DNA sequences as well as the radioactively labelled sequence. If the unlabelled "competitor" sequence can also bind the factor, it will do so and since it is in excess it will prevent the labelled sequence from binding the factor. However, since the competitor DNA is unlabelled, no radioactive band will be detected (Fig. 7.6, track C). This method can therefore be used to determine whether a novel DNA binding activity has a similar binding specificity to a known protein and is therefore likely to be identical or closely related to it. This relationship between the binding activity and known proteins can also be probed if an antibody to the known protein is available. Thus, the antibody can be added to the assay to see whether it binds to the DNA binding factor forming a complex of even lower mobility, which is known as a "supershifted" complex (Fig. 7.6, track D).

DNaseI footprinting assay

The DNA mobility shift assay is therefore of great value in characterising the proteins binding to a specific DNA sequence. However, it does not provide any information as to where the protein binds within the DNA sequence or its position relative to other proteins binding to adjacent DNA sequences in the regulatory region of a particular gene. This is achieved by the DNaseI footprinting assay. In this technique, a DNA fragment covering the region of interest is radiolabelled at one end and is then incubated with a nuclear or whole cell extract. It is then exposed to digestion with DNaseI under conditions where the DNA is reduced to a series of fragments, each labelled at one end and differing by one base pair in size. However, if a protein is bound to a particular region of the DNA, this region will be protected from the digestion by DNaseI. Hence, when the DNA fragment is run on a gel capable of resolving one base pair differences, the fragments corresponding to cuts at the protected site will be missing. Sites of protein binding therefore appear as gaps in the ladder of DNA fragments and this is known as a "footprint" (Fig. 7.7).

Although technically more difficult to carry out, this method thus offers advantages over the DNA mobility shift assay. Firstly, it delineates the bases which are actually bound by the binding protein. Secondly, it is possible to visualise multiple footprints on a single DNA molecule, thereby elucidating the pattern of proteins bound to the promoter or regulatory region of a particular gene.

As with the DNA mobility shift assay, it is also possible to determine whether a particular binding activity/footprint is observed with extracts prepared from cells exposed to different treatments or from different cell types and relate this to the pattern of expression of the gene being studied. Similarly, the DNA sequence specificity of any binding activity can be studied by adding an excess of an unlabelled oligonucleotide containing a sequence related to that of a particular footprint. Thus, if the DNA binding activity producing the footprint can bind to the unlabelled sequence, it will do so and hence the footprint will disappear since that region of DNA will no longer be protected from digestion. DNaseI footprinting is therefore a valuable technique for examining the interaction of proteins with particular DNA sequences.

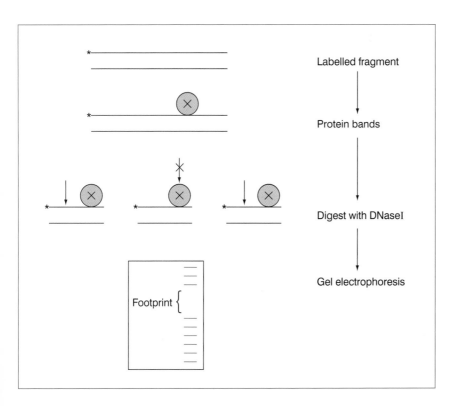

Figure 7.7

DNaseI footprinting assay. A protein (X) that binds to the test sequence, which is radioactively labelled at one end, protects it from digestion by DNaseI. Hence, the DNA fragments corresponding to digestion in this region will not be formed and will appear as a "footprint" in the ladder of bands formed by DNaseI digestion at other, unprotected, parts of the DNA. Hence, protein binding to specific sequences can be detected and localised as a footprint.

193

7.2.5 Mechanism of action of promoter regulatory elements

It is clear, therefore, that short DNA sequence elements located near the start site of transcription play an important role in regulating gene expression in eukaryotes. As indicated in Table 7.2 and discussed above, such sequences mediate transcriptional activation by binding a specific protein. This binding may give rise to gene activity in one of two ways (Fig. 7.8). First, as discussed in Chapter 6 and illustrated by the glucocorticoid receptor, binding of a specific protein may result in displacement of a nucleosome and generation of a DNaseI hypersensitive site, allowing easy access to the

Figure 7.8

Roles of short sequence elements in gene activation. These elements can either bind a factor that displaces a nucleosome and unmasks a binding site for another factor (a) or can bind a factor that directly activates transcription (b).

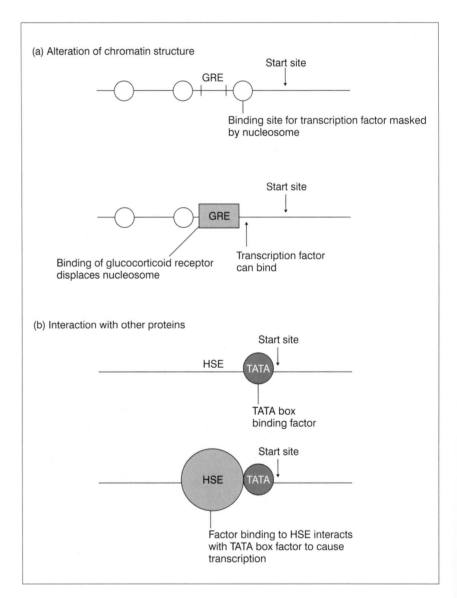

gene for other transcription factors. The direct activation of transcription by such factors constitutes the second mechanism of gene induction, and is illustrated both by the binding of other non-regulated factors to glucocorticoid-regulated genes following binding of the receptor and by the binding of the heat-shock factor to its consensus sequence in the heat-inducible genes. These factors are likely to act by interacting directly with proteins necessary for transcription, such as TBP or RNA polymerase itself. This interaction facilitates the formation of a stable transcription complex, which may enhance the binding of RNA polymerase to the DNA or alter its structure in a manner which increases its activity (for further discussion see Section 8.3.3; Latchman, 1998).

It should be noted that these two mechanisms of action are not exclusive. Thus the glucocorticoid receptor, whose binding to a specific DNA sequence displaces a nucleosome, also contains an activation domain capable of interacting with other bound transcription factors (see Section 8.3.2). Hence, following binding of the receptor, transcription is increased by inter-factor interactions between the receptor and other bound factors.

Short DNA sequence elements act in a similar manner to gene regulatory elements in prokaryotes, and are located at a similar position in the promoter region close to the start site of transcription. Unlike the situation in prokaryotes, however, important elements involved in the regulation of eukaryotic gene expression are also found at very large distances from the site of initiation of transcription. The nature of such regulatory elements will now be discussed.

7.3 ENHANCERS

7.3.1 Regulatory sequences that act at a distance

The first indication that sequences located at a distance from the start site of transcription might influence gene expression in eukaryotes came with the demonstration that sequences over 100 bases upstream of the transcriptional start site of the histone H2A gene were essential for its high-level transcription (Grosschedl and Birnsteil, 1980). Moreover, although this sequence was unable to act as a promoter and direct transcription, it could increase initiation from an adjacent promoter element up to one hundredfold when located in either orientation relative to the start site of transcription.

Subsequently, a vast range of similar elements have been described in both cellular genes and those of eukaryotic viruses, and have been called enhancers because, although they lack promoter activity and are unable to direct transcription themselves, they can dramatically enhance the activity of promoters (for reviews see Hatzopoulos *et al.*, 1988; Muller *et al.*, 1988; Thompson and McKnight 1992). Hence, if an enhancer element is linked to a promoter, such as that derived from the β-globin gene, the activity of the promoter can be increased several hundredfold. Variation in the position and orientation at which the enhancer element was placed relative to the promoter has led to three conclusions with regard to the action of enhancers. These are: (a) an enhancer element can activate a promoter when placed up to several thousand bases from the promoter; (b) an enhancer can activate a promoter when placed in either orientation relative to the promoter; or (c) an enhancer can activate a promoter when placed upstream or downstream of the transcribed region, or within an intervening sequence which is removed from the RNA by splicing (see Section 5.2). These characteristics constitute the definition of an enhancer and are summarised in Fig. 7.9.

7.3.2 Tissue-specific activity of enhancers

Following the discovery of enhancers, it was rapidly shown that many genes expressed in specific tissues also contained enhancers. Such enhancers frequently exhibited a tissue-specific activity, being able to enhance the activity of other promoters only in the tissue in which the gene from which they were derived is normally active and not in other tissues (Fig. 7.10). The tissue-specific activity of such enhancers acting on their normal promoters is

Figure 7.9

Characteristics of an enhancer element which can activate a promoter at a distance (a), in either orientation relative to the promoter (b), and when positioned upstream, downstream, or within a transcription unit (c).

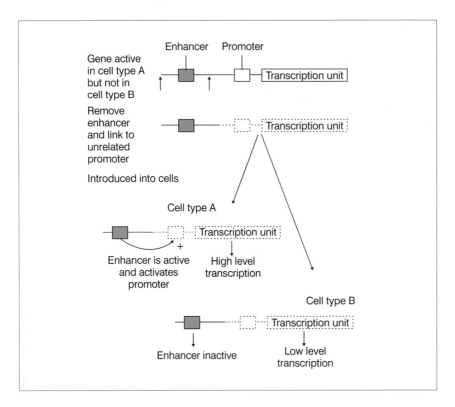

Figure 7.10

A tissue-specific enhancer can activate the promoter of its own or another gene only in one particular tissue and not in others.

likely, therefore, to play a critical role in mediating the observed pattern of gene regulation.

Thus, as previously discussed (Section 2.4), the genes encoding the heavy and light chains of the antibody molecule contain an enhancer located within the large intervening region separating the regions encoding the joining and constant regions of these molecules. When this element is linked to another promoter, such as that of the β-globin gene, it increases its activity dramatically when the hybrid gene is introduced into B-cells. In contrast, however, no effect of the enhancer on promoter activity is observed in other cell types, such as fibroblasts, indicating that the activity of the enhancer is tissue-specific.

Similar tissue-specific enhancers have also been detected in genes expressed specifically in the liver (α-foetoprotein, albumin, α-1-antitrypsin), the endocrine and exocrine cells of the pancreas (insulin, elastase, amylase), the pituitary gland (prolactin, growth hormone), and many other tissues. The tissue-specific activity of these enhancer elements is likely to play a crucial role in the observed tissue-specific pattern of expression of the corresponding gene. Thus, in the case of the insulin gene, early experiments involving linkage of different upstream regions of the gene to a marker gene, and subsequent introduction into different cell types, identified a region approximately 250 bases upstream of the transcriptional start site as being of crucial importance in producing high-level expression in pancreatic endocrine cells. This position corresponds exactly to the position of the tissue-specific enhancer, indicating the importance of this element in gene regulation. Similarly, mutation of conserved sequences within the tissue-specific enhancers of genes expressed in the exocrine cells of the pancreas, such as elastase and chymotrypsin, abolishes the tissue-specific pattern of expression of these genes.

In the case of the insulin gene, the importance of the enhancer element in producing tissue-specific gene expression was further demonstrated by experiments in which this enhancer (together with its adjacent promoter) was linked to the gene encoding the large T antigen of the eukaryotic virus SV40, whose production can be measured readily using a specific antibody. The resulting construct was introduced into a fertilised mouse egg and the expression of large T antigen analysed in all tissues of the transgenic mouse that developed following the return of the egg to the oviduct. Expression of large T was detectable only in the pancreas and not in any other tissue (Fig. 7.11) and was observed specifically in the β cells of the pancreatic islets which produce insulin (Fig. 7.12; Hanahan 1985). The enhancer of the insulin gene is therefore capable of conferring the specific pattern of insulin gene expression on an unrelated gene *in vivo*.

Hence, enhancer elements constitute another type of DNA sequence which is involved in the activation of genes in a particular tissue or in response to a particular stimulus, as suggested by the Britten and Davidson model. As envisaged by the model, the binding of cell-type specific proteins to the enhancer element has been demonstrated for many different enhancers, including those in the immunoglobulin and insulin genes. This has been achieved using the techniques described in Section 7.2.4 which can be used to determine the protein(s) binding to any specific DNA sequence, regardless of whether it is located adjacent to the promoter or within an enhancer.

Figure 7.11

Assay for expression of a hybrid gene in which the SV40 T-antigen protein-coding sequence is linked to the insulin gene enhancer and promoter. The gene was introduced into a fertilised egg and a transgenic mouse containing the gene in every cell of its body isolated. Expression of the T antigen is assayed by immunoprecipitation of protein from each tissue with an antibody specific for T antigen. Note that expression of the T antigen is detectable only in the pancreas (p) and not in other tissues, indicating the tissue-specific activity of the insulin gene enhancer. The track labelled "cos" contains protein isolated from a control cell line expressing T antigen. The Ig band in all tracks is derived from the immunoglobulin antibody used to precipitate the T antigen.

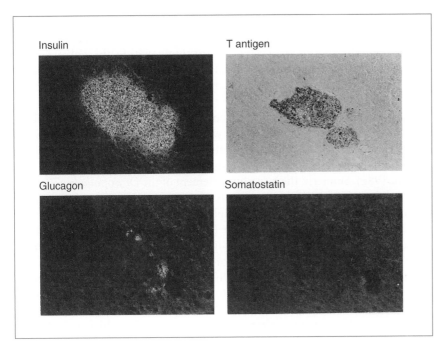

Insulin

T antigen

Glucagon

Somatostatin

Figure 7.12

Immunofluorescence assay of
pancreas preparations from the
transgenic mice described in the
legend to Fig. 7.11 with
antibodies to the indicated
proteins. Note that the
distribution of T antigen parallels
that of insulin and not that of the
other pancreatic proteins.

Therefore, in many cases the tissue-specific expression of a gene will be
determined both by the enhancer element and sequences adjacent to the
promoter. In the liver-specific pre-albumin gene, for example, gene activity
is controlled both by the promoter itself, which is active only in liver cells,
and by an upstream enhancer element, which activates any promoter
approximately tenfold in liver cells and not at all in other cell types.
Similarly, in the immunoglobulin genes, when the enhancer and the
promoter itself are separated, both exhibit B-cell-specific activity in isolation
but the maximal expression of the gene is observed only when the two
elements are brought together.

The importance of enhancer elements in the regulation of gene
expression therefore necessitates consideration of the mechanism by which
they act.

7.3.3 Mechanism of action of enhancers

In considering the nature of enhancers, we have drawn a distinction between
these elements, which act at a distance, and the sequences discussed in
Section 7.2, which are located immediately adjacent to the start site of transcription. In fact, however, closer inspection of the sequences within
enhancers indicates that they are often composed of the same sequences
found adjacent to promoters. For example, the immunoglobulin heavy-
chain enhancers contain the octamer motif (ATGCAAAT) which, as
discussed previously, is also found in the immunoglobulin promoters. Use of
DNA mobility shift and DNaseI footprinting assays (see Section 7.2.4)
has shown that these promoter and enhancer elements bind the identical

B-cell-specific transcription factor (as well as a related protein found in all cell types) and play an important role in the B-cell specific expression of the immunoglobulin heavy-chain gene. Interestingly, within the enhancer the octamer motif is found within a modular structure containing binding sites for several different transcription factors, which act together to activate gene expression (Fig. 7.13).

The close relationship of enhancer and promoter elements is further illustrated by the *Xenopus hsp70* gene, in which multiple copies of the HSE are located at positions far upstream of the start site and function as a heat-inducible enhancer element when transferred to another gene (Bienz and Pelham, 1986). Similarly, the heat-shock element of the *Drosophila hsp70* gene, which we have used as the basic example of a promoter motif, has been shown to function as an enhancer when multiple copies are placed at a position well upstream of the transcriptional start site (Bienz and Pelham, 1986).

Enhancers therefore appear to consist of sequence motifs, which are also present in similarly regulated promoters and may be present within the enhancer associated with other control elements or in multiple copies. In many cases, an enhancer will therefore consist of an array of different sequence motifs which function together to produce strong activation of transcription. The array of different regulatory proteins which assemble on such an enhancer has been called an enhanceosome (for a review see Carey, 1998).

It seems likely, therefore, that enhancers may activate gene expression by either or both of the mechanisms described previously for promoter elements, namely a change in chromatin structure leading to nucleosome displacement or by direct interaction with the proteins of the transcriptional apparatus. Indeed, the enhanceosome of the interferon β promoter has been shown to function by both these mechanisms. Thus, the enhanceosome complex of regulatory proteins which assembles on the enhancer recruits both histone acetylase and chromatin remodelling complexes, which will alter chromatin structure, as well as the complex of RNA polymerase II and associated proteins, which will direct enhanced transcription (Agalioti *et al.*, 2000; Lomvardas and Thanos, 2001) (for a review see Struhl, 2001).

In the case of chromatin structure changes, it is readily apparent that such changes caused by a protein binding to the enhancer could be propagated over large distances in both directions, causing the observed distance, position, and orientation independence of the enhancer. In agreement with this possibility, DNaseI hypersensitive sites have been mapped within a number of enhancer elements, including the immunoglobulin enhancer, and the

Figure 7.13

Protein-binding sites in the immunoglobulin heavy-chain gene enhancer. O indicates the octamer motif.

nucleosome-free gap in the DNA of the eukaryotic virus SV40 (see Section 6.7.1) is located at the position of the enhancer.

At first sight, models involving the binding of protein factors to the enhancer followed by direct interaction with proteins of the transcriptional apparatus are more difficult to reconcile with the action at a distance characteristic of enhancers. None the less, the binding to enhancers of very many proteins crucial for transcriptional activation (for a review see Hatzopoulos *et al.*, 1988) suggests that enhancers can indeed function in this manner. Models to explain this postulate that the enhancer serves as a site of entry for a regulatory factor. The factor would then make contact with the promoter-bound transcriptional apparatus either by sliding along the DNA, or via a continuous scaffold of other proteins, or by the looping out of the intervening DNA (for a review of the mechanisms of long distance gene activation see Bulger and Groudine, 1999) (Fig. 7.14).

Of these possibilities, both the sliding model and the continuous scaffold model are difficult to reconcile with the observed large distances over which

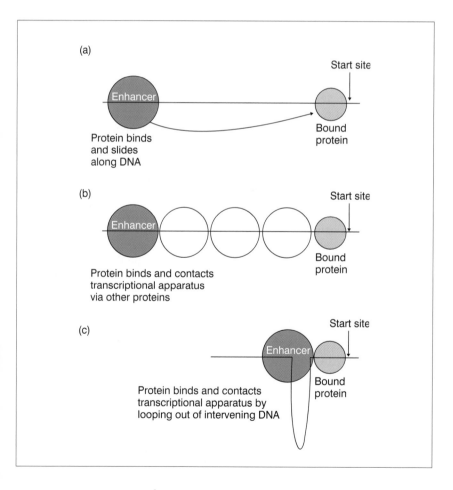

Figure 7.14

Possible models for the action of enhancers located at a distance from the activated promoter.

enhancers act. Similarly, these models cannot explain the observations of Atchison and Perry (1986) who found that the immunoglobulin enhancer activates equally two promoters placed 1.7 kb and 7.7 kb away on the same DNA molecule, since they would postulate that sliding or scaffolded molecules would stop at the first promoter (Fig. 7.15). Moreover, it has been shown that an enhancer can act on a promoter when the two are located on two separate DNA molecules linked only by a protein bridge, which would disrupt sliding or scaffolded molecules (Muller *et al.*, 1989).

Such observations are explicable, however, via a model in which proteins bound at the promoter and enhancer proteins have an affinity for one another and make contact via looping out of the intervening DNA. Moreover, such a model can explain readily the critical importance of DNA structure on the action of enhancers. Thus it has been shown that removal of precise multiples of ten bases (one helical turn) from the region between the SV40 enhancer and its promoter has no effect on its activity, but deletion of DNA corresponding to half a helical turn disrupts enhancer function severely.

Interestingly, some proteins which bind to enhancers actually bend the DNA so that interactions can occur between regulatory proteins bound at distant sites on the DNA (Fig. 7.16) (for a review see Werner and Burley, 1997). This is seen in the case of the T cell receptor α chain gene enhancer where the LEF-1 factor binds to a site at the centre of the enhancer and bends the DNA so that other regulatory transcription factors can interact with one another. Indeed, such effects are not confined to enhancers since binding of TBP to the TATA box in gene promoters (see Section 3.2.4) also bends the DNA, suggesting that this may be a widespread mechanism for producing interactions between regulatory proteins.

7.4 NEGATIVELY ACTING SEQUENCE ELEMENT

7.4.1 Silencers

Thus far we have assumed that enhancers act in an entirely positive manner. In a tissue containing an active enhancer-binding protein, the enhancer will activate a promoter, whereas in other tissues where the protein is absent or

Figure 7.15

An enhancer activates an adjacent promoter and a more distant one equally well.

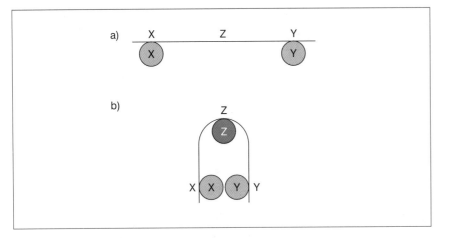

Figure 7.16

Binding of a factor (Z) which bends the DNA can produce interactions between other regulatory proteins (X and Y) which bind at distant sites on the DNA.

inactive, the enhancer will have no effect. Such a mechanism does indeed appear to operate for the majority of enhancers which, when linked to promoters, activate gene expression in one or a few cell types and have no effect in other cell types. In contrast, however, some sequences appear to act in an entirely negative manner, inhibiting the expression of genes which contain them. Following the initial identification of such an element in the cellular oncogene c-*myc*, similar elements, which are referred to as silencers, have been defined in genes encoding proteins as diverse as collagen type II, growth hormone and glutathione transferase P (for a review see Jackson, 1991). Like enhancers, silencers can act on distant promoters when present in either orientation but have an inhibitory rather than a stimulatory effect on the level of gene expression.

As with enhancer elements it is likely that these silencers act either at the level of chromatin structure by recruiting factors which direct the tight packing of adjacent DNA or by binding a protein which then directly inhibits transcription by interacting with RNA polymerase and its associated factors. Examples of silencers which appear to act in each of these ways have been observed (Fig. 7.17). Thus the silencer element located approximately 2 kb from the promoters of the repressed mating type loci in yeast plays a crucial role in organising this region into the tightly packed structure characteristic of non-transcribed DNA (for reviews see Jackson, 1991; Dillin and Rine, 1995).

Interestingly the silencer element appears to represent a site for attachment of the DNA to the nuclear matrix. Hence, it may act by promoting the further condensation of the chromatin solenoid to form a loop of DNA attached to the nuclear matrix which is the most condensed structure of chromatin (see Section 6.3; Fig. 6.9). This could be achieved by recruiting protein factors which direct tight packing of the adjacent DNA, in the same way as the SWI/SNF complex and the GAGA factor direct the unpacking of the adjacent DNA (see Section 6.7.2). Indeed, specific factors such as polycomb have been identified in *Drosophila* which have exactly this effect,

Figure 7.17

A silencer element (S) can inhibit activity of the promoter (P) either (a) by binding a protein which organises the DNA into a tightly packed chromatin structure or (b) by binding an inhibitory transcription factor which represses promoter activity.

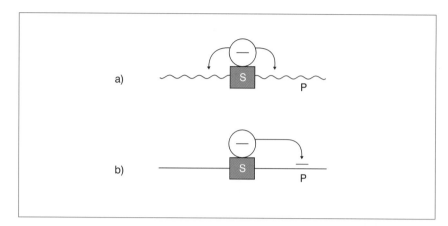

promoting tighter packing of chromatin and whose inactivation by mutation results in aberrant activation of homeotic genes. Hence, at least in this organism, chromatin structure is controlled by the antagonistic effects of proteins such as polycomb which direct tight packing of the chromatin and others such as GAGA/trithorax and brahma (see Section 6.7.2) which direct its opening (for reviews see Paro, 1995; Simon, 1995).

As well as acting at the level of chromatin structure, it is also possible for silencers to act by binding negatively acting transcription factors which interact with the basal transcriptional complex of RNA polymerase II and associated factors (Fig. 7.17b). Thus, the silencer in the gene encoding lysozyme appears to act, at least in part, by binding the thyroid hormone receptor which, in the absence of thyroid hormone, has a directly inhibitory effect on gene activity (see Section 8.3.4 for further discussion of the mechanisms of gene repression).

Hence, just as enhancer elements can act at a distance to enhance gene expression by opening up the chromatin or by directly stimulating transcription, silencers can inhibit gene expression either by promoting a more tightly packed chromatin structure or by directly repressing transcription.

7.4.2 Insulators

The existence of elements such as enhancers or silencers which can act over large distances to alter gene expression leads to the question of how the action of such elements is confined to the appropriate gene or genes. Thus, some mechanism must exist to prevent the effect of enhancers or silencers spreading to other genes in adjacent regions of the chromosome and producing an inappropriate pattern of gene expression.

This problem is solved by the existence of insulator sequences which, as their name implies, block the spread of enhancer or silencer activity so that it is confined to the appropriate gene or genes and other adjacent genes are "insulated" from these effects (Fig. 7.18) (for reviews see Sun and Elgin, 1999; Bell *et al.*, 2001). Such insulators are seen, for example, in the region of human chromosome 11 which contains several closely linked genes

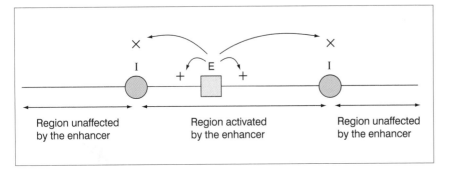

Figure 7.18

An insulator element (I) limits the action of an enhancer (E) so that it only activates gene expression in a specific region and not in adjacent regions.

encoding β-globin-type proteins (see Section 7.1.2). Thus, as discussed in Section 7.5.1, this region contains a strong enhancer-like element, known as a locus control region (LCR), which stimulates the expression of the β-globin genes in erythroid cells. As shown in Fig. 7.19, the region containing the LCR and the β-globin genes is flanked by two insulator sequences which prevent the LCR inappropriately activating the flanking genes encoding the folate receptor and the odorant receptor gene (Fig.7.19).

Interestingly, the activity of insulator elements can be regulated so that they are active in one situation and inactive in another. An example of this was discussed in Section 6.8.2 where the imprinting control region prevents the enhancer of one H19 gene from activating the Igf2 gene on the maternal chromosome but allows it to do so on the paternal chromosome. Thus, the imprinting control region is acting as an insulator element, blocking the ability of an enhancer to exert its stimulatory effect on an adjacent region on the maternal chromosome but having no effect on the paternal chromosome (see Fig. 6.39).

As discussed in Section 6.8.2 the imprinting control region/insulator is only active when it binds a protein, known as CTCF, which is essential for its insulator activity. Hence, as with the other DNA sequences discussed in this chapter, it appears that insulator sequences achieve their effect by binding

Figure 7.19

Insulator elements (I) flanking the β-globin gene cluster prevent the locus control region (LCR) from stimulating the expression of adjacent genes and confine its effect to the β-globin genes.

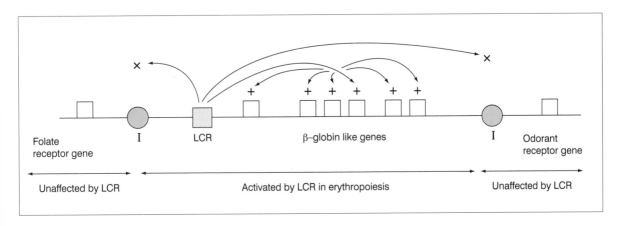

specific proteins. Evidently, the manner in which such protein binding allows the insulator to block the action of an enhancer or silencer will depend on the mechanism by which the enhancer or silencer itself is actually acting to alter gene expression (see Sections 7.3.3 and 7.4.1).

Thus, for example, if an enhancer or a silencer acts to induce changes in chromatin structure which are propagated over large distances, then an insulator with a protein bound to it would presumably act by preventing this change in chromatin structure from spreading past the point where the protein was bound. In agreement with this idea, it has been shown that the polycomb and trithorax proteins which are known to act by altering chromatin structure (see Section 6.7.2) bind to specific insulator elements in *Drosophila* (Gerasimova and Corces, 1998).

Similarly, if looping of the DNA takes place so as to bring proteins bound to the promoter and to the enhancer into contact, the silencer might change this looping pattern so that an enhancer and a promoter, which were separated by the insulator, were not brought into contact. This could involve, for example, the DNA being thrown into smaller loops so that the promoter and the distant enhancer were on separate loops which would not allow contact between the bound proteins (Fig. 7.20) (for a discussion see Bell *et al.*, 2001).

Figure 7.20

An insulator (I) may affect the looping pattern of the DNA so that when it is active proteins (X) bound at a promoter (P) cannot interact with proteins (Y) bound at an enhancer so preventing the enhancer from stimulating the promoter. In contrast, when the insulator is inactive, the looping pattern allows such interactions to occur and the enhancer can activate the promoter (compare Fig. 7.16).

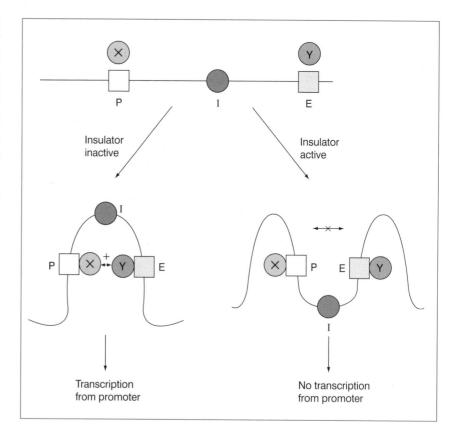

It is clear therefore that insulators play a key role in gene expression and its regulation by acting to ensure that the long distance effects of enhancer and silencer sequences are confined to the appropriate DNA region.

7.5 LOCUS CONTROL REGIONS

7.5.1 The locus control region

When specific genes are introduced into fertilised mouse eggs and used to create transgenic mice, the introduced gene is frequently expressed at very low levels and this expression is not increased when multiple copies of the genes are introduced. Similarly different copies of the introduced gene which integrate into the host chromosomes at different positions are expressed at very different levels, suggesting that gene activity is being influenced by adjacent chromosomal regions. This effect occurs even when the gene is introduced together with its adjacent promoter and enhancer elements. This has led to the concept that some sequences, which are necessary for high level gene expression independent of the position of the gene in the genome, are absent from the introduced gene.

This idea has been supported by studies in the β-globin gene cluster (see Section 7.1.2) which contains four functional β-globin-like genes and a non-functional pseudogene (Fig. 7.21). Thus a region located 10–20 kb upstream of the β-globin genes has been shown to confer high level, position-independent expression when linked to a single globin gene and introduced into transgenic mice. Moreover, this element acts in a tissue-specific manner since its presence allows the globin gene to be expressed in the correct pattern with high level expression in erythroid cells and not in other cell types.

Most importantly this element is not only required for expression of the β-globin genes in transgenic animals but also plays a role in the natural expression of the globin genes. Thus its deletion in human individuals leads to a lack of expression of any of the genes in the cluster. This occurs even in cases where all the genes, together with their promoter and enhancer regions, remain intact leading to a lethal disease known as Hispanic thalassaemia in which no functional haemoglobin is produced (Fig. 7.21).

The role of this element in stimulating activity of all the genes in the β-globin cluster has led to its being called a locus control region (LCR: for

Figure 7.21

Organisation of the β-globin gene locus showing the position of the locus control region (LCR) and the five DNaseI hypersensitive sites within it (arrows). The region showing enhanced DNaseI sensitivity in the reticulocyte lineage is indicated together with the functional genes (shaded boxes) encoding epsilon globin (ε), the two forms of gamma globin (Gγ and Aγ), delta globin (δ) and beta globin (β) and the non-functional β-globin-like pseudogene (open box). The extent of the deletion in patients with Hispanic thalassaemia is indicated by the line. Note that this deletion removes the LCR but leaves the genes themselves intact.

reviews see Bulger and Groudine, 1999; Li *et al.*, 1999). As with enhancers, following its original definition in the β-globin cluster, similar LCR elements have been identified in other genes including the α-globin cluster, the major histocompatibility locus and the CD2 and lysozyme genes. It is clear therefore that LCRs constitute another important element essential for the correct regulation of gene expression.

7.5.2 Mechanism of action of LCRs

DNA sequence analysis of LCRs has indicated that they are rich in sequence motifs which are also found in promoter and enhancer elements and which could bind transcription factors. As with promoters and enhancers therefore it is in principle possible that the LCR could function either by direct interaction with the transcriptional apparatus or by affecting chromatin structure. In practice, however, the second of these is likely to be correct. Thus in many cases regions with LCR activity do not affect gene activity when introduced transiently into cells under conditions where the exogenous DNA is not packaged into chromatin but can do so when the same gene construct integrates into the host chromosome.

It is likely therefore that the LCR functions by affecting chromatin structure of the adjacent genes so that they can be transcribed. In agreement with this the region defined as the LCR on the basis of its functional activity contains five DNaseI hypersensitive sites (see Section 6.7.1) which appear early in erythroid development prior to the expression of any of the genes in the β-globin cluster (Fig. 7.21). Moreover, the LCR is able to induce the adjacent region containing the β-globin genes to assume the enhanced DNaseI sensitivity which is characteristic of active or potentially active genes (see Section 6.4) and the LCR marks the boundary of the region of the genome which shows this preferential sensitivity in erythroid cells (Fig. 7.21).

Thus the presence of this region would render the adjacent DNA capable of being expressed in a position-independent manner, allowing high level, tissue-specific expression of the gene in transgenic mice, regardless of the position in the genome into which it integrated. In contrast a gene lacking this sequence would lack this stimulatory action and would also be subject to the influence of adjacent regulatory elements which might inhibit its expression (Fig. 7.22). In agreement with this idea, a transgene containing the CD2 locus control region is not inactivated when inserted into a tightly packed region of the DNA known as heterochromatin, whereas this does occur when the same gene is inserted into this region without the LCR.

As with other DNaseI hypersensitive sites (see Section 6.7.1) the sites in the LCR are likely to be free of nucleosomes and may be in a highly supercoiled form subject to torsional stress. This may allow them to propagate a wave of altered chromatin structure to the adjacent region, causing the formation of the "beads on a string" structure characteristic of active or potentially active DNA throughout the adjacent region (see Section 6.4).

Interestingly several LCRs contain MAR (matrix attachment region) sequences which are involved in the attachment of large domains or loops of

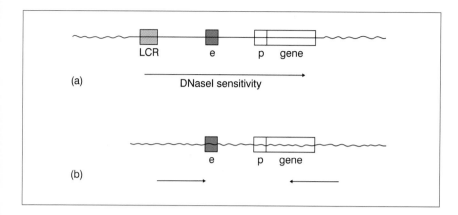

Figure 7.22

An inserted gene containing an LCR will be organised into an open chromatin conformation characterised by enhanced DNaseI sensitivity compared to flanking DNA (panel a). In contrast an inserted gene lacking the LCR will be subject to repression by adjacent regions which direct its organisation into a closed chromatin organisation (panel b).

chromatin to the nuclear matrix or scaffold (see Section 6.3). Thus a region controlled by a single LCR, such as that containing the β-globin genes, may constitute a single large domain or loop which is attached to the nuclear scaffold and whose chromatin structure is regulated as a single unit. In agreement with this idea, the MAR sequences contained within the LCR of the immunoglobulin gene have been shown to enhance the accessibility of the adjacent chromatin in B lymphocytes but not in other cell types (Jenuwein *et al.*, 1997).

Hence the LCR controls the structure of a specific region of DNA ensuring an appropriate pattern of gene regulation. Once an entire region of chromatin has been opened up in this way by the LCR, the promoter and enhancer elements of each individual gene in the region would direct its specific expression pattern. This is well illustrated in the β-globin gene cluster in which the DNaseI hypersensitive sites within the LCR and the overall DNaseI sensitivity appear very early in erythroid development prior to the expression of any of the β-globin genes. Subsequently the ε-globin, γ-globin and β-globin genes are expressed successively in embryonic, foetal and adult erythroid cells, respectively, with expression being preceded in each case by the appearance of DNaseI hypersensitive sites in the promoter/enhancer regions of the individual gene (Fig. 7.23). The LCR thus constitutes a control element which directs the chromatin structure of a large region of DNA, allowing the subsequent activation of individual genes within the cluster via the formation of DNaseI hypersensitive sites and transcription factor binding at their individual promoter and enhancer regions.

7.6 REGULATION OF TRANSCRIPTION BY RNA POLYMERASES I AND III

The very wide variety of patterns of expression in genes transcribed by RNA polymerase II is thus paralleled by a wide variety of DNA sequences which can produce such expression patterns by binding specific transcription factors which will be discussed in detail in Chapter 8. The much simpler patterns of expression of genes transcribed by RNA polymerases I and III

Figure 7.23

The appearance of multiple DNaseI hypersensitive sites within the LCR (shown as a single arrow for simplicity) in early erythroid cells precedes the appearance of other hypersensitive sites adjacent to each individual gene which occurs later in development as these genes are sequentially expressed.

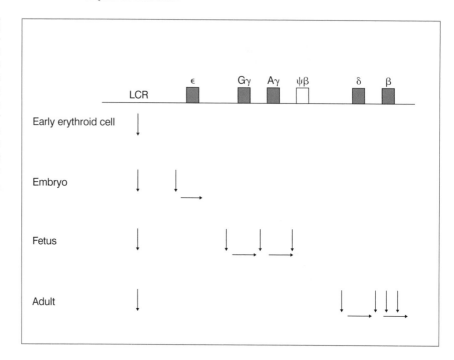

(see Sections 3.2.1 and 3.2.2) is paralleled by their simpler regulation which is generally achieved by altering the level of activity of basal transcription factors which are involved in the expression of all the genes transcribed by a particular polymerase.

Thus, for example, the fact that the TBP factor is involved in transcription by all three RNA polymerases (see Section 3.2.4) appears to be used to regulate the balance between transcription by the different RNA polymerases. Thus, the Dr1 inhibitory factor binds to TBP and inhibits its ability to recruit other transcription factors to RNA polymerase II and III gene promoters. It has no effect, however, on the activity of TBP within the RNA polymerase I transcriptional complex (White *et al.*, 1994). Hence Dr1 may regulate the balance between the transcription of the ribosomal genes which are the only genes transcribed by RNA polymerase I and all the other genes in the cell (Fig. 7.24).

The activity of RNA polymerase III is also altered in specific situations by altering the level of its specific basal transcription factors (see Section 3.2.2). Thus the activation of genes transcribed by RNA polymerase III following treatment of cells with serum is due to enhanced levels of active TFIIIC whereas the down regulation of polymerase III transcription during the differentiation of embryonal carcinoma cells is due to a decrease in TFIIIB levels (for a review see White, 1994).

As with RNA polymerase II, these regulatory processes involve factors which exert their effects by binding at the DNA level. In addition, however, in the case of RNA polymerase III the presence of transcriptional control elements within the transcribed region (see Section 3.2.2), and hence in both

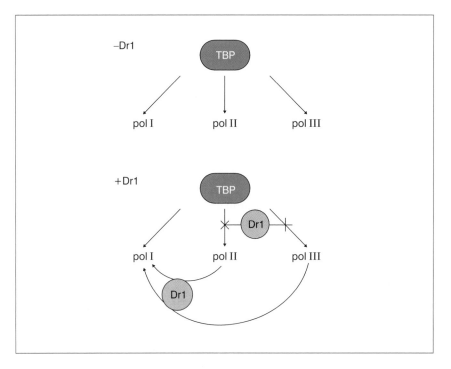

Figure 7.24

The Dr1 factor inhibits the activity of TBP within the basal transcription complex of genes transcribed by RNA polymerase II and III but not in that of the ribosomal genes which are transcribed by RNA polymerase I. Hence Dr1 may regulate the balance between the transcription of ribosomal genes by RNA polymerase I and that of all other genes in the cell by RNA polymerases II and III.

the DNA and the RNA product, is exploited in a unique regulatory mechanism involving the basal transcription factor TFIIIA which binds to the internal regulatory sequence of the 5S ribosomal RNA genes.

Thus, the toad, *Xenopus laevis*, contains two types of 5S genes: the oocyte genes, which are transcribed only in the developing oocyte before fertilisation, and the somatic genes, which are transcribed in cells of the embryo and adult. The internal control region of both these types of genes binds TFIIIA, whose binding is necessary for their transcription (see Section 3.2.2). However, sequence differences between the two types of gene (Fig. 7.25) result in a higher affinity of the TFIIIA factor for the somatic compared to the oocyte genes. Hence, the oocyte genes are only transcribed in the oocyte, where there are abundant levels of TFIIIA, and not in other cells, where the levels are only sufficient for activity of the somatic genes. In the developing oocyte TFIIIA is synthesised at high levels and transcription of the oocyte genes begins. As more and more 5S RNA molecules containing the TFIIIA binding site accumulate in the maturing oocyte, they bind the transcription factor. This factor is thus sequestered with the 5S RNA into storage particles and is unavailable for transcription of the 5S rRNA genes. Hence the level of free transcription factor falls below that necessary for the transcription of the oocyte genes and their transcription ceases, while transcription of the higher affinity somatic genes is unaffected (Fig. 7.26).

A simple mechanism involving differential binding of a transcription factor to two related genes is thus used in conjunction with the binding of the same factor to its RNA product to modulate expression of the 5S genes.

Figure 7.25

Sequence of the internal
control element of the
Xenopus somatic 5S genes.
The two base changes in the
oocyte 5S genes, which result
in a lower affinity for TFIIIA, are
indicated below the somatic
sequence.

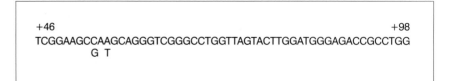

+46   +98
TCGGAAGCCAAGCAGGGTCGGGCCTGGTTAGTACTTGGATGGGAGACCGCCTGG
        G T

7.7 CONCLUSIONS

In this chapter we have discussed a number of sequence elements, which can confer a particular pattern of gene regulation, causing increased expression in response to a particular signal or in a particular tissue. Such sequences are present in a variety of locations close to or distant from the start site of transcription, in intervening sequences or, in the case of polymerase III transcription units, within the region encoding the RNA product itself. The interaction of the different positive and negative elements within these various regions controls the pattern of expression of a particular gene (for a review see Bonifer, 2000). In all cases, however, such regions act by binding a protein which recognises specific sequences within the control region. The activity of these proteins is crucial to the increased transcription mediated by the sequences to which they bind. The nature and function of these proteins is discussed in the next chapter.

Figure 7.26

Sequestration of the TFIIIA factor
by binding to 5S RNA made in
the oocyte results in a fall in the
level of the factor, switching off
transcription of the oocyte-
specific 5S genes. The somatic
5S genes, which have a higher
affinity binding site for TFIIIA,
continue to be transcribed.

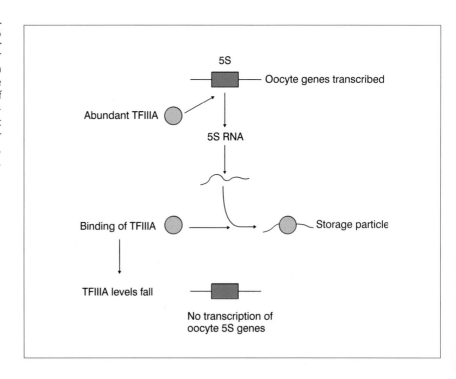

Agalioti, T., Lomvardas, S., Parekh, B., Yie, J., Maniatis, T. and Thanos, D. (2000). Ordered recruitment of chromatin modifying and general transcription factors to the IFN-β promoter. *Cell* **103**, 667–678.

Ashburner, M. and Bonner, J.J. (1979). The induction of gene activity in *Drosophila* by heat shock. *Cell* **17**, 241–254.

Atchison, M.L. and Perry, R.P. (1986). Tandem kappa immunoglobulin promoters are equally active in the presence of the kappa enhancer: implications for models of enhancer function. *Cell* **46**, 253–262.

Bell, A.C., West, A.G. and Felsenfeld, G. (2001). Insulators and boundaries: versatile regulatory elements in the eukaryotic genome. *Science* **291**, 447–450.

Bienz, M. and Pelham, H.R.B. (1986). Heat shock regulatory elements function as an inducible enhancer when linked to a heterologous promoter. *Cell* **45**, 753–760.

Bonifer, C. (2000). Developmental regulation of eukaryotic gene loci which *cis*-regulatory information is required? *Trends in Genetics* **16**, 310–315.

Britten, R.J. and Davidson, E.H. (1969). Gene regulation for higher cells: a theory. *Science* **165**, 349–358.

Bulger, M. and Groudine, M. (1999). Looping versus linking: toward a model for long-distance gene activation. *Genes and Development* **13**, 2465–2477.

Carey, M. (1998). The enhanceosome and transcriptional synergy. *Cell* **92**, 5–8.

Davidson, E.H., Jacobs, H.T. and Britten, R.J. (1983). Very short repeats and co-ordinate induction of genes. *Nature* **301**, 468–470.

Dillin, A. and Rine, J. (1995). On the origin of a silencer. *Trends in Biochemical Sciences* **20**, 231–235.

Gerasimova, T.I. and Corces, V.G. (1998). Polycomb and trithorax group proteins mediate the function of a chromatin insulator. *Cell* **92**, 511–521.

Gronemeyer, H. and Laudet, V. (1995). Nuclear receptors. *Protein Profile* **2**, 1173–1308.

Gronemeyer, H. and Moras, D. (1995). Nuclear receptors: how to finger DNA. *Nature* **375**, 190–191.

Grosschedl, R. and Birnsteil, M.L. (1980). Spacer DNA sequences up-stream of the TATAAATA sequence are essential for promotion of histone H2A transcription *in vivo*. *Proceedings of the National Academy of Sciences of the USA* **77**, 7102–7106.

Hanahan, D. (1985). Heritable formation of pancreatic beta-cell tumours in transgenic mice expressing recombinant insulin/simian virus 40 oncogenes. *Nature* **315**, 115–122.

Hatzopoulos, A.K., Schlokat, U. and Gruss, P. (1988). Enhancers and other cisacting sequences. In: *Transcription and Splicing*, (eds. Hames, B.D. and Glover, D.M.), pp 43–96. IRL Press, Oxford.

Jackson, M.E. (1991). Negative regulation of eukaryotic transcription. *Journal of Cell Science* 1–7.

Jenuwein, T., Forrester, W.C., Fernandez-Herrero, L.A. Laible, G., Dull, M. and Grosschedl, R. (1997). Extension of chromatin accessibility by nuclear matrix attachment regions. *Nature* **385**, 269–272.

Jones, N.C., Rigby, P.W.J. and Ziff, E.B. (1988). *Trans*-acting protein factors and the regulation of eukaryotic transcription. *Genes and Development* **2**, 267–281.

Latchman, D.S. (1998). *Eukaryotic Transcription Factors*. 3rd edition. Academic Press, London.

Latchman, D.S. (ed.). (1999). *Transcription Factors: A Practical Approach*. Oxford University Press, Oxford.

Li, Q., Harju, S. and Peterson, K.R. (1999). Locus control regions coming of age at a decade plus. *Trends in Genetics* **15**, 403–408.

Lomvardas, S. and Thanos, D. (2001). Nucleosome sliding via TBP binding *in vivo*. *Cell* **106**, 685–696.

Mangelsdorf, D.J. and Evans, R.M. (1995). The RXR heterodimers and orphan receptors. *Cell* **83**, 841–850.

Maniatis, T., Goodbourn, S. and Fischer, J.A. (1987). Regulation of inducible and tissue-specific gene expression. *Science* **236**, 1237–1245.

Morimoto, R.I. (1998). Regulation of the heat shock transcriptional response: cross talk between a family of heat shock factors, molecular chaperones, and negative regulators. *Genes and Development* **12**, 3788–3796.

Muller, H.-P., Soga, J.W. and Schaffner, W. (1989). An enhancer stimulates transcription in *trans* when linked to the promoter via a protein bridge. *Cell* **58**, 767–777.

Muller, M.M., Gerster, T. and Schaffner, W. (1988). Enhancer sequences and the regulation of gene transcription. *European Journal of Biochemistry* **176**, 485–495.

Parker, C.S. and Topol, J. (1984). A *Drosophila* RNA polymerase II transcription factor binds to the regulatory site of an *hsp70* gene. *Cell* **37**, 273–283.

Paro, R. (1995). Propagating memory of transcriptional states. *Trends in Genetics* **11**, 295–297.

Pelham, H.R.B. (1982). A regulatory upstream promoter element in the *Drosophila hsp70* heat-shock gene. *Cell* **30**, 517–528.

Schmitz, A. and Galas, D. (1979). The interaction of RNA polymerase and *lac* repressor with the *lac* control region. *Nucleic Acids Research* **6**, 111–137.

Simon, J. (1995). Locking in stable states of gene expression: transcriptional control during *Drosophila* development. *Current Opinion in Cell Biology* **7**, 376–385.

Struhl, K. (2001). A paradigm for precision. *Science* **293**, 1054–1055.

Sun, F.-L. and Elgin, S.C.R. (1999). Putting boundaries on silence. *Cell* **99**, 459–462.

Thompson, C.C. and McKnight, S.L. (1992). Anatomy of an enhancer. *Trends in Genetics* **8**, 232–236.

Werner, M.H. and Burley, S.K. (1997). Architectural transcription factors: proteins that remodel DNA. *Cell* **88**, 733–736.

White, R.J. (1994). *RNA Polymerase III Transcription*. p 147. R.G Landes, Austin, TX.

White, R.J., Khoo, B.C.-E., Inostroza, J.A. Reinberg, D. and Jackson, S.P. (1994). Differential regulation of RNA polymerase I, II and III by the TBP repressor Dr1. *Science* **266**, 448–450.

Williams, G.T., McClanahan T.K. and Morimoto, R.I. (1989). Ela transactivation of the human *hsp70* promoter is mediated through the basal transcriptional complex. *Molecular and Cellular Biology* **9**, 2574–2587.

Wirth, T., Staudt, L. and Baltimore, D. (1987). An octamer oligonucleotide upstream of a TATA motif is sufficient for lymphoid specific promoter activity. *Nature* **329**, 174–178.

Zimarino, V. and Wu, C. (1987). Induction of sequence-specific binding of *Drosophila* heat-shock activator protein without protein synthesis. *Nature* **327**, 727–730.

TRANSCRIPTIONAL CONTROL – TRANSCRIPTION FACTORS

8

8.1 INTRODUCTION

As discussed in Chapter 7, the expression of specific genes in particular cell types or tissues is regulated by DNA sequence motifs present within promoter or enhancer elements which control the alteration in chromatin structure of the gene that occurs in a particular lineage, or the subsequent induction of gene transcription. It was assumed for many years that such sequences would act by binding a regulatory protein which was only synthesised in a particular tissue or was present in an active form only in that tissue. In turn, the binding of this protein would result in the observed effect on gene expression. Indeed, as described in Section 7.2.4 cell extracts can be used in DNA mobility shift or DNaseI footprinting assays to show that they contain protein(s) able to bind to a specific sequence.

The isolation and characterisation of such factors proved difficult, however, principally because they were present in very small amounts. Hence, even if they could be purified, the amounts obtained were too small to provide much information as to the properties of the protein.

This obstacle was overcome by the cloning of the genes encoding a number of different transcription factors. Two general approaches were used to achieve this (for a review, see Latchman, 1999). In one approach (Fig. 8.1), exemplified by the work of Kadonaga and Tjian (1986), the transcription factor Spl was purified by virtue of its ability to bind to its specific DNA binding site. The partial amino acid sequence of the protein was then obtained from the small amount of material isolated, and was used in conjunction with the genetic code to predict a set of DNA oligonucleotides, one of which would encode this region of the protein. The oligonucleotides were then hybridised to a complementary DNA library prepared from Spl-containing HeLa cell mRNA. A cDNA clone derived from the Spl mRNA must contain the sequence capable of encoding the protein and hence will hybridise to the probe. In this experiment one single clone derived from the Spl mRNA was isolated by screening a library of 1 million recombinants prepared from the whole population of HeLa cell mRNAs (Kadonaga *et al.*, 1987).

An alternative, more direct, approach to the cloning of transcription factors is exemplified by the work of Singh *et al.* (1988) on the NFκB protein, which is involved in regulating the expression of the immunoglobulin genes in B cells (Fig. 8.2). As in the previous method, a cDNA library was constructed containing copies of all the mRNAs in a specific cell type. However, the library was constructed in such a way that the sequences within it would be translated into their corresponding proteins. This was achieved by inserting the cDNA into the coding region of the bacteriophage β-galactosidase gene, resulting in the translation of the eukaryotic insert as

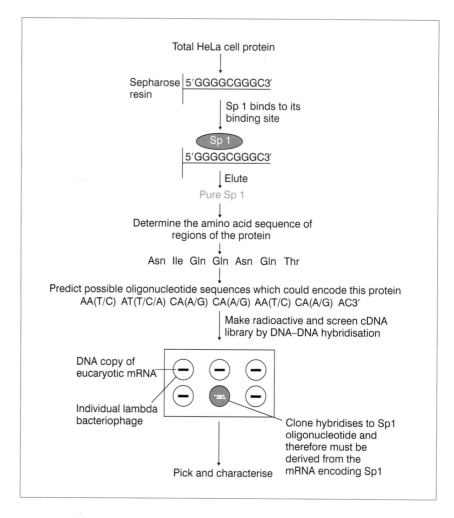

Figure 8.1

Isolation of cDNA clones for the Spl transcription factor by screening with short oligonucleotides predicted from the protein sequence of Spl. Because several different triplets of bases can code for any given amino acid, multiple oligonucleotides that contain every possible coding sequences are made. Positions at which these oligonucleotides differ from one another are indicated by the brackets containing more than one base.

Total HeLa cell protein

Sepharose resin | 5′GGGGCGGGC3′

Sp 1 binds to its binding site

Sp 1

5′GGGGCGGGC3′

Elute

Pure Sp 1

Determine the amino acid sequence of regions of the protein

Asn Ile Gln Gln Asn Gln Thr

Predict possible oligonucleotide sequences which could encode this protein
AA(T/C) AT(T/C/A) CA(A/G) CA(A/G) AA(T/C) CA(A/G) AC3′

Make radioactive and screen cDNA library by DNA–DNA hybridisation

DNA copy of eucaryotic mRNA

Individual lambda bacteriophage

Clone hybridises to Sp1 oligonucleotide and therefore must be derived from the mRNA encoding Sp1

Pick and characterise

part of the bacteriophage protein. Most interestingly, these fusion proteins were capable of binding DNA with the same specificity as the original transcription factor encoded by the cloned mRNA. Hence the library could be screened directly with the radiolabelled DNA binding site for a particular transcription factor. A clone containing the mRNA for this factor, and hence expressing it as a fusion protein, bound the labelled DNA and could be identified readily and isolated.

Unlike the previous method, this procedure involves DNA–protein rather than DNA–DNA binding and can be used without prior purification of the transcription factor, provided its binding site is known. Since most factors are identified on the basis of their binding to a particular site, this is not a significant problem and the use of these two methods has resulted in the isolation of the genes encoding a wide variety of transcription factors.

In turn, this has resulted in an explosion of information on these factors (for a general review, see Latchman, 1998). Thus, once the gene for a factor

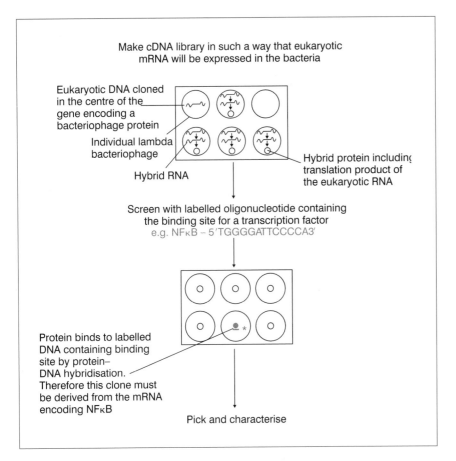

Make cDNA library in such a way that eukaryotic
mRNA will be expressed in the bacteria

Eukaryotic DNA cloned
in the centre of the
gene encoding a
bacteriophage protein

Individual lambda
bacteriophage

Hybrid RNA

Hybrid protein including
translation product of
the eukaryotic RNA

Screen with labelled oligonucleotide containing
the binding site for a transcription factor
e.g. NFκB – 5′TGGGGATTCCCCA3′

Protein binds to labelled
DNA containing binding
site by protein–
DNA hybridisation.
Therefore this clone must
be derived from the mRNA
encoding NFκB

Pick and characterise

Figure 8.2

Isolation of cDNA clones for the
NFκB transcription factor by
screening an expression library
with a DNA probe containing
the binding site for the factor.

has been cloned, Southern blotting (Section 2.2.4) can be carried out to study the structure of the gene, Northern blotting (Section 1.3.1) can be used to search for RNA transcripts derived from it in different cell types and related genes expressed in other tissues or other species can be identified.

More importantly, considerable information can be obtained from the cloned gene about the corresponding protein and its activity. Thus, not only can the DNA sequence of the gene be used to predict the amino acid sequence of the corresponding protein, but the existence of functional domains within the protein with particular activities can also be defined. As described above, if the gene encoding a transcription factor is expressed in bacteria, it continues to bind DNA in a sequence-specific manner. Hence, if the gene is broken up into small pieces and each of these is expressed in bacteria (Fig. 8.3), the abilities of each portion to bind to DNA, to other proteins, or to a potential regulatory molecule can be assessed using, for example DNA mobility shift assays, exactly as with cellular extracts (see Section 7.2.4). This mapping can also be achieved by transcribing and translating pieces of the DNA into protein fragments in the test tube and testing their activity in the same way.

Figure 8.3

Mapping of the DNA binding region of a transcription factor by testing the ability of different regions to bind to the appropriate DNA sequence when expressed in bacteria.

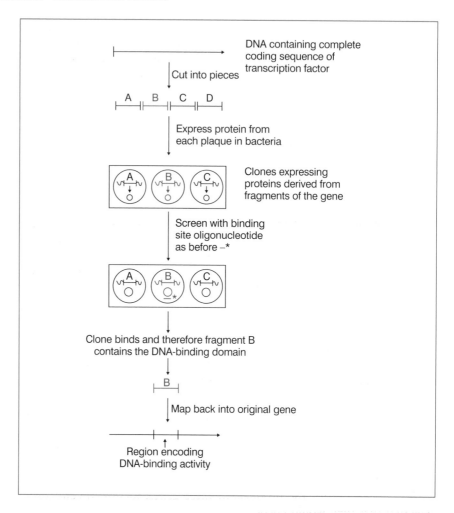

Each of the domains identified in this way can be altered by mutagenesis of the DNA and subsequent expression of the mutant protein as before. The testing of the effect of these mutations on the activity mediated by the particular domain of the protein will thus allow the identification of the amino acids that are critical for each of the observed properties of the protein.

In this way large amounts of information have accumulated on individual transcription factors. Rather than attempt to consider each factor individually, we will focus on the properties necessary for such a factor and illustrate our discussion by referring to the manner in which these are achieved in individual cases.

It should be evident from the foregoing discussion that the first property such a factor requires is the ability to bind to DNA in a sequence-specific manner, and this is discussed in Section 8.2. Subsequently, the bound factor must influence transcription either positively or negatively by interacting with other transcription factors or with the RNA polymerase itself. Section

8.3 therefore considers the means by which DNA-bound transcription factors actually regulate transcription. Finally, in the case of transcription factors which activate a particular gene in one tissue only, some means must be found to ensure that the transcription factor is active only in that tissue. Section 8.4 discusses how this is achieved, either by the expression of the gene encoding the factor only in one particular tissue or by a tissue-specific modification which results in the activation of a factor present in all cell types.

8.2 DNA BINDING BY TRANSCRIPTION FACTORS

8.2.1 Introduction

Extensive studies of eukaryotic transcription factors have identified several structural elements, which either bind directly to DNA or which facilitate DNA binding by adjacent regions of the protein (for reviews see Travers, 1993; Latchman, 1998). These motifs will be discussed in turn, using transcription factors that contain them to illustrate their properties.

8.2.2 The helix-turn-helix motif

The homeobox

The small size and rapid generation time of the fruit fly *Drosophila melanogaster* has led to it being one of the best-characterised organisms genetically, and a number of mutations which affect various properties of the organism have been described. These include mutations which affect the development of the fly, resulting, for example, in the production of additional legs in the position of the antennae (Fig. 8.4). Genes of this type are likely to play a crucial role in the development of the fly and, in particular, in determining the body plan, and are known as homeotic genes (for a review see Lawrence and Morata, 1994).

The critical role for the products of these genes, identified genetically, suggested that they would encode regulatory proteins which would act at particular times in development to activate or repress the activity of other genes encoding proteins required for the production of particular structures. This idea was confirmed when the genes encoding these proteins were cloned. Thus, these proteins were shown to be able to bind to DNA in a sequence-specific manner and to be able to induce increased transcription of genes which contained this binding site. Thus, in the case of the homeotic gene *fushi tarazu* (*ftz*), mutation of which produces a fly with only half the normal number of segments, the protein has been shown to bind specifically to the sequence TCAATTAAATGA. When the gene encoding this protein is introduced into *Drosophila* cells with a marker gene containing this sequence, transcription of the marker gene is increased. This up-regulation is entirely dependent on binding of the Ftz protein to this sequence in the promoter of the marker gene, since a 1 bp change in this sequence, which abolishes binding, also abolishes the induction of transcription (Fig. 8.5).

The product of another homeotic gene, the engrailed protein, binds to the identical sequence to that bound by Ftz. Its binding does not produce

219

Figure 8.4

Effect of a homeotic mutation, which produces a middle leg (b) in the region that would contain the antenna of a normal fly (a). al, all and alll: first, second and third antennal segments; ar, arista; ta, tarsus; ti, tibia; fe, femur; ap, apical bristle.

increased transcription of the marker gene, however, and indeed it prevents the activation by Ftz. Hence, the expression of Ftz alone in a cell would activate particular genes, whereas Ftz expression in a cell also expressing the engrailed product would have no effect (Fig. 8.6). In this way interacting homeotic gene products expressed in particular cells could control the developmental fate of the cells by regulating the expression of specific target genes (for a review see Edelman and Jones, 1993).

Interestingly, there is evidence that the homeotic genes may be necessary not only for the actual production of a specific cell type but also for the long-term process of commitment to a particular cellular phenotype which was discussed in Section 6.2. Thus, in the case of the imaginal discs of *Drosophila* (Section 6.2), commitment to the production of a particular adult structure was maintained through many cell generations in the absence

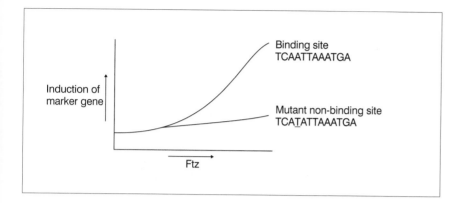

Figure 8.5

Effect of expression of the Ftz protein on the expression of a gene containing its binding site, or a mutated binding site containing a single base pair change which abolishes binding of Ftz.

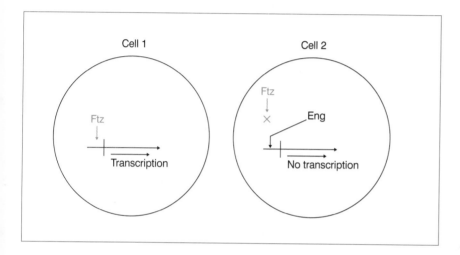

Figure 8.6

Blockage of gene induction by Ftz in cells expressing the engrailed (Eng) protein which binds to the same sequence as Ftz but does not activate transcription.

of differentiation. If, during this time, however, a mutation is introduced into one of the homeotic genes in the disc cells, inactivating it, when eventually the cells are allowed to differentiate they will produce the wrong structure. Thus, for example, if the homeotic gene *ultrabithorax* (*ubx*) is inactivated in a disc cell which normally gives rise to the haltere (balancer), these cells will produce wing tissue when allowed to differentiate. Thus the continual expression of homeotic genes within the cells is essential for their commitment to a particular pathway of differentiation.

A possible molecular mechanism for this is provided by the demonstration that the Ubx protein binds to its own promoter and up-regulates its own transcription. Hence once production of this protein has been induced, presumably during the commitment process, it will continue indefinitely and thus maintain this commitment (Fig. 8.7).

In addition to this mechanism, other processes maintain the chromatin structure of the homeotic genes in an active state so that transcription continues once the gene has initially been activated. This is achieved by members of the trithorax group of proteins, which include the GAGA and

Figure 8.7

The Ubx protein activates its own promoter, producing a positive feedback loop maintaining high-level production of Ubx.

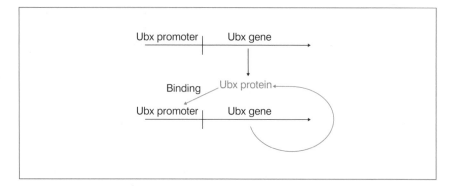

brahma factors discussed in Section 6.7.2. These factors bind to active homeotic genes and maintain them in an open chromatin structure, allowing transcription to continue. Conversely, other proteins of the polycomb group bind to inactive homeotic genes and maintain their inactive chromatin structure so preventing their inappropriate activation (for reviews see Simon, 1995; Pirrotta, 1997). As might be expected, mutations of the genes encoding trithorax or polycomb group proteins result in gross abnormalities in the fly due to, respectively, a failure to activate the appropriate homeotic genes or their inappropriate activation (Fig. 8.8).

Interestingly, as noted in Section 6.2, the changes in commitment which occur in imaginal discs when the process of commitment breaks down after the discs are cultured for long periods are precisely those which occur in homeotic mutations. Hence a change in the chromatin structure and expression of a specific homeotic gene in the imaginal disc will result in a change in the pattern of commitment similar to that which occurs when this gene is mutated.

The clear evidence that homeotic gene products regulate both their own genes and other genes by binding specifically to DNA has led to extensive investigation of their structure in order to identify the region that mediates this sequence-specific DNA binding. When the genes encoding these proteins were first cloned, it was found that they each contained a short

Figure 8.8

(a) Members of the trithorax group of proteins (T) maintain the active chromatin structure (solid line) of transcriptionally active homeotic genes. Inactivation of these proteins by mutation results in an inactive chromatin structure (wavy line) leading to a failure of transcription. (b) Members of the polycomb family (p) maintain the inactive chromatin structure (wavy line) of transcriptionally inactive homeotic genes. Their inactivation by mutation results in an active chromatin structure (solid line) leading to the inappropriate transcription of these genes.

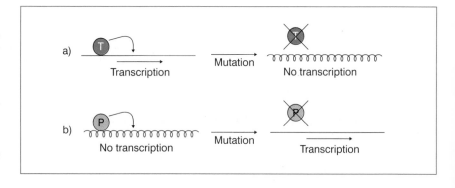

related DNA sequence of about 180 bp, capable of encoding 60 amino acids (Fig. 8.9), which was flanked on either side by sequences that differed dramatically between the different genes. The presence of this sequence, which was named the homeobox or homeodomain (for reviews see Gehring *et al.*, 1994; Kornberg, 1993), in all these genes suggested that it plays a critical role in mediating their regulatory function. This suggestion was confirmed subsequently by the use of the homeobox as a probe to isolate other previously uncharacterised *Drosophila* regulatory genes.

The role of the homeodomain in DNA binding has been confirmed directly by the synthesis of the homeobox region of the Antennapedia protein without the remainder of the protein, either by expression in bacteria or by chemical synthesis, and showing that it can bind to DNA in the identical sequence-specific manner to that exhibited by the intact protein.

The localisation of this DNA binding to a short region of the protein, only 60 amino acids in length, allows a detailed structural prediction of the corresponding protein, which reveals that it contains a so-called helix-turn-helix motif which is highly conserved between the different homeobox-containing proteins (for a review see Gehring *et al.*, 1994). In this motif, a short region which can form an α-helical structure is followed by a β-turn and then another α-helical region. The position of these elements in the homeodomain is shown in Fig. 8.9 and a diagram of the helix-turn-helix motif is given in Fig. 8.10.

The prediction that this structure exists in the DNA-binding homeodomain has been directly confirmed by X-ray crystallographic analysis. Moreover, by carrying out this analysis on the homeodomain bound to its DNA binding site, it has been shown that the helix-turn-helix motif does indeed contact DNA, with the second helix lying partly within the major groove where it can make specific contacts with the bases of the DNA (Fig. 8.11). This second helix (labelled the recognition helix in the homeobox sequence in Fig. 8.9) can thus mediate sequence-specific binding.

The presence of this structure therefore indicates how the homeobox proteins can bind specifically to particular DNA binding sequences, which is the first step in transcriptional activation of their target genes. The role of the helix-turn-helix motif in the recognition of specific sequences in the DNA has been demonstrated directly. Thus a mutation which changes a

Figure 8.9

Amino acid sequences of several *Drosophila* homeodomains, showing the conserved helical motifs. Differences between the sequences of the Ubx and Ftz homeodomains from that of Antp are indicated, a blank denotes identity in the sequence. The helix-turn-helix region is indicated.

Antp	Arg	Lys	Arg	Gly	Arg	Gln	Thr	Tyr	Thr	Arg	Tyr	Gln	Thr	Leu	Glu	Leu	Glu	Lys	Glu	Phe	His	Phe	Asn	Arg	Tyr	Leu	Thr	Arg	Arg	Arg
Ubx			Arg																					Thr		His				
Ftz	Ser				Thr																						Ile			

Helix — Turn — Recognition helix

Antp	Arg	Ile	Glu	Ile	Ala	His	Ala	Leu	Cys	Leu	Thr	Glu	Arg	Gln	Ile	Lys	Ile	Trp	Phe	Gln	Asn	Arg	Arg	Met	Lys	Trp	Lys	Lys	Glu	Asn
Ubx				Met		Tyr									Glu										Leu				Ile	
Ftz			Asp		Asn			Ser		Ser															Ser				Asp	Arg

Figure 8.10

The helix-turn-helix motif.

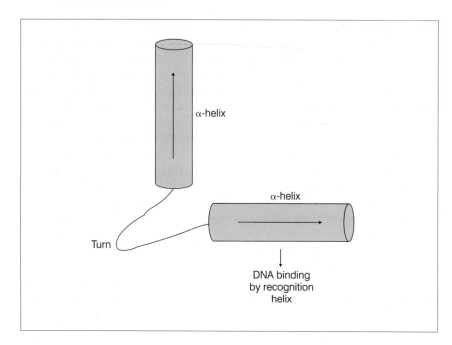

α-helix

α-helix

Turn

DNA binding
by recognition
helix

Figure 8.11

Binding of the helix-turn-helix
motif to DNA, with the
recognition helix in the major
groove of the DNA.

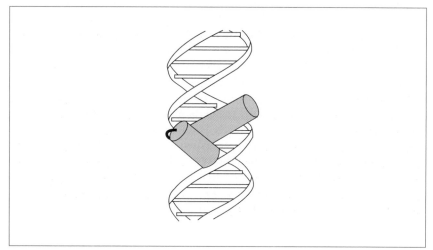

lysine at position 9 of the recognition helix in the Bicoid protein to the glutamine found in the equivalent position of the Antennapedia protein results in the protein binding to DNA with the sequence specificity of an Antennapedia rather than a Bicoid protein. Hence, not only does the helix-turn-helix motif mediate DNA binding, but differences in the precise sequence of this motif in different homeoboxes control the precise DNA sequence to which these proteins bind. Clearly, further structural and genetic studies of how this is achieved will throw considerable light on the way in which these proteins function.

The POU domain

The obvious importance of the homeobox and its presence in organisms as diverse as *Drosophila* and yeast prompted a search for proteins containing this element in other organisms. Both *Xenopus* and mammals have been shown to contain proteins of this type, which are expressed in specific cell types in the early embryo and are likely to play a regulatory role (for reviews see Krumlauf, 1994; Duboule, 2000). Hence these proteins may provide a link between the processes regulating development in a variety of different organisms.

In addition, however, another class of regulatory proteins has been identified which contains the homeobox as one part of a much larger, 150–160 amino acid, conserved region known as the POU domain (for reviews see Verrijzer and Van der Vliet 1993; Ryan and Rosenfeld, 1997). Unlike the homeobox proteins, these regulatory proteins were not identified by mutational analysis or by homology to other regulatory proteins, but were characterised as transcription factors having a particular pattern of activity.

Thus the mammalian Oct-1 and Oct-2 proteins both bind to the octamer sequence (consensus ATGCAAATNA) in the promoters of genes such as the histone H2B gene and those encoding the immunoglobulins, and mediate the transcriptional activation of these genes. When the genes encoding these proteins were cloned they were found to possess a 150–160 amino acid sequence that was also found in the mammalian Pit-1 protein, which regulates gene expression in the pituitary by binding to a sequence related to, but distinct from, the octamer and in the protein encoded by the nematode gene *unc-86*, which is involved in sensory neuron development.

This POU (Pit–Oct–Unc) domain contains both a homeobox-like sequence and a second conserved domain, the POU-specific domain (Fig. 8.12). Although there are some differences between different POU proteins, in general the isolated homeodomain of the POU proteins alone is sufficient for sequence-specific DNA binding, but unlike the classical homeobox, the binding is of relatively low affinity in the absence of the POU-specific domain. Hence both parts of the POU domain are required for high affinity sequence-specific DNA binding, indicating that the POU homeodomain and the POU-specific domain form two parts of a DNA-binding element which are held together by a flexible linker sequence.

Figure 8.12

Amino acid sequences of the POU proteins. The homeodomain and the POU-specific domain are indicated. Solid boxes indicate regions of identity between the different POU proteins. The final line shows a consensus sequence obtained from the four proteins.

Interestingly, like the POU homeodomain, the POU-specific domain can form a helix-turn-helix motif. It has therefore been suggested that the recognition helix from the POU-specific domain and that from the POU home-odomain bind to adjacent regions within the major groove of the DNA.

It is clear, therefore, that the POU proteins represent a family of proteins, related to the homeobox proteins, which are likely to play a critical role in development. Thus inactivation of the Pit-1 gene leads to a failure of pituitary gland development resulting in dwarfism in mouse and humans while the *unc-86* mutation results in a failure to form specific neurones in the nematode.

8.2.3 The zinc finger motif

The two cysteine–two histidine zinc finger

As discussed in Section 7.6, one of the earliest gene regulatory systems to be characterised was that of the gene encoding the 5S RNA of the ribosome, in which a transcription factor, TFIIIA, binds to the internal control region of the gene (see Section 3.2.2). This transcription factor was among the first to be purified (Miller *et al.*, 1985). The pure protein was shown to have a periodic repeated structure and to contain between 7–11 atoms of zinc associated with each molecule of the pure protein.

The basis for this repeated structure was revealed when the gene encoding this protein was cloned and used to predict the corresponding amino acid sequence. This protein sequence contained nine repeats of a 30 amino acid sequence of the form Tyr/Phe–X–Cys–X–Cys–X_{2-4}–Cys–X_3–Phe–X_5–Leu–X_2–His–X_{3-4}–His–X_5, where X is a variable amino acid. This repeating structure therefore contains two invariant pairs of cysteine and histidine residues which were predicted to bind a single zinc atom, accounting for the multiple zinc atoms bound by the purified protein.

This 30 amino acid repeating unit is referred to as a zinc finger, on the basis that a loop of 12 amino acids, containing the conserved leucine and phenylalanine residues as well as several basic residues, projects from the surface of the protein and is anchored at its base by the conserved cysteine and histidine residues, which directly coordinate an atom of zinc (Fig. 8.13). The binding of zinc by the cysteine and histidine residues has been confirmed directly by X-ray crystallographic analysis of the TFIIIA protein. Such structural studies have also revealed that the finger region consists of two antiparallel β-sheets and an α-helix packed against one of the β- sheets with the α-helix contacting the major groove of the DNA as occurs for the recognition helix of the homeodomain proteins (Fig. 8.14) (for reviews see Rhodes and Klug, 1993; Klug and Schwabe, 1995) (colour plate 11).

It is clear, therefore, that, like the helix-turn-helix motif, the zinc finger represents a protein structure capable of mediating the DNA binding of transcription factors. Although originally identified in the RNA polymerase III transcription factor TFIIIA, this motif has now been identified in a number of RNA polymerase II transcription factors and shown to play a critical role in their ability to bind to DNA and thereby influence transcription (Table 8.1; for reviews see Evans and Hollenberg, 1988; Klug and Schwabe, 1995; Turner and Crossley, 1999).

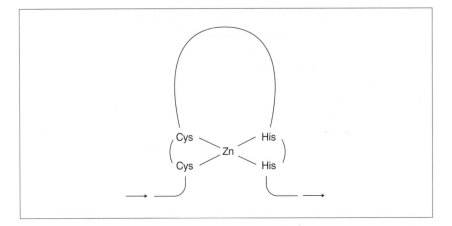

Figure 8.13

Schematic structure of the two cysteine–two histidine zinc finger.

Figure 8.14

Detailed structure of the zinc finger in which two antiparallel β sheets (solid lines) are packed against an adjacent α-helix (wavy line). The straight line indicates the region which contacts the major groove of the DNA.

Thus three contiguous copies of the 30 amino acid zinc finger motif are found in the transcription factor Spl whose cloning was discussed earlier (Section 8.1). The sequence-specific binding pattern of the intact Spl protein can be reproduced by expressing in *E. coli* a truncated protein containing only the zinc finger region, confirming the importance of this region in DNA binding (Kadonaga *et al.*, 1987). Similarly, the *Drosophila* Krüppel protein, which is vital for proper thoracic and abdominal development, contains four zinc finger motifs. A single mutation, which results in the replacement of the conserved cysteine in one of these fingers by a serine which could not bind zinc, leads to the complete abolition of the function of the protein, resulting in a mutant fly whose appearance is indistinguishable from that produced by complete deletion of the gene.

The zinc finger therefore represents a DNA binding element which is present in variable numbers in many regulatory proteins. Indeed, the linkage between the presence of this motif and the ability to regulate gene

Table 8.1

Transcriptional regulatory proteins containing Cys$_2$ His$_2$ zinc fingers

Organism	Gene	Number of fingers
Drosophila	krüppel	4
	hunchback	6
	snail	4
	glass	5
Yeast	ADRI	2
	Swi5	3
Xenopus	TFIIIA	9
	Xfin	37
Mammal	NGF-Ia (Egr1)	3
	MKI	7
	MK2	9
	Evi I	10
	Sp I	3

expression is now so strong that, as with the homeobox, it has been used as a probe to isolate the genes encoding new regulatory proteins. The Krüppel zinc finger, for example, has been used in this way to isolate Xfin, a 37 finger protein expressed in the early *Xenopus* embryo (colour plate 11) (for a review of Krüppel factors see Bieker, 2001).

The possible involvement of zinc finger proteins in controlling development in vertebrates is mirrored in *Drosophila*, where numerous proteins involved in regulating development, such as Krüppel, Hunchback and Snail, contain zinc fingers. The interactions of these proteins with the homeobox proteins, which contain the alternative DNA binding helix-turn-helix motif, are of central importance in the development of *Drosophila* and possibly other organisms.

The multi-cysteine zinc finger

Throughout this work we have noted that the effect of steroid hormones on mammalian gene expression is one of the best characterised examples of gene regulation. Thus the steroid-regulated genes were among the first to be shown to be regulated at the level of gene transcription (Chapter 4) by means of the binding of a specific receptor to a specific DNA sequence (Section 7.2), resulting in a change in chromatin structure and the generation of a DNase I-hypersensitive site (Section 6.7). When the genes encoding the DNA binding receptors for the various steroid hormones, such as glucocorticoid and oestrogen, were cloned, they were found to constitute a family of proteins encoded by distinct but related genes. In turn, these proteins were related to other receptors which mediated the response of the cell to hormones such as thyroid hormone, retinoic acid or vitamin D, leading to the idea of an evolutionarily related family of genes encoding nuclear receptors, known as the steroid–thyroid hormone receptor gene superfamily (for reviews see Gronemeyer and Laudet, 1995; Mangelsdorf *et al.*, 1995; Weatherman *et al.*, 1999).

When the detailed structures of the members of this family were compared (Fig. 8.15) it was found that each had a multi-domain structure, which included a central highly conserved domain. On the basis of experiments in which truncated versions of the receptors were introduced into cells and their activities measured, it was shown that this conserved domain mediated the DNA binding ability of the receptor, while the C-terminal region was involved in the binding of the appropriate hormone and the N-terminal region was involved in producing maximal induction of transcription of target genes.

Sequence analysis of the DNA binding domain in a variety of receptors showed that it conformed to a consensus sequence of the form $Cys–X_2–Cys–X_{13}–Cys–X_2–Cys–X_{15-17}–Cys–X_5–Cys–X_9–Cys–X_2–Cys–X_4–Cys$. Like the cysteine–histidine finger described in the previous section, the DNA binding of this element is dependent upon the presence of zinc or a related heavy metal such as cadmium. Moreover, this element can be drawn as two conventional zinc fingers, in which four cysteines replace the two cysteine–two histidine structure of the conventional finger in binding zinc and which are separated by a linker region containing the 15–17 variable amino acids (Fig. 8.16). Such a structure is supported by spectrographic analysis of this region of the receptor, which clearly demonstrates the presence of two zinc atoms, each coordinated by four cysteines in a tetrahedral array (colour plate 12).

However, such structural analysis also indicates that the two fingers in the steroid receptors constitute a single structural element, unlike the situation

Figure 8.15

Domain structure of individual members of the steroid–thyroid hormone receptor superfamily. The proteins are aligned on the DNA binding domain, which shows the most conservation between different receptors. The percentage homologies in each domain of the receptors to that of the glucocorticoid receptor are indicated.

in the cysteine–histidine fingers where each finger forms a separate structural unit. Moreover, the multi-cysteine finger cannot be converted to a cysteine–histidine finger by substituting two of its cysteines with histidines. Thus, while the multi-cysteine domain is clearly similar to the cysteine–histidine domain in its coordination of zinc, it is distinct in its lack of histidines and conserved phenylalanine and leucine residues as well as in its structure and the two elements are unlikely to be evolutionarily related (for reviews see Schwabe and Rhodes, 1991; Rhodes, and Klug, 1993).

Whatever its precise relationship to the cysteine–histidine finger, it is clear that, like this type of finger, the multi-cysteine domain in the hormone receptors is involved in mediating DNA binding (colour plate 13). Similar single domains containing multiple cysteines separated by non-conserved residues have also been identified in other DNA binding proteins, such as the yeast transcription factors GAL4, PPRI, LAC9, etc., which all contain a cluster of six invariant cysteines, and in the adenovirus transactivator, E1A, which has a cluster of four cysteines within the region that mediates *trans*-activation (Table 8.2; for a review see Evans and Hollenberg, 1988).

The existence of a short DNA binding region in a number of different steroid-receptor proteins which bind distinct but related sequences (Section 7.2.3) has allowed a dissection of the elements in this structure that are important in sequence-specific DNA binding. Thus, as illustrated in Table 7.3, the sequences that confer responsiveness to glucocorticoid or oestrogen treatment are distinct but related to one another. If the cysteine-rich region of the oestrogen receptor is replaced by that of the glucocorticoid receptor, a chimeric receptor is obtained which has the DNA binding specificity of the glucocorticoid receptor but, because all the other regions of the protein are derived from the oestrogen receptor, it continues to bind oestrogen. Hence this hybrid receptor induces the expression of glucocorticoid-responsive genes (which carry its DNA binding site) in response to treatment with oestrogen (to which it binds) (Fig. 8.17). Further so-called fingerswop experiments using smaller parts of this region have shown that this change in specificity can also be achieved by the exchange of the N-terminal, four-

Figure 8.16

Structure of the four-cysteine zinc finger.

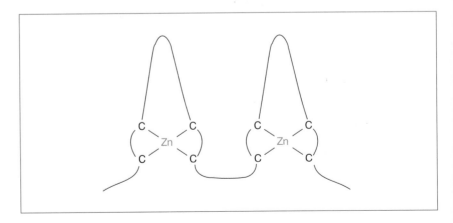

Finger type	Factor	Species
Cys_4Cys_5	Steroid, thyroid receptors	Mammals
Cys_4	EIA	Adenovirus
Cys_6	GAL4, PPRI, LAC9	Yeast

Table 8.2

Transcriptional regulatory proteins with multiple cysteine fingers

cysteine finger, together with the region immediately following it, which are therefore critical for determining the sequence-specific binding to the DNA.

These findings have been further refined by exchanging individual amino acids in this region of the glucocorticoid receptor for their equivalents in the oestrogen receptor. As shown in Fig. 8.18, the alteration of the two amino acids between the third and fourth cysteines of the N-terminal finger to their oestrogen receptor equivalents results in a glucocorticoid receptor which switches on oestrogen-responsive genes. Hence the change of only two critical amino acids within a protein of 777 amino acids can completely change the DNA binding specificity of the receptor.

The specificity of the hybrid receptor for oestrogen-responsive genes can be further enhanced by changing another amino acid, which is located in the linker region between the two fingers (Fig. 8.18), indicating that this region also plays a role in controlling the specificity of binding to DNA. Interestingly, this region following the finger can form an α-helical structure similar to the recognition helix seen in the helix-turn-helix motif. Thus, the DNA binding specificity of the steroid receptors appears to involve the cooperation of a zinc finger motif and an adjacent helical motif.

In contrast to the effect of mutations in the first finger and adjacent region, further alteration of five amino acids in the second finger is sufficient

Figure 8.17

Effect of exchanging the DNA binding domain (shaded) of the oestrogen receptor with that of the glucocorticoid receptor on the binding of hormone and gene induction by the hybrid receptor.

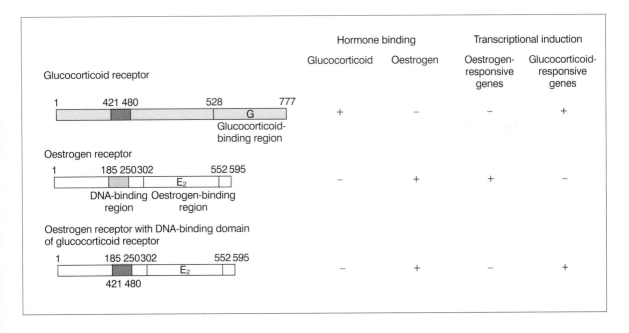

to change the binding specificity of the receptor such that it now recognises the thyroid-hormone-receptor binding sites (Umesono and Evans, 1989; Fig. 8.18). Thyroid-hormone-binding sites do not differ from those of the oestrogen receptor in sequence but only in the spacing between the two halves of the palindromic DNA recognition sequence (Table 7.3a). This indicates that the second finger is critical for mediating protein–protein interactions between the two copies of the receptor that bind to the two halves of the palindromic sequence (see Section 7.2.3), and thus for controlling the optimal spacing of these halves for binding of the particular receptor.

Hence, by studying the multiple related steroid receptors and their relationship with the related DNA sequences to which they bind, it has been possible to determine the critical role of both the first zinc finger and its adjacent helix in controlling the sequence to which these receptors bind, and of the second zinc finger in determining the spacing of adjacent sequences which is optimal for the binding of each receptor.

As well as determining the nature of the receptor homeodimers which bind to the palindromic repeats illustrated in Table 7.3a, the DNA binding domain also plays a critical role in determining the receptors which bind to the repeat sequences illustrated in Table 7.3b. Thus when these direct repeats are separated by only one base, they can bind a homeodimer of the retinoid–X–receptor (RXR) and hence confer response to 9-*cis* retinoic acid which binds to this receptor (Fig. 8.19). In contrast, the RXR homeodimer cannot bind to the response elements when the direct repeats are separated by two, three, four or five base pairs. On these response elements the DNA binding domain of RXR interacts with the DNA binding domain of another member of the nuclear receptor family to form heterodimers which can bind to each of these sites. In these heterodimer combinations, the effect of RXR is suppressed and the response of the heterodimer is determined by the other component. Thus a spacing of two or five bases binds a heterodimer of RXR and the retinoic acid receptor (RAR) and results in a response to all *trans*-retinoic acid which binds to RAR, a spacing of four base pairs binds a

Figure 8.18

Effect of amino acid substitutions in the zinc finger region of the glucocorticoid receptor on its ability to bind to and activate genes which are normally responsive to different steroid hormones.

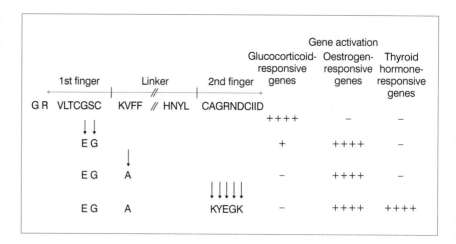

heterodimer of RXR and the thyroid hormone receptor and therefore responds to thyroid hormone, and a spacing of three base pairs binds a heterodimer of RXR and the vitamin D receptor leading to a response to vitamin D (Fig. 8.19) (for reviews see Gronemeyer and Moras, 1995; Mangelsdorf and Evans, 1995).

Hence, the different DNA binding domains of the various nuclear receptors produce different patterns of homeodimer and heterodimer binding to different binding sites, thereby allowing the diverse members of the nuclear receptor family to produce a wide variety of different responses.

8.2.4 The leucine zipper, the helix-loop-helix motif and the basic DNA binding domain

In earlier sections of this chapter, we have examined how the presence of unusual structural motifs, such as the zinc finger, in several different regulatory proteins led to the identification of the crucial role of these elements in DNA binding. A similar approach has led to the identification of another such motif, the leucine zipper (for reviews see Kerppola and Curan, 1995; Hurst, 1996). Thus, in studies of the gene encoding the transcription factor C/EBPα which is involved in stimulating the expression of several liver-specific genes, it was noted that it contained a region of 35 amino acids, in which every seventh amino acid was a leucine. Similar runs of leucine residues were also noted in the yeast transcriptional regulatory protein GCN4, as well as in the proto-oncogene proteins Myc, Fos and Jun, which were originally identified on the basis of their ability to transform cultured cells to a cancerous phenotype (see Chapter 9) and are believed to act by regulating the transcription of other cellular genes (Fig. 8.20).

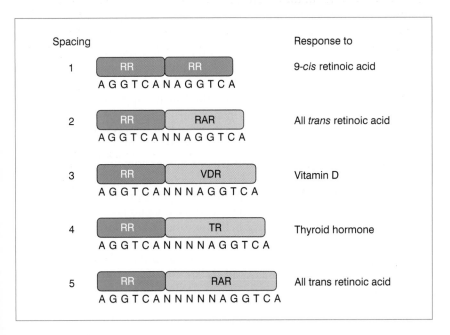

Figure 8.19

Patterns of nuclear receptor heterodimers which form on various directly repeated response elements with different spacings (N) between the two halves of the repeat. Note that the response of the element is determined by the nature of the receptor which associates with RXR.

Figure 8.20

Alignment of the leucine-rich region in several cellular transcription factors. Note the conserved leucine residues (L) which occur every seven amino acids.

C/EBP	L T S D N D R L R K R V E Q L S R E L D T L R G I F R Q L
Jun B	L E D K V K T L K A E N A G L S S A A G L L R E Q V A Q L
Jun	L E E K V K T L K A Q N S E L A S T A N M L R E Q V A Q L
GCN 4	L E D K V E E L L S K N Y H L E H E V A R L K K L V G E R
Fos	L Q A E T D Q L E D E K S A L Q T E I A N L L K E K E K L
Fra 1	L Q A E T D K L E D E K S G L Q R E I I E L Q K Q K E R L
C-Myc	V Q A E E Q K L I S E E D L L R K R R E Q L K H K L E Q L
N-Myc	L Q A E E H Q L L L E K E K L Q A R Q Q Q L L K K I E H A
I-Myc	L V G A E K K M A T E K R Q L R C R Q Q Q L Q K R I A Y L

It was proposed that the leucine-rich region would form an α-helix in which the leucines would occur every two turns on the same side of the helix. These leucine residues would then facilitate the dimerisation of two molecules of the transcription factor by promoting the interdigitation of two such helices, one on each of the individual molecules (Fig. 8.21). In agreement with this idea, replacement of individual leucine residues in C/EBPα with valine or isoleucine residues abolishes the ability of the protein to form a dimer. In turn, these mutations also prevent the binding of the protein to its specific recognition sequence.

Unlike the helix-turn-helix motif, however, the leucine zipper does not bind directly to DNA. Rather, by facilitating the dimerisation of the protein it provides the correct protein structure for DNA binding by the adjacent region of the protein, which is rich in basic amino acids that can interact directly with the acidic DNA (Fig. 8.22) (for a review see Ellenberger, 1994). In agreement with this idea, mutations in the basic DNA binding domain abolish the ability of the protein to bind to the DNA without abolishing its ability to dimerise. Similar juxtapositions of a basic DNA binding domain and the leucine zipper are also found in the Fos and Jun oncogene proteins and in the yeast transcription factor GCN4, where a single 60 amino acid region contains all the information needed for both dimerisation and sequence-specific DNA binding. Hence the leucine zipper has a role similar to that of the second zinc finger in the steroid receptors (Section 8.2.3) which modulates the activity of the DNA binding region rather than being involved directly in binding.

Although originally identified in leucine zipper-containing proteins, the basic DNA binding domain has also been identified by homology comparisons in a number of other transcriptional regulatory proteins which lack the leucine zipper. In this case, however, the basic domain is associated with an adjacent region that can form a helix-loop-helix structure. This motif is distinct from the helix-turn-helix motif described in Section 8.2.2 and consists of two amphipathic helices (containing all the charged amino acids on one side of the helix) separated by an intervening non-helical loop. Although originally thought to be the DNA binding domain of these proteins, this helix-loop-helix motif is now known to play a similar role to the leucine zipper in mediating protein dimerisation and facilitating DNA

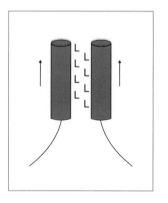

Figure 8.21

Model of the leucine zipper and its role in the dimerisation of two molecules of a transcription factor.

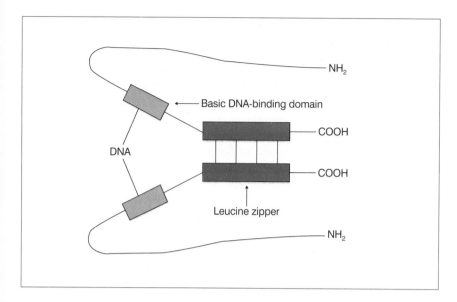

Figure 8.22

Model for the structure of the leucine zipper and the adjacent DNA binding domain following dimerisation of the transcription factor C/EBP.

binding by the adjacent basic DNA binding motif (for reviews see Ellenberger, 1994; Littlewood and Evan, 1995).

The helix-loop-helix motif with its adjacent basic DNA binding region is present in a number of different transcription factors expressed in different tissues. Thus, for example, it is found in several of the factors which control muscle development such as the MyoD transcription factor, whose artificial expression in an undifferentiated fibroblast cell line can induce it to differentiate to skeletal muscle cells by activating the expression of muscle-specific genes (for reviews see Buckingham, 1994; Olson and Klein, 1994). Thus, the MyoD gene is likely to be the critical regulatory locus which is activated by treatment of these cells with 5-azacytidine, allowing this agent to induce these cells to differentiate into muscle cells (see Section 6.5.2).

The leucine zipper and helix-loop-helix structures therefore act to mediate the dimerisation of the transcription factors which contain them, so forming a dimeric molecule which is able to bind to DNA via the adjacent basic DNA binding domain. This ability provides an additional aspect to gene regulation by such proteins. Thus, in addition to the formation of a dimer by two identical factors, it is possible to envisage the formation of a heterodimer between two different factors, which might have different properties in terms of sequence-specific binding and gene activation compared to homodimers of one or other of the two factors.

Such homo- and heterodimerisation resulting in binding to different response elements occurs in the case of the nuclear receptors as discussed in Section 8.2.3. In the case of basic domain-containing factors, an example of this type is seen in the case of the related oncoproteins Fos and Jun (see Section 9.4.1; Kerppola and Curran, 1995). Thus Jun can bind as a homodimer to the AP1 recognition sequence, TGAGTCAG, which mediates transcriptional induction by phorbol esters (Table 7.2). In contrast, Fos cannot bind to DNA alone but can form a heterodimer with the Jun protein. This heterodimer binds to the AP1 recognition site with a 30-fold greater affinity than the Jun homodimer, and is considerably more effective in enhancing transcription of genes containing the binding site. Both hetero- and homodimer formation and DNA binding are dependent on the leucine zipper motif which is found in both proteins. Hence dimerisation by the leucine zipper motif allows two different complexes with different binding affinities and different activity to form on the identical DNA binding site (Fig. 8.23).

The failure of the Fos protein to form homodimers and its inability to bind to DNA in the absence of Jun has been shown to be due to differences in its leucine zipper region from that of Jun. Thus, if the leucine zipper region of Fos is replaced by that of Jun the resulting protein can dimerise. This dimerisation allows the chimeric protein to bind to DNA through the basic region of Fos which is therefore a fully functional DNA binding domain.

As well as having a positive role, heterodimerisation can also have a negative role. Thus, as discussed in Section 8.3.4, the ability of the MyoD factor to stimulate gene expression is inhibited by heterodimerisation with the Id factor which has a helix-turn-helix motif but no basic domain. Since DNA binding requires the cooperation of two basic domains within the heterodimer, the MyoD–Id heterodimer cannot bind to DNA. Hence MyoD cannot activate the expression of muscle-specific genes and thereby promote the production of skeletal muscle cells in the presence of the Id factor (Fig. 8.24). It is clear, therefore, that the ability of the leucine zipper and the helix-loop-helix motif to facilitate the formation of different dimeric complexes between different transcription factors is likely to play a crucial

Figure 8.23

Model for DNA binding by the Jun homodimer and the Fos–Jun heterodimer.

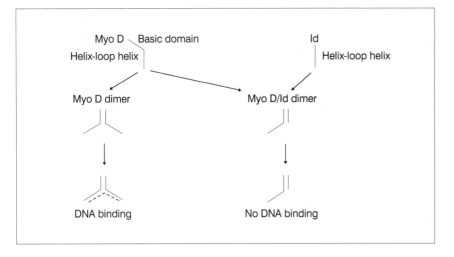

Figure 8.24

DNA binding by the MyoD protein is inhibited by the Id factor which contains the helix-loop-helix dimerisation motif but not the basic DNA binding domain.

role in the regulation of gene expression, by producing complexes with different binding affinities and different activities (for reviews see Jones, 1990; Lamb and McKnight, 1991).

8.2.5 Other DNA binding domains

As the genes encoding more and more transcription factors are isolated and characterised, it has become clear that while the DNA binding domains of many factors fall into the three classes discussed above, not all do so. Hence additional types of DNA binding motifs exist. Interestingly, however, as the structures of these DNA binding domains are progressively understood, relationships between the different classes of DNA binding domain have emerged. Thus structural analysis of the Ets DNA binding domain, which is present, for example, in the *ets-1* proto oncogene (see Section 9.4.3) and the mouse PU-1 gene has shown it to be identical to the winged helix-turn-helix or fork head DNA binding motif identified in the *Drosophila* fork head factor and the mammalian liver transcription factor HNF-3 (Donaldson *et al.*, 1996). Hence all these factors share a common winged-helix-turn-helix motif.

In turn, this motif, as its name suggests, contains a helix-turn-helix structure which is also present in the homeodomain proteins discussed in Section 8.2.2. These two DNA binding motifs are thus related to one another, although the winged helix-turn-helix motif also contain a β-sheet structure with two loops which appear as wings protruding from the factor, thereby giving it its name (for a review see Brennan, 1993). In the majority of winged helix-containing proteins, it is the helix-turn-helix motif which is responsible for DNA binding. However, in the winged helix-containing protein hRFX1, it is the β-sheet wing structure which binds to the DNA rather than the helix-turn-helix motif, indicating that members of the winged helix family can use one of two distinct structures to bind to DNA (Gajiwala *et al.*, 2000).

In another example of the relationship of different groups of factors, the existence of the POU DNA binding domain, containing a POU-specific domain in addition to a POU homeodomain, is paralleled in the Pax family of vertebrate transcription factors, many of which contain both a homeodomain and a so-called "paired" domain, both of which contribute to high affinity DNA binding (for a review see Mansouri *et al.*, 1996). Unlike the POU-specific domain, however, the paired domain can be found as an isolated DNA binding domain without the homeobox, both in some members of the PAX family and in several *Drosophila* factors, including the paired gene in which it was originally identified.

It is clear therefore that a number of different structures exist which can mediate sequence-specific DNA binding and many of these are related to one another. Each of these DNA binding motifs is common to a number of different transcription factors, with differences in the precise amino acid sequence of the motif in each factor controlling the precise DNA sequence which it binds and hence the target genes for the factor.

8.3 REGULATION OF TRANSCRIPTION

8.3.1 Introduction

Although binding to DNA is generally a necessary prerequisite for the activation of transcription, it is clearly not in itself sufficient for this to occur. Following binding, the bound transcription factor must somehow regulate transcription, for example by directly activating the RNA polymerase itself or by facilitating the binding of other transcription factors and the assembly of a stable transcriptional complex. Although some transcription factors can inhibit transcription, the majority of factors defined so far act to activate transcription. We will therefore discuss in turn the features of activating transcription factors that produce this activation (Section 8.3.2) and the manner in which they do so (Section 8.3.3) and then discuss the manner in which specific factors inhibit transcription (Section 8.3.4).

8.3.2 Activation domains

Identification of activation domains

It is clear from the preceding sections of this chapter that transcription factors have a modular structure in which a particular region of the protein mediates DNA binding while another may mediate binding of a co-factor, such as a hormone, and so on. It seems likely, therefore, that a specific region of each individual transcription factor will be involved in its ability to up-regulate transcription following DNA binding.

In the majority of cases, it is clear that such activation regions are distinct from those which produce DNA binding. This domain-type structure is seen clearly in the yeast transcription factor GCN4, which mediates the induction of the genes encoding the enzymes of amino acid biosynthesis in response to amino acid starvation. Thus, if a 60 amino acid region of this protein, containing the DNA binding region, is introduced into cells, it can bind to the DNA of GCN4-responsive genes but fails to activate transcription.

Hence, although DNA binding is necessary for transcriptional activation to occur, it is not sufficient, and gene activation must be dependent upon a region of the protein that is distinct from that mediating DNA binding.

Unlike the DNA binding regions, the region of a transcription factor that mediates gene activation cannot, therefore, be identified on the basis of a simple assay of, for example, the ability to bind to DNA or another protein. Rather, a functional assay of gene activation following binding to DNA is required. Activation regions have therefore been identified on the basis of so-called "domain-swap" experiments, in which the DNA binding region of one transcription factor is combined with various regions of another factor and the ability to activate transcription of a gene containing the binding site of the first factor is assessed (Fig. 8.25). Following binding of the hybrid factor to the target gene binding site, gene activation will occur only if the hybrid factor also contains an activation domain provided by the second factor, and hence the activation domain can be identified.

Thus, in the case of the yeast transcription factor GCN4 discussed above, if a 60 amino acid region, outside the DNA binding domain, is linked to the DNA binding region of the bacterial regulatory protein, Lex A, the hybrid factor will activate transcription in yeast from a gene containing the binding

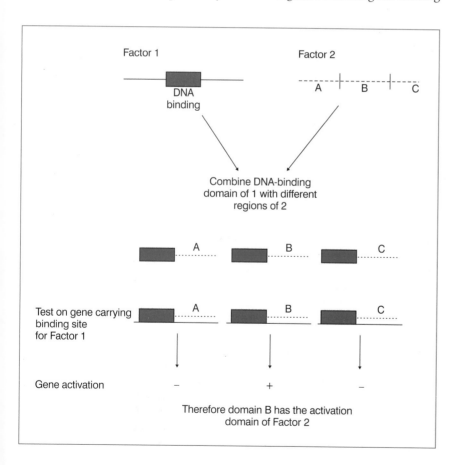

Figure 8.25

Domain swapping experiment, in which the activation domain of factor 2 is mapped by combining different regions of factor 2 with the DNA binding domain of factor 1 and assaying the hybrid proteins for the ability to activate transcription of a gene containing the DNA binding site of factor 1.

site for Lex A, whereas neither the Lex A DNA binding domain nor the GCN4 region will do so alone. Hence this region of GCN4 contains an activation domain, which can increase transcription following DNA binding and is separate from the region of the protein that normally mediates DNA binding (Fig. 8.26a).

Following its initial use in yeast, similar domain-swapping experiments have also been used to identify the activation domains of mammalian transcription factors. In the glucocorticoid receptor, for example, two independent regions, each able to produce gene activation, have been identified in this way (Fig. 8.26b).

The success of domain-swap experiments is further proof of the modular nature of transcription factors, allowing the DNA binding domain of one factor and the activation domain of another to cooperate together to produce gene activation. This is analogous to the exchange of the DNA binding domain of the glucocorticoid and oestrogen receptors, which allows the creation of a hybrid receptor that binds to oestrogen-responsive genes through the DNA binding domain but is responsive to the presence of glucocorticoid through the steroid binding domain of the protein.

An extreme example of this modularity is provided by the herpes simplex virus *trans*-activating protein VP16, which activates transcriptionally the viral immediate-early genes during lytic infection of mammalian cells. Thus, although this protein contains a very potent activation region, which can strongly induce gene transcription when fused to the DNA binding domain of GAL4, it contains no DNA binding domain and cannot bind to DNA itself. Rather, following infection, it forms a complex with the cellular octamer-binding protein, Oct-1 (see Section 8.2.2). Oct-1 provides the DNA

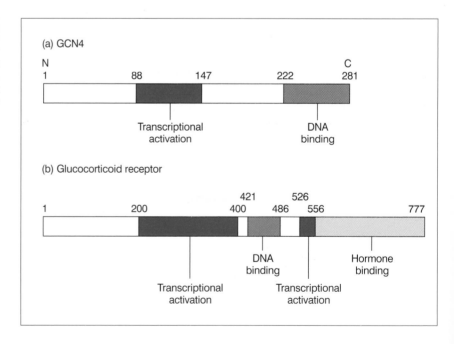

Figure 8.26

Structure of the yeast GCN4 factor (a) and the mammalian glucocorticoid receptor (b), indicating the distinct regions that mediate DNA binding or transcriptional activation.

binding domain which allows binding to the sequence TAATGARAT (R = purine) in the viral promoters, and activation is achieved by the activation domain of VP16 (for a review see Goding and O'Hare, 1989). Hence, in this instance, DNA binding and activation motifs actually reside on separate molecules (Fig. 8.27).

Nature of activating regions

Acidic domains

Although the activating regions identified in various transcription factors do not show strong amino acid sequence homology to each other, in many cases they have a very high proportion of acidic amino acids, resulting in a strong net negative change (for a review see Triezenberg, 1995). Thus, in the N-terminal activating region of the glucocorticoid receptor, 17 acidic amino acids are contained in an 82 amino acid region. Similarly, the activating region of the yeast factor GCN4 contains 17 negative charges in an activating region of only 60 amino acids. This has led to the idea that activation regions consist of so-called "acid blobs" or "negative noodles" which are necessary for the activation of transcription. However, although the net negative charge in acidic activation regions is likely to be of importance in their ability to stimulate transcription, it appears that other features, such as the presence of several conserved hydrophobic residues, are also necessary for transcriptional activation (Uesugi et al., 1997).

Other activating domains

Although acidic domains of the type discussed above have been identified in a wide range of transcriptional activators from yeast to man, it is clear that this type of structure is not the only one which can mediate transcriptional activation. Thus, of the two regions of the human Spl transcription factor that could mediate activation of transcription, neither was particularly acidic. Instead, each of these two domains was particularly rich in glutamine residues, and the intactness of the glutamine-rich region was essential for transcriptional activation. Similar sequences have also been identified in the homeotic proteins Antennapedia and Cut, in Zeste, another *Drosophila* transcriptional regulator, and in the POU proteins

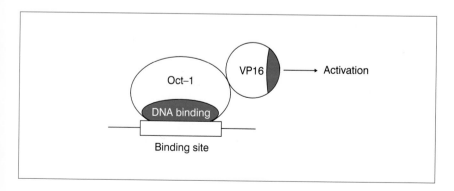

Figure 8.27

Activation of gene transcription by interaction of the cellular factor Oct-1, which contains a DNA binding domain, and the herpes simplex virus VP16 protein, which contains an activation domain but cannot bind to DNA.

Oct-1 and Oct-2, indicating that this type of activating region is not confined to a single protein.

A further type of activation domain has been identified in the transcription factor CTF/NF1 which binds to the CCAAT box present in many eukaryotic promoters (Table 7.2). The activation domain of this protein is not rich in acidic or glutamine residues but instead contains numerous proline residues, forming approximately one-quarter of the amino acids in this region. Similar proline-rich regions are found in other transcription factors, such as AP2 and Jun, indicating that, as with glutamine-rich domains, this element is not confined to a single protein.

Hence, as with DNA binding domains, it is clear that several distinct protein motifs are involved in the activation of transcription (for reviews see Mitchell and Tjian, 1989; Triezenberg, 1995).

8.3.3 How is transcription activated?

The widespread interchangeability of activation domains from yeast, *Drosophila* and mammalian transcription factors, discussed in the previous section, suggests that common mechanisms may mediate transcriptional activation in a wide range of organisms. This is supported by the observations that mammalian transcription factors, such as the glucocorticoid receptor, can activate a gene carrying its binding site in cells of *Drosophila*, tobacco plants and mammals. Indeed, yeast and mammalian factors can cooperate together in gene activation, a gene bearing binding sites for GAL4 and the glucocorticoid receptor being synergistically activated by the two factors in mammalian cells so that the activation observed with the two factors together is greater than the sum of that observed with each factor independently. This cooperation between two factors which come from widely different species and would therefore never normally interact, suggests that they both function by interacting with some highly conserved component of the basal transcriptional complex (see Section 3.2.3) which is involved in the basic process of transcription in all different species.

In fact, there is evidence that activators can interact with a variety of different targets within the basal transcriptional complex and these will be discussed in turn. Clearly such interactions could stimulate transcription by increasing the binding of a particular component of the basal transcriptional complex, so enhancing its assembly (Fig. 8.28a). Alternatively, stimulation could occur by the activator altering the conformation of an already bound factor so stimulating the activity and/or stability of the complex (Fig. 8.28b). It is clear that both of these mechanisms are actually used and they could evidently operate whether the complex actually assembles in a fully stepwise fashion, in which each component binds sequentially, or in a situation where several components bind together in a holoenzyme complex with RNA polymerase itself (see Section 3.2.3 for further discussion of these alternative models of complex assembly). In discussing the various targets within the complex, however, we shall consider them in the order in which they bind in the stepwise model (for reviews of activator mechanisms see Ranish and Hahn, 1996; Stargell and Struhl, 1996).

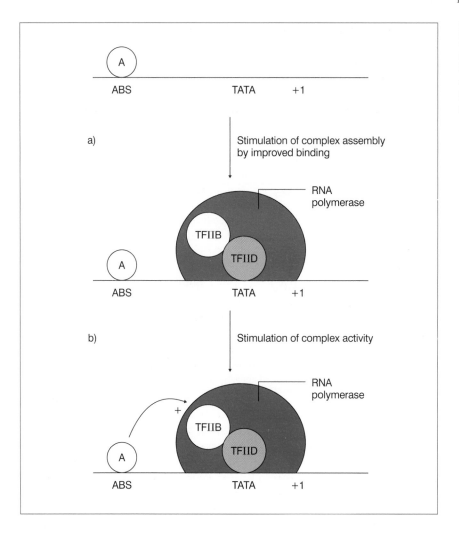

Figure 8.28

An activator (A) bound to its binding site (ABS) can stimulate either complex assembly (a) or the activity of the already assembled complex (b).

TFIID

As described in Section 3.2.3, the binding of the TFIID complex (containing TBP and associated proteins) to the TATA box is the initial stage of complex assembly in both the stepwise and holoenzyme assembly models. Hence TFIID constitutes an obvious target for activating molecules. Indeed early studies indicated that both the recruitment of TFIID to the promoter and its conformation when bound are affected by activator molecules. Increased binding of TFIID would evidently result in enhanced binding of the other components of the basal transcription complex which bind subsequently, while alterations in TFIID configuration might act by improving its ability to recruit these other factors or by directly enhancing its activity within the assembled basal transcription complex (Fig. 8.29). Such findings are complicated, however, by the fact that, as discussed in Section 3.2.4, TFIID is a multi-protein complex consisting of the DNA binding component TBP and

Figure 8.29

An activator (A) bound to its binding site (ABS) can enhance the binding of TFIID to the TATA box, thereby improving the rate of assembly of the basal transcriptional complex by facilitating the subsequent binding of other components which is dependent upon prior binding of TFIID (a). In addition, the activator can also alter the configuration of TFIID so stimulating its activity either by increasing its ability to recruit other components of the complex or by enhancing its ability to stimulate transcription (b).

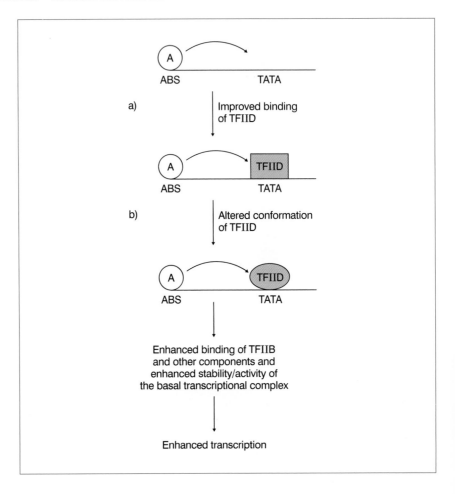

a variety of TAFs (TBP-associated factors) (for reviews see Hahn, 1998; Green, 2000).

Indeed it is likely that both TBP itself and one or more of the TAFs can be targets for transcriptional activators (Fig. 8.30). Thus acidic activators have been shown directly to interact with TBP, and single amino acid mutations in such activators, which abolish interaction with TBP, also abolish the ability to activate transcription, supporting an important role in this effect. It appears that such activators act by enhancing the rare of recruitment of TBP to the promoter, hence increasing the rate of assembly of the basal transcriptional complex and this effect of activators has now been demonstrated in intact cells as well as in cellular extracts (Li *et al.*, 1999; Kuras and Struhl, 1999).

Although there is thus evidence that activators can interact directly with TBP, it is clear that in some circumstances activation requires interaction with one or more of the TAFs. Thus, in many cases, stimulation of transcription *in vitro* does not occur with purified TBP alone but only when the full

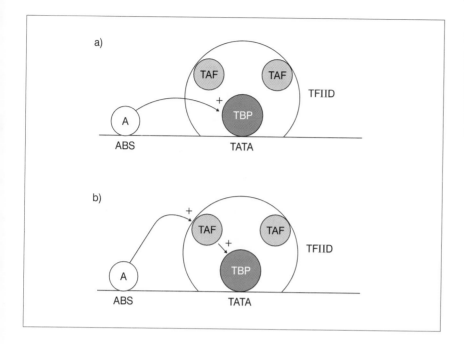

Figure 8.30

An activating molecule (A) can interact with TFIID either by interacting with the TATA box binding protein (TBP) directly (panel a) or by interacting indirectly with TBP via TBP associated factors (TAF) (panel b).

TFIID complex is present, indicating that such activation requires interaction with the TAFs. Most interestingly, different types of activation domain appear to contact different TAFs. Thus, for example, glutamine-rich activators such as Spl can bind $TAF_{II}110$ while acidic activators bind $TAF_{II}40$ and multiple activators including proline-rich activators bind $TAF_{II}55$. Hence different components within TFIID can be targeted by different classes of activator (Fig. 8.31). Because TAFs can function as intermediaries between activators and the basal transcription complex they are often referred to as co-activators (see below).

TFIIB

Although there is considerable evidence for the interaction of activators and TFIID, it is clear that activators can also act to stimulate the assembly/activity of the basal transcriptional complex after TFIID has bound. As discussed in Section 3.2.3 the binding of TFIID to the promoter is followed by the binding of TFIIB. As with TBP and the TAFs, there is clear evidence for the interaction of TFIIB with activators. Thus TFIIB can be purified from a mixture of proteins on a column containing a bound acidic activator. Moreover, this interaction is of importance for activation of transcription since mutations in acidic activators which abolish the interaction with TFIIB prevent the activator from stimulating transcription (for a review see Hahn, 1993).

As with TFIID, such interactions can both stimulate the binding of TFIIB to the promoter and alter its configuration when bound, so improving its ability to recruit the other components of the basal transcriptional complex, such as the RNA polymerase which binds subsequently (Fig. 8.32). Hence,

Figure 8.31

Different classes of activator can have distinct targets within TFIID with acidic activators contacting TAF$_{II}$40, glutamine-rich activators contacting TAF$_{II}$110 and proline-rich activators contacting TAF$_{II}$55.

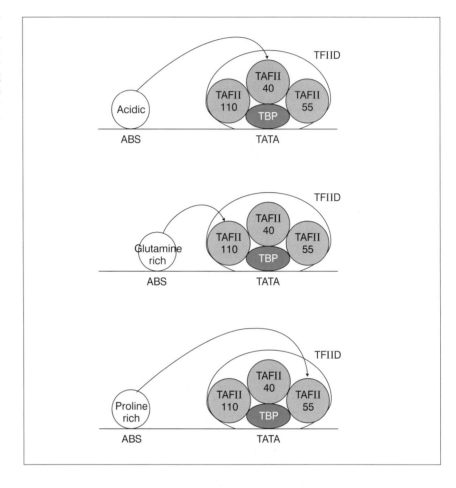

as with the various components of TFIID, the single TFIIB polypeptide is evidently a target for transcriptional activators.

RNA polymerase II and the mediator complex

As discussed in Section 3.2.3 binding of TFIIB allows the subsequent recruitment of the RNA polymerase itself together with the TFIIF factor.

The RNA polymerase would be an obvious target for activators, so allowing the activator to directly stimulate enzyme activity. Thus, as discussed in Section 3.2.3, the C-terminal region of the largest subunit of RNA polymerase II is known to contain a repeating motif of a sequence whose consensus is Tyr–Ser–Pro–Thr–Ser–Pro–Ser, which is conserved from yeast to mammals. As well as being phosphorylated by TFIIH, this region is also implicated in the response to transcriptional activators. Thus, deletion of this region from RNA polymerase II prevents enhanced transcription in response to transcriptional activators while increasing the number of repeated elements enhances the response to activators (for a review see Stargell and Struhl, 1996).

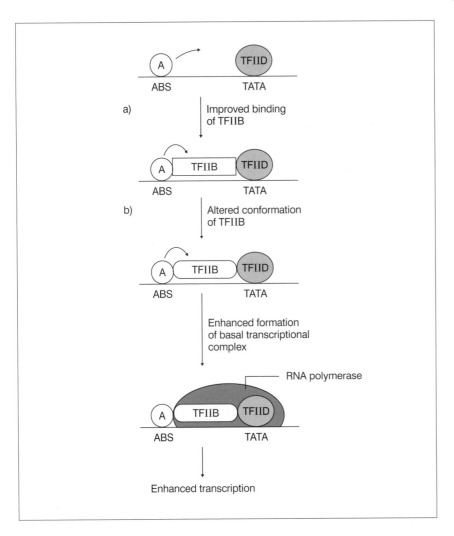

Figure 8.32

The binding of an activator (A) to its appropriate binding site (ABS) can enhance the rate of transcription by enhancing the binding of the TFIIB factor and/or by altering its configuration so that it recruits the other components of the basal transcriptional complex more efficiently. Compare with Fig. 8.29, which shows the same effects for TFIID.

Although these findings indicate that this region of the RNA polymerase is necessary for transcriptional activation, they do not prove that activators contact this region directly. Indeed, work in yeast and subsequently in mammalian cells has identified a mediator complex containing over 20 polypeptides which associates with this C-terminal region of the RNA polymerase II and is required for the response to transcriptional activators. It is clear therefore that activators can act by interacting with the mediator complex which in turn stimulates RNA polymerase activity (for reviews see Kingston, 1999; Malik and Roeder, 2000; Myers and Kornberg, 2000) (Fig. 8.33). Interestingly, electron microscope analysis of the mediator complex bound to RNA polymerase II suggests that the mediator partially envelopes the polymerase (Asturias *et al.*, 1999). This would allow it to receive signals from activators and transmit them to the polymerase (Fig. 8.34).

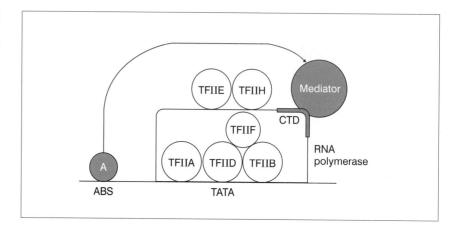

Interestingly, the mediator complex has also been shown to stimulate the
ability of TFIIH to phosphorylate the C-terminal region of RNA polymerase. It is possible therefore that activators may enhance the ability of the
mediator to stimulate this effect of TFIIH, thereby also stimulating the start
of transcriptional elongation which is dependent upon such phosphorylation (Fig. 8.35) (see Section 3.2.3).

The mediator complex has been shown to be associated with the RNA
polymerase holoenzyme prior to binding to the DNA. Hence, as discussed in
Section 3.2.3 this holoenzyme contains not only RNA polymerase II, and
basal transcription factors such as TFIIB, TFIIE, TFIIF and TFIIH, but also
contains mediator components which respond to transcriptional activators
and the SWI/SNF complex which can alter chromatin structure (see Section
6.7.2).

Co-activators

As discussed above, transcriptional activators often contact their ultimate
targets indirectly acting, for example, via TAFs (see above) or the mediator
complex (see above). This has led to the concept of co-activators which act
to transmit the signal from the DNA binding transcriptional activator to the
basal transcriptional complex.

Indeed, co-activators have been identified in a number of different situations involving specific transcriptional activators, where they are recruited
to the DNA by the DNA bound activator and then activate transcription.
Perhaps the best known co-activator is CBP (CREB binding protein: for
reviews see Shikama *et al.*, 1997; Goodman and Smolik, 2000). As its name
suggests, CBP was first defined as a protein which binds to the CREB transcription factor (for reviews of CREB see de Cesare *et al.*, 1999; Shaywitz
and Greenberg, 1999; de Cesare and Sassone-Corsi, 2000). Most importantly, CBP only binds to CREB when CREB has been phosphorylated on
the serine amino acid at position 133. Since such phosphorylation is necessary for CREB to activate transcription, this indicated that recruitment of
CBP was likely to play a key role in mediating transcriptional activation by

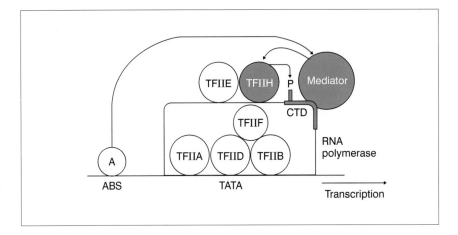

Figure 8.35

As well as affecting transcriptional initiation via the mediator, it is possible that activators act by stimulating the ability of the mediator to interact with TFIIH and stimulate it to phosphorylate the CTD (P), thereby promoting transcriptional elongation.

CREB (see Section 8.4.3). Hence, CBP does not bind to DNA itself but is recruited to the DNA by DNA-bound phosphorylated CREB and then activates transcription.

CBP has now been shown to play a key role in mediating transcriptional activation via a number of DNA binding transcription factors which are discussed in this book including the steroid–thyroid hormone receptor family (Section 8.2.3), NFκB (Section 8.4.1), p53 (Section 9.5.2) and MyoD (Section 8.2.4). Hence, it is clear that CBP, and its close relative p300, play a key role as co-activators for a number of transcriptional activators (Fig. 8.36). Indeed, the widespread use of CBP by transcription factors activated by different signalling pathways may result in competition between these pathways for limited amounts of CBP. This in turn would account for the phenomenon in which simultaneous stimulation of two different pathways,

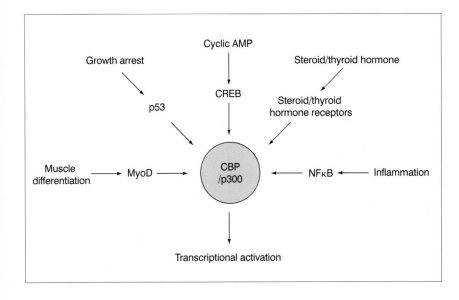

Figure 8.36

A variety of activating transcription factors stimulate transcription via the closely related CBP and p300 co-activators.

such as those stimulated by glucocorticoid and phorbol esters, results in each pathway inhibiting the other so that no transcriptional activation occurs.

Although CBP can interact with members of the steroid/thyroid hormone receptor family, it is not the only co-activator to do so. In several cases, specific co-activators can bind to the receptors only after hormone treatment. Hence, the binding of specific co-activators to these receptors only after hormone treatment allows the receptors to activate transcription in a hormone-dependent manner (for review see Freedman, 1999; Glass and Rosenfeld, 2000; Rosenfeld and Glass, 2001).

It is likely that co-activators can act to stimulate transcription via at least two distinct mechanisms, following recruitment to the DNA by a DNA bound activator. Both these mechanisms have been shown to operate in the case of CBP. Thus, CBP has been shown to bind to various components of the basal transcription complex such as TBP, TFIIB and the RNA polymerase holoenzyme, indicating that it can bridge the gap between CREB and the basal complex allowing CREB to activate transcription (Fig. 8.37a). Moreover, as discussed in Section 6.6.1, CBP also has histone acetyltransferase activity indicating that it can also stimulate transcription by acetylating histones and therefore opening up the chromatin structure (Ogryzkio *et al.*, 1996) (Fig. 8.37b). Hence, co-activators can act either via linking the activator to the basal complex or by modifying chromatin structure.

A multitude of targets for activators

There exists therefore a bewildering array of potential targets for activator molecules including the RNA polymerase/mediator complex, co-activators,

Figure 8.37

A co-activator (CA) may act (a) by linking an activator (A) to the basal transcriptional complex or (b) by promoting the conversion of a closed chromatin structure (wavy line) to a more open structure (solid line). In both cases, the co-activator will be recruited to the DNA by a DNA bound activator.

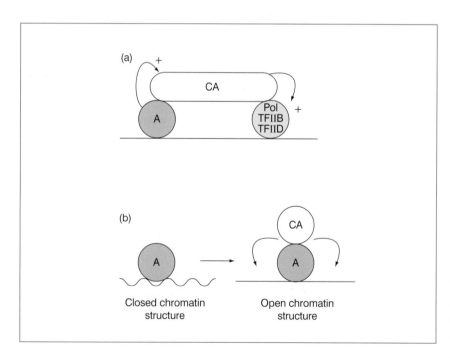

TFIIB and the different components of TFIID. Indeed, other components of the basal transcriptional complex such as TFIIA, TFIIE and TFIIH have also been observed to interact directly with transcriptional activators. All these various possibilities are not mutually exclusive, however. Thus, it is likely that all these components can act as a target either for the same activating factor or different activating factors.

Several possibilities may account for such a wide range of activator targets. Thus, it is possible that different organisms differ in the preferred target for transcriptional activators. Indeed, while TAFs associated with TBP appear to be of critical importance for transcriptional activation in higher organisms such as *Drosophila* they may be of much less importance in yeast, Alternatively, it has been suggested that the sole requirement for an activator is to be able to bind to DNA and then bind any component of the basal transcriptional complex and hence recruit it to the DNA (for a review see Ptashne and Gann, 1997). Thus, an activator can function simply by interacting with any component of the complex and thereby enhancing its binding to the promoter.

It is most probable, however, that a multiplicity of targets for activators is required to produce the strong synergistic activation of transcription which is observed when different activating factors are added together, compared to the level observed when each factor is added separately (see above), as well as the strong enhancement of transcription observed when multiple copies of a single factor can bind the target DNA. Thus, different activating factors or different molecules of the same activating factor could contact different targets within the basal transcriptional complex, so ensuring the great enhancement of transcriptional activity which is the ultimate aim of activating molecules (Fig. 8.38).

8.3.4 Repression of transcription

Although the majority of transcription factors described so far act in a positive manner, a number of cases have now been reported in which a transcription factor exerts an inhibiting effect on transcription and several possible mechanisms by which this can be achieved have been described (for

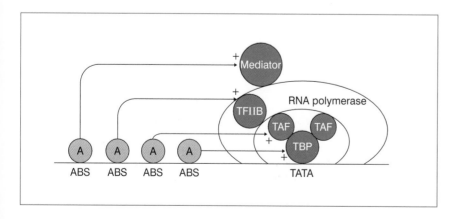

Figure 8.38

Synergistic activation of transcription by activator molecules (A) interacting with different components of the basal transcriptional complex, such as the mediator complex, TFIIB or the different components of TFIID (TBP and non-DNA-binding TAFs).

reviews see Hanna-Rose and Hansen, 1996, Latchman, 1996; Smith and Johnson, 2000; Fig. 8.39).

The simplest means of achieving repression is for a repressor to prevent an activating molecule from binding to DNA. This can occur by a negative factor promoting a tightly packed chromatin structure which does not allow activating molecules to bind (Fig. 8.39a) (see Chapter 6 for a discussion of the effect of chromatin structure on gene expression). An example of this effect was discussed in Section 8.2.2 in which the polycomb repressor binds to *Drosophila* homeotic genes and organises them into an inactive chromatin structure incapable of binding activators (for reviews see Simon, 1995; Pirrotta, 1997).

Alternatively, the negatively acting factor can bind specifically to the binding site of the activator so preventing it binding (Fig. 8.39b). This effect is seen in the β-interferon promoter where the binding of two positively acting factors is necessary for gene activation. Another factor acts negatively by binding to this region of DNA and simply preventing the positively acting factors from binding (Fig. 8.39b). In response to viral infection, the negative

Figure 8.39

Mechanisms by which an inhibitory factor (R) can repress transcription. These involve inhibiting an activator (A) binding to its binding site (ABS) in the DNA either by producing an inactive chromatin structure (panel a) or by competing for the DNA binding site (panel b) or by sequestering the activator in solution (panel c); inhibiting the ability of bound activator to stimulate transcription (panel d) or direct repression (panel e).

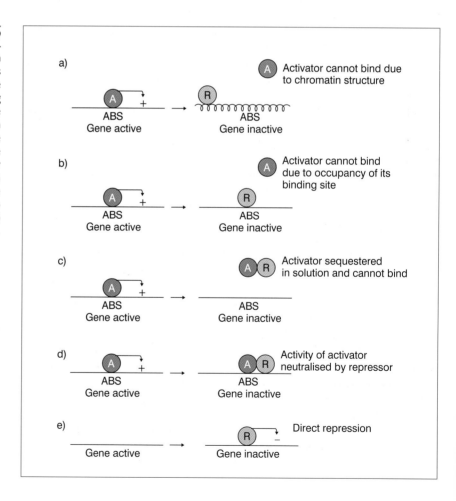

factor is inactivated, allowing the positively acting factors to bind and transcription occurs. A similar example was described in Section 8.2.2, whereby the DNA binding of the engrailed gene product inhibits gene activation by preventing the binding of the Ftz activator protein.

In a related phenomenon (Fig. 8.39c) repression is achieved by formation of a complex between the activator and the repressor in solution, preventing the activator binding to the DNA. This is seen in the case of the inhibitory factor Id which dimerises with the muscle determining factor MyoD via its helix-loop-helix motif and prevents it binding to DNA and activating transcription (see Section 8.2.4).

In addition to inhibiting DNA binding by these different means, a negative factor can also act by interfering with the activation of transcription mediated by a bound factor in a phenomenon known as quenching (Fig. 8.39d). Thus, in the case of the promoter driving expression of the c-*myc* gene, a negatively acting factor *myc*-PRF binds to a site adjacent to that occupied by a positively acting factor *myc*-CF1 and prevents it from activating c-*myc* gene expression.

In all cases discussed so far the negative factor exerts its inhibiting effect in an essentially passive manner by neutralising the action of a positively acting factor by preventing either its DNA binding or its activation of transcription. It is clear, however, that some factors can have an inherently negative action on transcription which does not depend upon the neutralisation of a positively acting factor (Fig. 8.39e) (for a review see Hanna-Rose and Hansen, 1996).

Thus, the *Drosophila* even-skipped protein is able to repress transcription from a promoter lacking DNA binding sites for any activating proteins, indicating that it has a directly negative effect on transcription. A similar direct inhibitory effect has been defined in the case of the mammalian c-*erb*A gene which encodes the thyroid hormone receptor, a member of the steroid–thyroid hormone receptor family (see Sections 7.2.3 and 8.2.2). Thus, the binding of this receptor to its specific DNA binding site in the absence of thyroid hormone results in the direct inhibition of transcription. This effect is mediated by a co-repressor which binds to the receptor in the absence of hormone and inhibits transcription (for reviews see Knoepfler and Eisenman, 1999; Glass and Rosenfeld, 2000; Rosenfeld and Glass, 2001). In the presence of thyroid hormone, however, the receptor undergoes a conformational change which results in the release of the co-repressor and allows co-activators to bind (see Section 8.3.3). This then allows the receptor to activate rather than repress transcription (Fig. 8.40).

This phenomenon indicates that the distinction between activators and repressors is not a precise one. Thus, in some cases, the same factor can directly stimulate or directly inhibit transcription, depending on the circumstances with this effect being controlled via the regulated binding of co-repressor and co-activator molecules.

In a number of directly acting transcriptional repressors such as even-skipped, the thyroid hormone repressor and the MeCP2 protein which binds to methylated DNA (see Section 6.5.3), a specific region of the protein

Figure 8.40

The thyroid hormone receptor
encoded by the c-erbA α gene
represses transcription in the
absence of thyroid hormone (T)
via a specific inhibitory domain
(white) which recruits an
inhibitory co-repressor (COR).
Following binding of hormone,
the protein undergoes a
conformational change which
releases the co-repressor and
leads to binding of a co-activator
(CoA) which allows the receptor
to activate transcription.

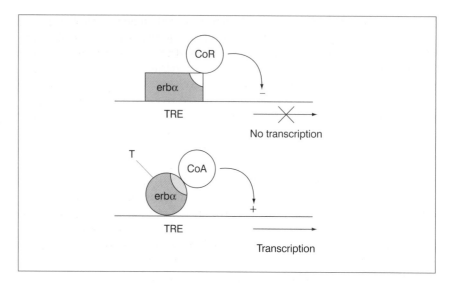

has been shown to be able to confer the ability to inhibit gene expression upon the DNA binding domain of another protein when the two are artificially linked, paralleling the similar behaviour of activation domains in such domain-swop experiments (Section 8.3.2). As noted in Section 6.6.1 the nuclear receptor co-repressor, which binds to the inhibitor domain of the thyroid hormone receptor, associates with the Sin3/RPD3 protein complex which can alter chromatin structure by deacetylating histones. A similar ability to recruit a histone deacetylase has also been observed in the case of MeCP2, as discussed in Section 6.5.3.

Hence, some inhibitory domains may function, at least in part, by recruiting molecules which can alter chromatin structure. In other cases, however, they are likely to inhibit transcription by reducing the formation and/or stability of the basal transcriptional complex either by interacting directly with the basal complex or by recruiting a co-repressor which then interacts with the basal complex. Such inhibitory domains therefore act in a manner similar to activation domains but reduce rather than enhance the rate of transcription.

Hence, inhibitory factors are likely to play a critical role in the regulation of transcription both by inhibiting the activity of positively acting factors and in some cases by having a direct inhibitory effect on transcription.

8.4 WHAT REGULATES THE REGULATORS?

8.4.1 Introduction

The crucial role of transcription factors in tissue-specific gene regulation and in the regulation of transcription in response to specific stimuli clearly leads to the question of how such factors are themselves regulated so that they produce gene activation or repression only in a specific tissue or in response to a particular signal.

As discussed in Section 7.1.3, the Britten and Davidson model envisaged that the products of regulatory or integrator genes, which controlled the activity of other genes, would be synthesised in response to a particular signal, or in a particular tissue, and would be absent in other tissues. Hence, the activity of the regulated genes would be directly correlated with the presence of the regulatory protein (Fig. 8.41a). An example of this type is provided by the MyoD factor discussed in Section 7.2.4 which is expressed only in skeletal muscle cells and whose over-expression in other cell types can induce them to differentiate into skeletal muscle cells.

A contrasting example is provided, however, by a factor involved in immunoglobulin gene transcription in B lymphocytes, namely NFκB (for reviews see Foo and Nolan, 1999; Karin and Ben-Neriah 2000; Perkins, 2000). This factor is detected in a form capable of stimulating immunoglobulin gene transcription only in extracts of mature immunoglobulin-producing B cells, no activity being detectable in immature B cells or in non-B cells, such as HeLa cells. The NFκB protein and its corresponding RNA are detectable in all cell types, however, suggesting that this protein exists in an inactive form in most cell types and is activated in mature B cells. In agreement with this, active NFκB capable of stimulating immunoglobulin gene transcription can be induced in other cell types, such as T cells or HeLa cells, by treatment with phorbol esters. This effect occurs even under conditions where new protein synthesis cannot occur, indicating that it takes place via the activation of pre-existing NFκB protein.

These two examples illustrate, therefore, how the ability of transcription factors to act only in a particular tissue can be controlled either by tissue-specific synthesis (Fig. 8.41a) or by tissue-specific activation of pre-existing protein (Fig. 8.41b). These two mechanisms will now be considered.

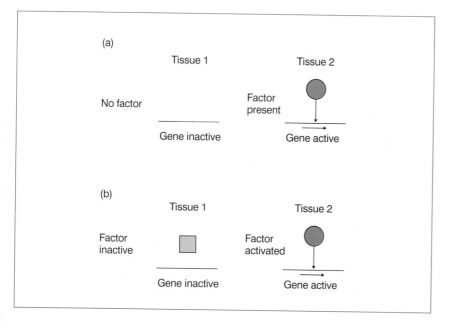

Figure 8.41

Gene activation mediated by the synthesis of a transcription factor only in a specific tissue (a) or its activation in a specific tissue (b).

8.4.2 Regulated synthesis of transcription factors

Clearly, all the various levels of gene regulation such as transcription, splicing, translation, etc., which were discussed in Chapter 3, could be used to regulate the expression of the genes encoding transcription factors, and there is evidence that several of these are used. These will be discussed in turn with illustrative examples.

Regulation of transcription

Although the low abundance of many transcription factors makes the demonstration of transcriptional control difficult, it has been demonstrated in the case of the C/EBPα protein which, as previously described (Section 8.2.4), regulates the transcription of several different liver-specific genes, such as transthyretin and α-1-antitrypsin. This transcription factor is made at high level only in the liver and, in agreement with this, significant transcription of its gene is detectable only in this tissue (Xanthopoulos *et al.*, 1989). Hence, the regulated transcription of the C/EBPα gene controls the production of the corresponding protein which, in turn, directly controls the liver-specific transcription of other genes. Interestingly, such transcriptional control is supplemented by control at the level of mRNA translation which produces activator and repressor forms of the protein (see below).

Regulation of splicing

As discussed in Section 5.2, in mammals alternative splicing is used widely to generate two different forms of a protein, with different functions, from a single gene. A similar theme is seen in the transcription factors, and has been well characterised in the c-*erb*A α gene, which encodes the receptor mediating gene expression in response to thyroid hormone. As discussed in Section 8.3.4, gene activation by this protein is dependent on the presence of thyroid hormone, and in the absence of hormone the receptor represses gene expression. Activation by the receptor is dependent upon the presence of a region in the receptor capable of binding thyroid hormone. Interestingly, however, two alternative forms of the protein exist which are encoded by alternatively spliced mRNAs. One of these (α-1) contains the hormone-binding domain and can mediate gene activation in response to thyroid hormone, while the other form (α-2) contains another protein sequence instead of part of this domain (Fig. 8.42a). Therefore the α-2 form cannot respond to the hormone, although since it contains the DNA binding domain it can bind to the binding site for the receptor in hormone-responsive genes. By doing so, it prevents binding of the α-1 form and hence the induction of the gene in response to thyroid hormone (Fig. 8.42b) (for a review see Foulkes and Sassone-Corsi, 1992). These two alternatively spliced forms of the transcription factor, which are made in different amounts in different tissues, therefore mediate opposing effects on thyroid-hormone-dependent gene expression.

Interestingly, as well as alternative splicing events which produce activators or repressors from the same gene, it is also possible to use alternative splicing to regulate the activity of a specific factor. This is seen in the case of

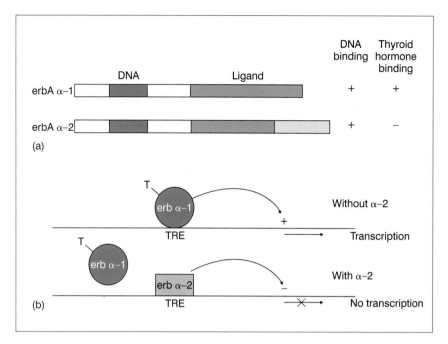

Figure 8.42

(a) Relationship of the erbA α-1 and α-2 proteins. Note that only the α-1 protein has a functional thyroid hormone binding-domain. (b) Inhibition of erbA α-1 binding and of gene activation in the presence of the α-2 protein.

the basic leucine zipper transcription factor Hac1p where alternative splicing produces two different forms of the protein with different C-termini, one of which is more resistant to degradation than the other (Cox and Walter, 1996). In response to the presence of unfolded proteins in the cell, the splicing of the Hac1p RNA is switched to favour the mRNA encoding the protein form which is more resistant to degradation. The level of active Hac1p therefore rises, allowing it to fulfil its role of activating genes encoding proteins which assist other proteins to fold properly.

Regulation of translation

As discussed in Section 5.6.2, the yeast transcription factor GCN4, which induces the expression of genes involved in amino acid biosynthesis in response to amino acid starvation, is itself regulated at the level of translation (for a review see Morris and Geballe, 2000). Thus, when amino acids are lacking, translation is initiated preferentially at the start point for GCN4 production, whereas when amino acids are abundant, translation initiates at short open reading frames upstream of the GCN4 start site and does not re-initiate at the GCN4 site (Fig. 5.34). Hence, GCN4 is synthesised in response to amino acid starvation and activates the genes encoding the enzymes required for the biosynthetic pathways necessary to make good this deficiency.

Such regulation of translation by alternative initiation is also observed in mammalian cells in the case of two transcription factor expressed in the liver (C/EBPα and C/EBPβ (see Section 5.6.2) and supplements the transcriptional control of C/EBPα gene expression described above. In this case, however, following transcription in the liver alternative initiation produces two different forms of each protein, known as liver activator protein (LAP) and liver inhibitor protein (LIP). Only LAP contains the N-terminal activation domain. In contrast, LIP production involves translational initiation at a downstream AUG residue so that this protein lacks the activation domain and contains only the DNA binding domain and the leucine zipper regions which are also found in LAP (Fig. 8.43). LIP is therefore capable of dimerising and binding to DNA but obviously cannot activate transcription (Fig. 8.43). It therefore acts as an inhibitory protein which interferes with the activation of transcription by LAP by competing with it for binding to the DNA. This use of alternative initiation codons to produce different forms of a transcription factor with opposite effects on transcription thus parallels the similar use of alternative splicing to do this in factors such as the thyroid hormone receptor (see above).

8.4.3 Regulated activity of transcription factors

Although, as discussed above, many transcription factors are regulated by controlling their synthesis, a number are controlled by regulating the activity of the protein, which is present in many different cell types. Such a system has obvious advantages, in that it allows a direct effect of the agent inducing gene expression on the activity of the factor, either by binding to it or by modifying the protein, for example, by phosphorylation and hence results in a rapid response. Such post-translational modifications therefore allow specific signal transduction pathways to alter the activity of cellular transcription factors and so produce alterations in gene expression (for reviews see Hill and Treisman, 1995; Treisman, 1996). Examples of these kinds of modifications will now be discussed (Fig. 8.44).

Activation by protein–ligand or protein–protein interaction

Perhaps the simplest mechanism for regulating the activity of transcription factor in response to a specific ligand is for it to bind the ligand directly and undergo a conformational change (Fig. 8.44a). An example of such a protein–ligand interaction activating a transcription factor is provided by the ACE1 factor in yeast, which mediates the induction of the metalloth-

Figure 8.43

Alternative use of different initiation codons results in the generation of the LAP protein which can activate transcription due to its possession of a functional activation domain and the LIP protein which lacks this domain and therefore acts as a repressor.

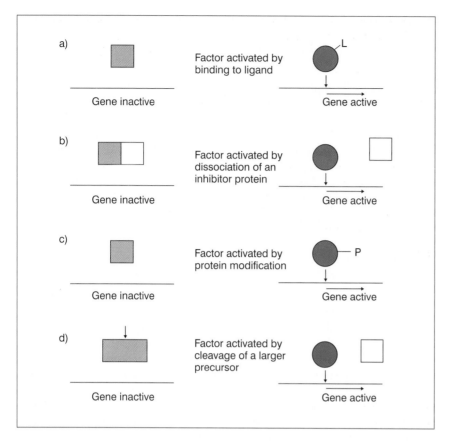

Figure 8.44

Mechanisms by which transcription factors can be activated from an inactive (square) to an active (circle) form by post-translational changes.

ionein gene in response to copper. This protein has been shown to undergo a major conformational change upon binding of copper, which allows it to bind to metallothionein gene regulatory sites and induce transcription. Hence, the activity of this transcription factor is directly modulated by copper, allowing it to mediate gene activation in response to the metal (Fig. 8.45). Similarly, in mammalian cells, the ability of the DREAM transcriptional repressor to bind to DNA is decreased by binding of calcium ions (for a review see Mandel and Goodman, 1999). Hence, in the presence of calcium, the DREAM repressor is removed from the DNA and its target genes are therefore activated in response to calcium.

A similar case of a ligand-induced regulation is seen with the thyroid hormone receptor (see Section 8.3.4). In this case, however, rather than affecting the ability of the receptor to bind to DNA, the ligand induces a conformational change in the receptor which allows it to activate gene expression in response to thyroid hormone treatment by binding a co-activator rather than a co-repressor (Fig. 8.40).

In contrast, however, other members of the steroid–thyroid hormone receptor family do not bind to DNA in the absence of the appropriate hormone. This is seen in the case of the glucocorticoid receptor, for

Figure 8.45

Activation of the ACE1 factor in response to copper results in transcription of the metallothionein gene.

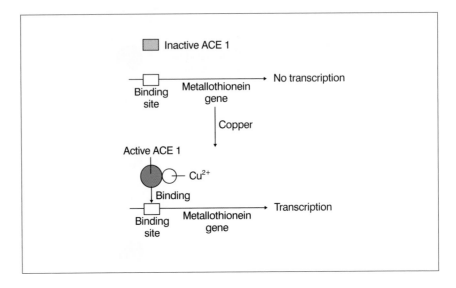

example, where the receptor is normally located in the cytoplasm and only binds to DNA and activates transcription when the hormone is added. Originally it was thought that, as with ACE1, binding of hormone to the receptor activated its ability to bind to DNA and switch on transcription of hormone-responsive genes. However, it has been shown that, although the receptor binds to DNA only in the presence of the hormone in the cell, in the test tube it will bind even when no hormone is present. This has led to the idea that in the cell the receptor is prevented from binding to DNA by its association with another protein, and that the hormone acts to release it from this association and allow it to fulfil its inherent ability to bind to DNA.

In agreement with this idea, the glucocorticoid receptor, unlike the thyroid hormone receptor, has been shown to be associated in the cytoplasm with a 90 000 molecular weight heat-inducible protein (hsp90). Upon steroid binding the receptor dissociates from the hsp90 and moves to the nucleus, where it activates gene transcription (Fig. 8.46). Hence, the transcription factor is activated by hormone not by a protein–ligand interaction but by disruption of a protein–protein interaction which inhibits the inherent DNA binding ability of the receptor (Fig. 8.44b).

Activation by protein modification or proteolytic cleavage
Many transcription factors are modified extensively by the addition, for example, of O-linked monosaccharide residues or by phosphorylation (for a review see Hunter and Karin, 1992). Such modifications represent obvious targets for agents that induce gene activation. Thus, such agents could act by altering the activity of a modifying enzyme, such as a kinase, which in turn would act on the transcription factor, resulting in its activation and the switching on of gene expression (Fig. 8.44c). Thus, for example, it has been shown that the activity of the STAT-1 transcription factor is modulated by methylation of a specific arginine residue (Mowen et al., 2001) whilst acetylation enhances

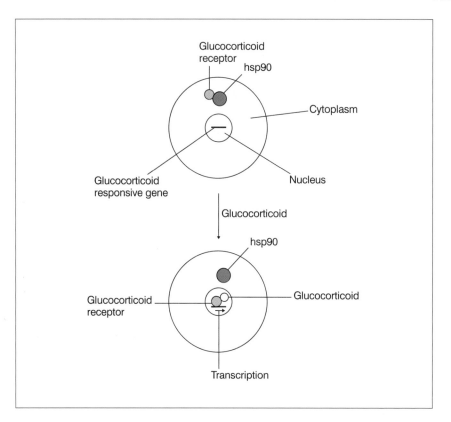

Figure 8.46

Binding of glucocorticoid to the glucocorticoid receptor results in its dissociation from hsp90 and movement of the hormone–receptor complex to the nucleus where it activates the transcription of glucocorticoid-responsive genes.

the DNA binding ability and hence the activity of the p53 transcription factor (Luo *et al.*, 2000) (see Section 9.5.2). Hence, processes such as methylation and acetylation can affect transcription factors directly as well as targeting the histones within the chromatin structure, as discussed in Section 6.6.

However, the most extensive modification which affects transcription factors is phosphorylation. One of the best-characterised examples of this involves the mammalian CREB factor, which mediates the induction of several genes in response to treatment with cyclic AMP and binds to the cyclic AMP response element in these genes (Table 7.2) (for reviews see de Cesare *et al.*, 1999; Shaywitz and Greenberg, 1999; de Cesare and Sassone-Corsi, 2000). The ability of CREB to stimulate transcription is enhanced greatly by its phosphorylation by protein kinase A on the serine residue located at amino acid 133 of the protein. Such phosphorylation stimulates the activity of the activation domain located in this region of the protein. Hence, the activation of gene expression by cyclic AMP is mediated by its activation of protein kinase A which, in turn, phosphorylates and activates CREB on serine residue 133. CREB is bound to the cyclic AMP response element (CRE) both before and after cyclic AMP treatment. However, as described in Section 8.3.3, phosphorylation of CREB allows it to bind the CBP co-activator and thereby recruit it to DNA and activate transcription in a cyclic AMP-dependent manner (Fig. 8.47).

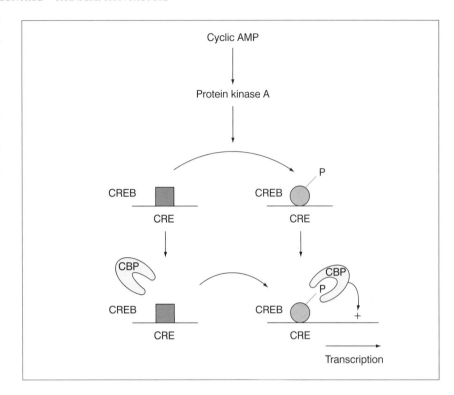

A more complex pattern of activation is seen in the case of the heat shock factor (HSF-1). As described in Section 7.2.2, HSF-1 is present in unstressed cells but becomes activated in response to exposure to stresses such as elevated temperature and then activates transcription of genes containing the HSE to which it binds. This activation proceeds in two stages (for a review see Morimoto, 1998) (Fig. 8.48). Firstly, HSF-1 is converted from a monomer to a trimer of three HSF-1 molecules which can bind to the HSE. As with the glucocorticoid receptor, the activation of HSF-1 involves the dissociation of hsp90 which binds to HSF-1 in unstressed cells and maintains it in an inactive form which cannot bind to DNA (Zou *et al.*, 1998).

Interestingly, hsp90, like many of the proteins inducible by heat or other stresses functions as a so-called "chaperone" protein, promoting the folding of unfolded or poorly folded proteins. Clearly, heat or other stresses will increase the amount of such proteins, with hsp90 being "called away" to deal with them and thus releasing HSF-1. Hence, this system neatly couples the effect of the inducing stimulus in producing unfolded proteins with the activation of HSF-1.

However, although trimerisation allows HSF1 to bind to the HSE, this is not sufficient to promote transcriptional activation. As in the case of CREB, HSF-1 must be phosphorylated (on serine 230) to allow it to activate transcription (Holmberg *et al.*, 2001). This system therefore combines the disruption of a protein–protein complex which is observed in the case of the glucocorticoid receptor with regulation by phosphorylation as in the case of CREB.

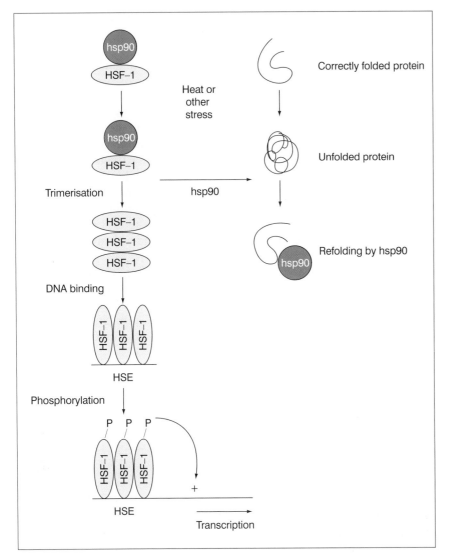

Figure 8.48

In non-stressed cells, HSF-1 exists in a non-DNA binding form associated with hsp90. In response to stresses such as elevated temperatures, hsp90 dissociates from HSF-1 and binds to unfolded proteins produced as a result of the stress. This allows HSF-1 to form a trimer which can bind to DNA but which needs to be further modified by phosphorylation before it can activate transcription.

As well as regulating the CREB/CBP interaction and the activation of HSF-1 the regulation of phosphorylation is also likely to be involved in the activation of NFκB in response to phorbol ester treatment, as described above (Section 8.4.1). In this case, however, the target for phosphorylation is not the transcription factor itself but rather an associated protein, IκB, which inhibits NFκB activity. Prior to phorbol ester treatment, NFκB is bound to IκB as an inactive complex which is located in the cytoplasm. Following treatment, however, the IκB is phosphorylated and this renders it a target for proteolytic enzymes which rapidly degrade it. The free NFκB then moves to the nucleus where it activates gene expression (for reviews see Foo and Nolan, 1999; Karin and Ben-Neriah, 2000; Perkins, 2000). Such activation is clearly similar to the dissociation of the glucocorticoid receptor from hsp90 following steroid treatment, although in this case phosphorylation rather than steroid binding dissociates the protein–protein complex.

It is clear therefore that transcription factors can be regulated by differences in the rate of degradation of the protein such as occurs for phosphorylated IκB or for the two different forms of the Hac1p factor generated by alternative splicing of its mRNA (see Section 8.4.2). Similarly, as will be discussed in Section 9.5.2, the p53 protein is rapidly degraded following binding to the MDM2 protein. In addition however, transcription factor activity can also be regulated by proteolytic cleavage of an inactive precursor (Fig. 8.44d). Thus an NFκB-related protein p105 is synthesised as an inactive precursor in which the active part of the molecule is inhibited by an IκB-like region to which it is linked in a single protein. Following exposure to an activating stimulus, the IκB-like region is phosphorylated as for IκB itself and the precursor molecule is cleaved to release the NFκB-like region which is now able to move to the nucleus and activate gene expression (Fig. 8.49).

Such regulation by protein precursor cleavage is not confined to the p106 system (for a review see Pahl and Baeurele, 1996). It is also seen, for example, in the SREBP proteins which activate gene expression in response to elevated cholesterol (for a review see Brown and Goldstein, 1997). Thus, in the absence of cholesterol, SREBP is synthesised as an inactive precursor

Figure 8.49

Activation mechanisms in the NFκB family. (a) Prior to exposure to an activating signal, the NFκB factor exists as an inactive form complexed to the distinct IκB inhibitory protein. In response to an activating signal the IκB factor is phosphorylated (P), resulting in the release of free NFκB which activates transcription. (b) In the p105 protein the NFκB and IκB-like regions are part of a single large precursor which is inactive. Following exposure to an activating signal, the IκB-like region is phosphorylated and the protein is cleaved releasing the NFκB-like region which is free to activate transcription.

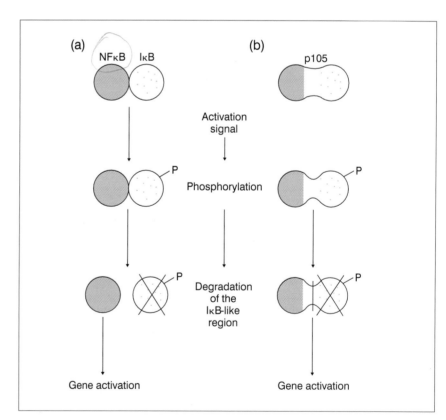

which is anchored within the cytoplasm to the endoplasmic reticulum membrane by a specific region of the protein. Following exposure to cholesterol, this region is cleaved off, freeing the active region and allowing it to move to the nucleus and activate its target genes.

Interestingly, as in the case of SREBP, in many of the cases we have discussed, the regulatory process results in the movement of the transcription factor into the nucleus where it can activate its target genes. This is seen, for example, for the glucocorticoid receptor following its release from hsp90 or for the NFκB factor following release from IκB. An interesting example of such regulated entry into the nucleus which does not involve either disruption of a protein–protein interaction or proteolytic cleavage is seen in the case of the Tubby transcription factor (for a review see Cantley, 2001).

Thus, the Tubby factor has been shown to bind to phospholipid in the plasma membrane and is therefore anchored in this position via a protein–lipid interaction rather than a protein–protein interaction (Fig. 8.50). Following the activation of G protein receptors by specific stimuli, a phospholipase enzyme is activated. In turn, this enzyme cleaves the anchoring

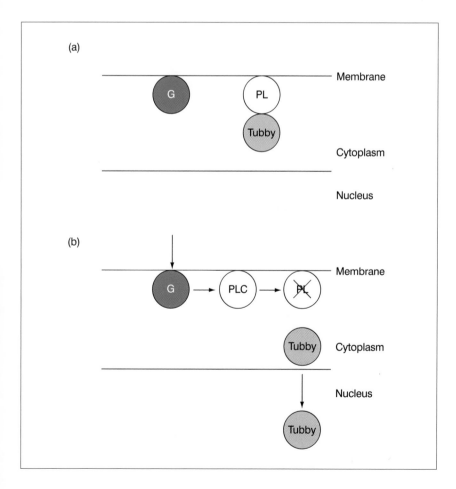

Figure 8.50

(a) The Tubby transcription factor is anchored in the cell membrane by interaction with a phospholipid (PL). (b) Following G protein-coupled receptor activation, the phospholipase C enzyme (PLC) is activated and cleaves the phospholipid. This releases Tubby and allows it to enter the nucleus and activate transcription.

phospholipid, releasing Tubby and allowing it to move to the nucleus and activate its target genes.

In this case nuclear localisation is therefore achieved by regulation of a protein–lipid interaction rather than by a protein–protein interaction or proteolytic cleavage. Hence, regulation of nuclear localisation by a number of different mechanisms is a widely used means of regulating transcription factor activity. Of course, as we have seen, it is not the only mechanism with, for example, direct regulation of DNA binding activity by the ligand being seen in the case of the ACE1 factor, regulation of the ability of an already DNA bound transcription factor to activate transcription being observed in the case of CREB and regulation of protein degradation being seen in the case of the Hac1p factor (see Section 8.4.2) (Fig. 8.51).

Hence, a variety of mechanisms exist for the activation of transcription factors (Fig. 8.44) with many cases involving several of these processes so that, for example, the activation of the glucocorticoid receptor involves both ligand binding and protein–protein interaction while the NFκB cases involve phosphorylation and protein–protein interaction or proteolytic cleavage. Moreover, these mechanisms can affect the transcription factor in a number

Figure 8.51

Regulatory processes can target the degradation of a transcription factor (a), its location in the cell (b), its ability to bind to DNA (c) or its ability to activate transcription following DNA binding (d).

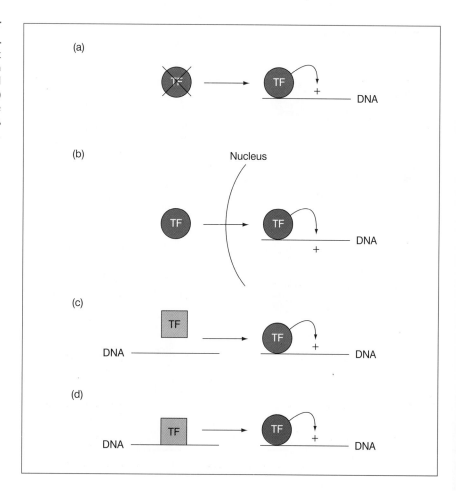

of different ways, regulating, for example, its degradation/stabilisation, its location in the cell, its DNA binding ability or the ability of a DNA bound factor to activate transcription (Fig. 8.51).

8.5 CONCLUSIONS

Although the existence of transcription factors had been known or inferred for some time, it is clear that the cloning of the genes encoding these factors has increased our knowledge of their functional properties greatly. A number of different motifs which mediate DNA binding, protein dimerisation and transcriptional activation have been identified and these are listed in Table 8.3.

Interestingly, unlike most cases of gene regulation, the regulation of transcription factors is often modulated by post-transcriptional processes, including protein modification. Such post-transcriptional regulation is

Domain	Role	Factors containing domain	Comments
Homeobox	DNA binding	Numerous *Drosophila* homeotic genes, related genes in other organisms	DNA binding mediated via helix-turn-helix motif
Cysteine–histidine zinc finger	DNA binding	TFIIIA, Krüppel, SP1 etc.	Multiple copies of finger motif
Cysteine–cysteine zinc finger	DNA binding	Steroid–thyroid hormone receptor family	Single pairs of fingers, related motifs in Adenovirus E1A and yeast GAL4, etc.
Basic element	DNA binding	C/EPB, c-fos, c-jun, GCN4	Often found in association with leucine zipper
Leucine zipper DNA	Protein dimerisation	C/EPB, c-fos, c-jun GCN4, c-myc	Mediates dimerisation which is essential for binding by adjacent domain
Helix-loop-helix DNA	Protein dimerisation	c-myc, *Drosophila* daughterless, MyoD, E12, E47	Mediates dimerisation which is essential for binding by adjacent domain
Acidic region	Gene activation	Yeast GCN4, GAL4, steroid-thyroid receptors, etc.	May form α-helical structure
Glutamine-rich region	Gene activation	SP1	Related regions in Oct-1, Oct-2, AP2, etc.
Proline-rich region	Gene activation	CTF/NF1	Related regions in AP2, c-jun, Oct-2

Table 8.3

Transcription factor domains

sensible in that transcriptional regulation of the genes encoding these factors would only set the process one step back, leading to the problem of how these genes are themselves transcriptionally regulated. In contrast, regulation at the post-transcriptional level allows a direct response to particular stimuli and results, for example, in the rapid activation of the glucocorticoid receptor by dissociation from hsp90 in response to a hormone, and the rapid activation of CREB by phosphorylation in response to cyclic AMP.

Our understanding of the functional domains of these factors and of their activation now opens the way for the detailed molecular modelling of how these factors interact with each other, with co-activators and co-repressors and with the RNA polymerase itself, as well as other components of the basal transcriptional complex. Ultimately, this will allow an elucidation of the way in which they stimulate transcription.

REFERENCES

Asturias, F.J., Jiang, Y.W., Myers, L.C., Gustafsson, C.M. and Kornberg, R.D. (1999). Conserved structures of mediator and RNA polymerase II holoenzyme. *Science* **283**, 985–987.

Bieker, J J. (2001). Krüppel-like factors: three fingers in many pies. *Journal of Biological Chemistry* **276**, 34355–34358.

Brennan, R.G. (1993). The winged-helix DNA-binding motif: another helix-turn-helix take off. *Cell* **74**, 773–776.

Brown, M.S. and Goldstein, J.L. (1997). The SREBP pathway: regulation of cholesterol metabolism by proteolysis of a membrane-bound transcription factor. *Cell* **89**, 331–340.

Buckingham, M. (1994). Which myogenic factors make muscle? *Current Biology* **4**, 61–63.

Cantley, L.C. (2001). Translocating Tubby. *Science* **292**, 2019–2021.

Cox, J.S. and Walter, P. (1996). A novel mechanism for regulating the activity of a transcription factor that controls the unfolded protein response. *Cell* **87**, 391–404.

De Cesare, D., Fimia, G.M. and Sassone-Corsi, P. (1999). Signalling routes to CREM and CREB: plasticity in transcriptional activation. *Trends in Biochemical Sciences* **24**, 281–285.

De Cesare, D. and Sassone-Corsi, P. 2000. Transcriptional regulation by cyclic AMP-responsive factors. *Progress in Nucleic Acids Research and Molecular Biology* **64**, 343–369.

Donaldson, W., Peterson, J.M., Graves, B.J. and McIntosh, L.P. (1996). Solution structure of the ETS domain from murine Ets-1: a winged helix-turn-helix DNA-binding motif. *EMBO Journal* **15**, 125–134.

Duboule, D. (2000). A Hox by any other name. *Nature* **403**, 607–610.

Edelman, G.M. and Jones, F.S. (1993). Outside and downstream of the homeobox. *Journal of Biological Chemistry* **268**, 20683–20686.

Ellenberger, T. (1994). Getting a grip on DNA recognition – structures of the basic region leucine zipper and the basic region helix-loop-helix DNA-binding domains. *Current Opinion in Structural Biology* **4**, 12–21.

Evans, R.M. and Hollenberg, S.M. (1988). Zinc fingers: gilt by association. *Cell* **52**, 1–3.

Foo, S.Y. and Nolan, G.P. (1999). NF-κB to the rescue RELs, apoptosis and cellular transformation. *Trends in Genetics* **15**, 229–235.

Foulkes, N.S. and Sassone-Corsi, P. (1992). More is better: activator and repressor from the same gene. *Cell* **68**, 411–414.

Freedman, L.P. (1999). Increasing the complexity of co-activation in nuclear receptor signalling. *Cell* **97**, 5–8.

Gajiwala, K.S., Chen, H., Cornille, F., Roques, B.P., Reith, W., Mach, B. and Burley, S.K. (2000). Structure of the winged-helix protein hRFX1 reveals a new mode of DNA binding. *Nature* **403**, 916–921.

Gehring, W.J., Affolter, M. and Burglin, T. (1994). Homeodomain proteins. *Annual Review of Biochemistry* **63**, 487–526.

Glass, C.K. and Rosenfeld, M.G. (2000). The coregulator exchange in transcriptional functions of nuclear receptors. *Genes and Development* **14**, 121–141.

Goding, C.R. and O'Hare, P. (1989). Herpes simplex virus Vmw65-octamer binding protein interaction: a paradigm for combinatorial control of transcription. *Virology* **173**, 363–367.

Goodman, R.H. and Smolik, S. (2000). CBP/p300 in cell growth, transformation and development. *Genes and Development* **14**, 1553–1577.

Green, M.R. (2000). TBP-associated factors (TAF$_{II}$s): multiple, selective transcriptional mediators in common complexes. *Trends in Biochemical Sciences* **25**, 59–63.

Gronemeyer, H. and Laudet, V. (1995). Nuclear receptors. *Protein Profile* **2**, 1173–1308.

Gronemeyer, H. and Moras, D. (1995). How to finger DNA. *Nature* **375**, 190–191.

Hahn, S. (1993). Efficiency in activation. *Nature* **363**, 672–673.

Hahn, S. (1998). The role of TAFs in RNA polymerase II transcription. *Cell* **95**, 579–582.

Hanna-Rose, W. and Hansen, U. (1996). Active repression mechanisms of eukaryotic transcription repressors. *Trends in Genetics* **12**, 229–234.

Hill, C.S. and Treisman, R. (1995). Transcriptional regulation by extracellular signals: mechanisms and specificity. *Cell* **80**, 199–211.

Holmberg, C.I., Hietakangas, V., Mikhailov, A., Rantanen, J.O., Kallio, M., Meinander, A., Hellman, J., Morrice, N., MacKintosh, C., Morimoto, R.I., Eriksson, J.E. and Sistonen, L. (2001). Phosphorylation of serine 230 promotes inducible transcriptional activity of heat shock factor 1. *EMBO Journal* **20**, 3800–3810.

Hunter, T. and Karin, M. (1992). The regulation of transcription by phosphorylation. *Cell* **70**, 375–387.

Hurst, H. (1996). Leucine zippers. *Protein Profile* **3**, 1–72.

Jones, N. (1990). Transcriptional regulation by dimerization: two sides to an incestuous relationship. *Cell* **61**, 9–11.

Kadonaga, J.T., Carner, K.R., Masiarz, F.R. and Tjian, R. (1987). Isolation of cDNA encoding the transcription factor Sp1 and functional analysis of the DNA-binding domain. *Cell* **51**, 1079–1090.

Kadonaga, J.T. and Tjian, R. (1986). Affinity purification of sequence-specific DNA-binding proteins. *Proceedings of the National Academy of Sciences USA* **83**, 5889–5893.

Karin, M. and Ben-Neriah, Y. (2000). Phosphorylation meets Ubiquitination: the control of NF-κB activity. *Annual Reviews of Immunology* **18**, 621–663.

Kerppola, T. and Curran, T. (1995). Zen and the art of Fos and Jun. *Nature* **373**, 199–200.

Kingston, R.E. (1999). A shared but complex bridge. *Nature* **399**, 199–200.

Klug, A. and Schwabe, J.W.R. (1995). Zinc fingers. *FASEB Journal* **9**, 597–604.

Knoepfler, P.S. and Eisenman, R.N. (1999). Sin meets NuRD and other tails of repression. *Cell* **99**, 447–450.

Kornberg, T.B. (1993). Understanding the homeodomain. *Journal of Biological Chemistry* **268**, 26813–26816.

Krumlauf, R. (1994). Hox genes in vertebrate development. *Cell* **78**, 191–201.

Kuras, L. and Struhl, K. (1999). Binding of TBP to promoters *in vivo* is stimulated by activators and requires Pol II holoenzyme. *Nature* **399**, 609–613.

Lamb, P. and McKnight, S.L. (1991). Diversity and specificity in transcriptional regulation: the benefits of heterotypic dimerization. *Trends in Biochemical Sciences* **16**, 417–422.

Latchman, D.S. (1996). Inhibitory transcription factors. *International Journal of Biochemistry and Cell Biology* **28**, 965–974.

Latchman, D.S. (1998). *Eukaryotic Transcription Factors*. 3rd edition. Academic Press, London.

Latchman, D.S. (ed.). (1999). *Transcription Factors: A Practical Approach*. 2nd edition. Oxford University Press, Oxford.

Lawrence, P.A. and Morata, G. (1994). Homeobox genes: their function in *Drosophila* segmentation and pattern formation. *Cell* **78**, 181–189.

Li, X.-Y., Virbasius, A., Zhu, X. and Green, M.R. (1999). Enhancement of TBP binding by activators and general transcription factors. *Nature* **399**, 605–609.

Littlewood, T.D. and Evan, G.I. (1995). Helix-loop helix. *Protein Profile* **2**, 621–702.

Luo, J., Chen, D., Shiloh, A. and Gu, W. (2000). Deacetylation of p53 modulates its effect on cell growth and apoptosis. *Nature* **408**, 377–381.

Malik, S. and Roeder, R.G. (2000). Transcriptional regulation through mediator-like coactivators in yeast and metazoan cells. *Trends in Biochemical Sciences* **25**, 277–283.

Mandel, G. and Goodman, R.H. (1999). DREAM on without calcium. *Nature* **398**, 29–30.

Mangelsdorf, D.J. and Evans, R.M. (1995). The RXR heterodimers and orphan receptors. *Cell* **83**, 841–850.

Mangelsdorf, D.J., Thummel, C., Beato, M., Herrlich, P., Schutz, G., Umesono, K., Blumberg, B., Castner, P., Mark, M., Shambon, P. and Evans, R.M. (1995). The nuclear receptor super family; the second decade. *Cell* **83**, 835–839.

Mansouri, A., Hallonet, M. and Gruss, P. (1996). Pax genes and their roles in cell differentiation and development. *Current Opinion in Cell Biology* **8**, 851–857.

Miller, J., McLachlan, A.D. and Klug, A. (1985). Repetitive zinc-binding domains in the protein transcription factor III A from *Xenopus* oocytes. *EMBO Journal* **4**, 1609–1614.

Mitchell, P.J. and Tjian, R. (1989). Transcriptional regulation in mammalian cells by sequence specific DNA binding proteins. *Science* **245**, 371–378.

Morimoto, R.I. (1998). Regulation of the heat shock transcriptional response: cross talk between a family of heat shock factors, molecular chaperones and negative regulators. *Genes and Development* **12**, 3788–3796.

Morris, D.R. and Geballe, A.P. (2000). Upstream open reading frames as regulators in mRNA translation. *Molecular and Cellular Biology* **20**, 8635–8642.

Mowen, K.A., Tang, J., Zhu, W., Schurter, B.T., Shuai, K., Herschman, H.R. and David, M. (2001). Arginine methylation of STAT1 modulates IFNα/β-induced transcription. *Cell* **104**, 731–741.

Myers, L.C. and Kornberg, R.D. (2000). Mediator of transcriptional regulation. *Annual Reviews of Biochemistry* **69**, 729–749.

Ogryzkio, V.V., Schiltz, R.L., Russanova, V., Howard, B.H. and Nakatani, Y. (1996). The transcriptional coactivators p300 and CBP are histone acetyltransferases. *Cell* **87**, 953–959.

Olson, E.N. and Klein, W.H. (1994). bHLH factors in muscle development: dead lines and commitments what to leave in and what to leave out. *Genes and Development* **8**, 1–8.

Pahl, H.L. and Baeurele, P.A. (1996). Control of gene expression by proteolysis. *Current Opinion in Cell Biology* **8**, 340–347.

Perkins, N.D. (2000). The Rel/NF-κB family: friend and foe. *Trends in Biochemical Sciences* **25**, 434–440.

Pirrotta, V. (1997). PcG complexes and chromatin silencing. *Current Opinion in Genetics and Development* **7**, 249–258.

Ptashne, M. and Gann, A. (1997). Transcriptional activation by recruitment. *Nature* **386**, 569–577.

Ranish, J.A. and Hahn, S. (1996). Transcription factors and activators. *Current Opinion in Genetics and Development* **6**, 151–158.

Rhodes, D. and Klug, A. (1993). Zinc finger structure. *Scientific American* **268**, 32–39.

Rosenfeld, M.G. and Glass, C.K. (2001). Coregulator codes of transcriptional regulation by nuclear receptors. *Journal of Biological Chemistry* **276**, 36865–36868.

Ryan, A.K. and Rosenfeld, M.G. (1997). POU domain family values:- flexibility, partnerships and developmental codes. *Genes and Development* **11**, 1207–1225.

Schwabe, J.W.R. and Rhodes, D. (1991). Beyond zinc fingers: steroid hormone receptors have a novel structural motif for DNA recognition. *Trends in Biochemical Sciences* **16**, 291–296.

Shaywitz, A.J. and Greenberg, M.E. (1999). CREB: A stimulus-induced transcription factor activated by a diverse array of extracellular signals. *Annual Reviews of Biochemistry* **68**, 821–861.

Shikama, N., Lyon, J. and La Thangue, N.B. (1997). The p300/CBP family:integrating signals with transcription factors and chromatin. *Trends in Cell Biology* **7**, 230–236.

Simon, J. (1995). Locking in stable states of gene expression: transcriptional control during *Drosophila* development. *Current Opinion in Cell Biology* **7**, 376–385.

Singh, H., LeBowitz, J.H, Baldwin, A.S. and Sharp, P.A. (1988). Molecular cloning of an enhancer binding protein isolation by screening of an expression library with a recognition site DNA. *Cell* **52**, 415–429.

Smith, R.L. and Johnson, A.D. (2000). Turning genes off by Ssn6-Tup1: a conserved system of transcriptional repression in eukaryotes. *Trends in Biochemical Sciences* **25**, 325–330.

Stargel, L.A. and Struhl, K. (1996). Mechanisms of transcriptional activation *in vivo*: two steps forwards. *Trends in Genetics* **12**, 311–315.

Travers, A. (1993). *DNA-Protein Interactions*. Chapman and Hall, London.

Treisman, R. (1996). Regulation of transcription by MAP kinase cascades. *Current Opinion in Cell Biology* **8**, 205–215.

Triezenberg, S.J. (1995). Structure and function of transcriptional activation domains. *Current Opinion in Genetics and Development* **5**, 190–196.

Turner, J. and Crossley, M. (1999). Mammalian Krüppel-like transcription factors: more than just a pretty finger. *Trends in Biochemical Sciences* **24**, 236–240.

Uesugi, M., Nyanguile, O., Lu, H., Levine, A.J. and Verdine, G.L. (1997). Induced alpha helix in the VP16 activation domain upon binding to a human TAF. *Science* **277**, 1310–1313.

Umesono, K. and Evans, R.M. (1989). Determinants of target gene specificity for steroid/thyroid hormone receptors. *Cell* **57**, 1139–1146.

Verrijzer, C.P. and Van der Vliet, P.C. (1993). POU domain transcription factors. *Biochimica et Biophysica Acta* **1173**, 1–21.

Weatherman, R.V., Fletterick, R.J. and Scanlan, T.S. (1999). Nuclear-receptor ligands and ligand-binding domains. *Annual Reviews of Biochemistry* **68**, 559–581.

Xanthopoulos, K.G., Mirkovitch, J., Decker, T., Kuo, C.F. and Darnell, J.E. Jr. (1989). Cell-specific transcriptional control of the mouse DNA-binding protein mc/EBP. *Proceedings of the National Academy of Sciences USA* **86**, 4117–4121.

Zou, J., Guo, Y., Guettouche, T., Smith, D.F. and Voellmy, R. (1998). Repression of heat shock transcription factor HSF1 activation by HSP90 (HSP90 complex) that forms a stress-sensitive complex with HSF1. *Cell* **94**, 471–480.

GENE REGULATION AND HUMAN DISEASE 9

9.1 GENE REGULATION AND HUMAN DISEASE

As we have discussed in the preceding chapters, the regulation of gene expression in higher eukaryotes is a highly complex process. It is not surprising, therefore, that this process can go wrong, and the identification of the molecular basis of many human diseases has shown some to be due to defects in gene regulation. A number of different processes have been shown to be affected in different human diseases and these will be discussed in turn (for reviews see Engelkamp and Van Heyningen, 1996; Latchman, 1996) (Fig. 9.1). In discussing these diseases, it should be noted that only gene mutations which are compatible with life will manifest as human diseases, however severe these may be. Mutations in factors which are incompatible with survival at least to birth will evidently not be detected and it is likely that many mutations affecting gene regulation fall into this category.

Transcription factors (Fig. 9.1a)
A number of congenital diseases which are detectable at birth or shortly after involve mutations in the genes encoding specific transcription factors. Thus, for example, mutations in individual members of the PAX transcription factor family discussed in Section 8.2.5 have been shown to be involved in a number of congenital eye disorders, while mutations in the gene encoding the Pit-1 member of the POU family of transcription factors (see Section 8.2.2) result in a failure of pituitary gland development and consequent dwarfism in both mice and humans.

As well as such development defects, caused by mutations affecting transcription factors, mutations in the genes encoding the members of the steroid/thyroid hormone receptor family discussed in Section 8.2.3 can result in a failure to respond to the specific hormone which normally binds to the receptor and thereby regulates gene expression. Thus, for example, mutations in the gene encoding the glucocorticoid receptor result in a syndrome of steroid resistance in which the patients do not respond to glucocorticoid.

Interestingly, mutations in the peroxisome proliferator-activated receptor gamma (PPARγ), which is also a member of the steroid/thyroid hormone receptor family, have been found in a few human individuals with insulin resistance leading to type 2 diabetes. Although such cases are rare, they suggest that PPARγ plays a key role in insulin responses and that it may be a valuable therapeutic target to enhance such responses in other cases of diabetes and in obesity (for reviews see Schwartz and Kahn, 1999; Kersten *et al.*, 2000).

Figure 9.1

Mutations can affect a number of components of the gene expression processes including (a) transcriptional activators (A); (b) co-activators (CA); (c) the basal transcriptional complex (BTC); (d) chromatin modelling complexes (CMC); or (e) factors involved in RNA processing.

Co-activators (Fig. 9.1b)

As described in Section 8.3.3, co-activators play a key role in the action of transcriptional activators. In particular, the CBP co-activator is of vital importance in a number of different signalling pathways, linking the transcription factor which is activated by the pathway to either the basal transcriptional complex or altering chromatin structure. It is not surprising therefore that mutations which result in individuals lacking functional CBP are incompatible with life. Indeed, even having a single inactive CBP gene whilst retaining a single functional gene results in the severe human disease Rubinstein–Taybi syndrome which is characterised by mental retardation and physical abnormalities (for a review see D'Arcangelo and Curran, 1995).

Hence, even if a single CBP gene remains functional, the inactivation of the second copy produces disease. This is likely to be due to the amount of CBP in normal cells being low with various transcription factors competing for it (see Section 8.3.3). Hence, the reduction in the level of CBP due to loss of one gene copy results in disease (Fig. 9.2).

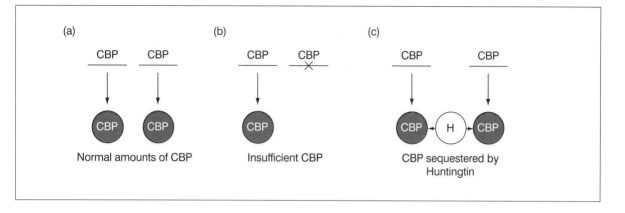

(a)   (b)   (c)

Normal amounts of CBP   Insufficient CBP   CBP sequestered by Huntingtin

Interestingly, CBP can also be involved in the disease processes even when it is intact and unmutated. Thus, it has been shown that individuals with the neurodegenerative disease Huntington's chorea produce an abnormal form of a protein known as Huntingtin. This abnormal form of Huntingtin binds to the normal CBP protein and prevents it from regulating transcription by depositing it in insoluble aggregates (Nucifora *et al.*, 2001) (Fig. 9.2).

Basal transcriptional complex (Fig. 9.1c)

One of the major targets for activators and co-activators is the basal transcriptional complex (see Section 8.3.3). Given the critical role for this complex (see Section 3.2.3), it is likely that most mutations in components of the complex would be incompatible with survival to birth and would not therefore be observed as post-natal human diseases. However, mutations in components of TFIIH have been observed in the human disease Xeroderma pigmentosum, which results in skin defects and a higher risk of cancer and several of these mutations have been shown to result in defective responses to transcriptional activators and repressors (Liu *et al.*, 2001).

Chromatin remodelling factors (Fig. 9.1d)

As discussed in Chapter 6, changes in chromatin structure are an essential prerequisite for activation of transcription. It is not surprising therefore that human diseases result when these processes are disrupted. Thus, mutations in the MeCP2 factor which prevent it from specifically recognising methylated DNA result in the neurological disease Rett syndrome (see Section 6.5.3).

In addition, as well as resulting from alteration in methylation recognition proteins, disease can also result from changes in methylation patterns on the DNA. Thus, in the human fragile X syndrome a triplet sequence CGG which is present in 10–50 tandem copies in the first exon of the FMR-1 gene is amplified so that over 230 tandem copies are present in patients with this syndrome. This amplified sequence is then heavily methylated on the C residues (see Section 6.5) resulting in transcriptional silencing of the FMR-1 gene which in turn produces the mental retardation and other symp-

Figure 9.2

To function effectively each cell needs to have two intact genes encoding the CBP co-activator protein (panel a). Specific diseases result when one gene is inactive resulting in reduced levels of CBP (panel b) or when CBP is intact but is sequestered by an abnormal Huntingtin protein (H) so that it cannot fulfil its co-activator role (panel c).

toms characteristic of the fragile X syndrome (for a review see Trottier *et al.*, 1993). Such altered methylation patterns have also been observed in other human diseases such as cancer (for a review see Baylin and Herman, 2000).

Other aspects of chromatin remodelling can also be affected in human diseases. Thus, a mutation in the SNF2 factor, which is part of the chromatin remodelling SWI/SNF complex discussed in Section 6.7.2, results in a severe form of α-thalassaemia that not only results in lack of globin gene expression as in other thalassaemias (see, for example, Section 7.5.1) but also leads to other symptoms, such as mental retardation, indicating that the activity of this factor is necessary for opening the chromatin structure of both the α-globin genes and a number of other genes, so preventing their transcription if it is absent (Gibbons *et al.*, 1995).

Post-transcriptional events (Fig. 9.1e)

As well as affecting transcriptional processes, mutations can also affect gene regulation at post-transcriptional stages. Thus, for example, the failure to produce one of the two alternatively spliced mRNAs derived from the porphobillinogen deaminase gene is the cause of one form of the disease acute intermittent porphyria (Grandchamp *et al.*, 1989).

Similarly, the SMN gene which was originally defined on the basis of its mutation in the disease spinal muscular atrophy has now been shown to encode a protein which is essential for RNA splicing (Pellizzoni *et al.*, 1998). Hence diseases can result both from mutations in RNA splicing proteins and from mutations in specific target RNAs which affect their splicing pattern.

Although a wide variety of human diseases arise due to defects in gene regulation, the human disease that exhibits the most extensive malregulation of gene expression is cancer. Thus, not only does cancer often result from the over-expression of certain cellular genes (known as proto-oncogenes) due to errors in their regulation, but several of these genes themselves actually encode transcription factors and cause the disease by affecting the expression of other genes. The remainder of this chapter will therefore focus on the connection between cancer and the malregulation of gene expression, and will illustrate how our increasing knowledge of this connection has aided our understanding both of the disease and of the processes that regulate gene expression in normal and transformed cancer cells.

9.2 PROTO-ONCOGENES

In order to provide a background to our discussion of cancer and gene regulation, it is necessary to discuss briefly the nature of cancer-causing oncogenes and the process that led to their discovery (for reviews see Broach and Levine, 1997; Hunter, 1997).

As long ago as 1911, Peyton Rous showed that a connective tissue cancer in the chicken was caused by an infectious agent. This agent was subsequently shown to be a virus and was named Rous Sarcoma virus (RSV) after its discoverer. This cancer-causing, tumorigenic virus is a member of a class of viruses called retroviruses, whose genome consists of RNA rather than DNA, as in most other organisms.

The majority of viruses of this type do not cause cancers but simply infect a cell and produce a persistent infection with continual production of virus by the infected cell. In the case of RSV, however, such infection also results in the conversion of the cell into a cancer cell, capable of indefinite growth and eventually killing the organism containing it.

In the case of non-tumorigenic retroviruses, the genome contains only three genes, which are known as *gag*, *pol* and *env*, and which function in the normal life cycle of the virus (Fig. 9.3). Following cellular entry, the viral RNA is converted into DNA by the action of the Pol protein, and this DNA molecule then integrates into the host chromosome. Subsequent transcription and translation of this DNA produces the viral structural proteins Gag and Env which coat the viral RNA genome, yielding viral particles that leave the cell to infect other susceptible cells.

Inspection of the RSV genome (Fig. 9.4) reveals an additional gene, known as the *src* gene, which is absent in the other viruses, suggesting that this gene is responsible for the ability of the virus to cause cancer. This idea was confirmed subsequently by showing that, if the *src* gene alone was introduced into normal cells, it was able to transform them to a cancerous phenotype. This gene was therefore called an oncogene (from the Greek *onkas* for mass or tumour) or cancer-causing gene.

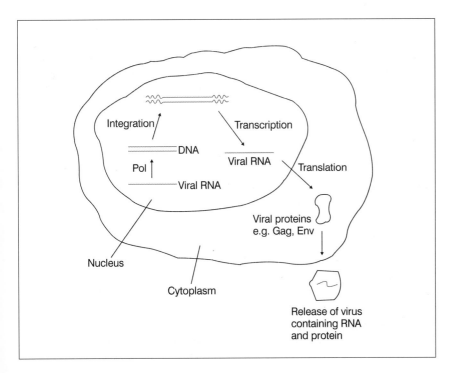

Figure 9.3

Life cycle of a typical non-tumorigenic retrovirus.

Figure 9.4

Comparison of the genome of a non-tumorigenic retrovirus with that of the tumorigenic retrovirus RSV.

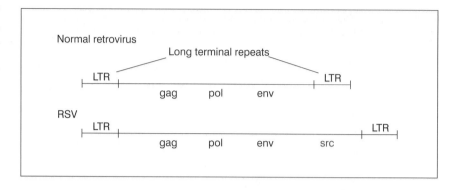

Following the identification of the *src* oncogene in RSV, a number of other oncogenes were identified in other oncogenic retroviruses infecting both chickens and mammals, such as the mouse or rat. Over 20 cellular proto-oncogenes which were originally identified in this way are now known (Table 9.1).

The identification of individual genes that are able to cause cancer obviously opened up many avenues of investigation for the study of this disease. From the point of view of gene regulation, however, the most exciting aspect of oncogenes was provided by the discovery that these cancer-causing genes are derived from genes present in normal cellular DNA. Thus, using Southern blotting techniques (Section 2.2.4) Takeya and Hanafusa (1985) detected a cellular equivalent of the viral *src* gene in the DNA of both normal and cancer cells (Fig. 9.5), and also showed that an mRNA capable of encoding the Src protein was produced in normal cells. Subsequent studies identified cellular equivalents of all the retroviral oncogenes. Many of these cellular equivalents of the viral oncogenes have now been cloned and shown to encode proteins identical or closely related to those present in the retroviruses. To avoid confusion, the viral oncogenes are given the prefix v, as in v-*src*, while their cellular equivalents are designated proto-oncogenes and given the prefix c, as in c-*src*.

The presence of cellular equivalents of the retroviral oncogenes suggests that the viral genes have been picked up from normal cellular DNA following integration of a non-tumorigenic virus next to the proto-oncogene. The subsequent incorrect excision of the viral genome resulted in its picking up the gene and converted it into a tumorigenic virus (Fig. 9.6). Hence the oncogene in the virus is derived from a cellular gene which is present in the DNA of normal cells.

Paradoxically the same gene can be present in normal cells without any apparent adverse effect and yet can cause cancer when incorporated into a virus. It is now clear that this conversion of the proto-oncogene into a cancer-causing oncogene is caused by one of two processes. Either the gene is mutated in some way within the virus, so that an abnormal product is formed, or alternatively the gene is expressed within the virus at much higher levels than are achieved in normal cells, and this high level of the normal product results in transformation (Fig. 9.7).

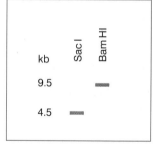

Figure 9.5

Southern blot showing hybridisation of the RSV *src* gene with normal chicken DNA cut with the restriction enzymes *Sac*I or *Bam*HI.

Oncogene	Species infected by virus	Normal function of protein
abl	Mouse	Tyrosine kinase
erbA	Chicken	Transcription factor and hormone receptor
erbB	Chicken	Receptor for epidermal growth factor, tyrosine kinase
ets	Chicken	Transcription factor
fes	Cat	Tyrosine kinase
fgr	Cat	Tyrosine kinase
fms	Cat	Tyrosine kinase, receptor for colony-stimulating factor
fos	Mouse	Transcription factor
jun	Chicken	AP1-related transcription factor
kit	Cat	Tyrosine kinase
lck	Chicken	Tyrosine kinase
mos	Mouse	Serine/threonine kinase
myb	Chicken	Transcription factor
myc	Chicken	Transcription factor
raf	Chicken	Serine/threonine kinase
ras	Rat	GTP-binding protein
rel	Turkey	Transcription factor
ros	Chicken	Tyrosine kinase
sea	Chicken	Tyrosine kinase
sis	Monkey	Platelet-derived growth factor B chain
ski	Chicken	Nuclear protein
src	Chicken	Tyrosine kinase
yes	Chicken	Tyrosine kinase

Table 9.1

Proto-oncogenes and their functions

Such conversion of a proto-oncogene into a cancer-causing oncogene is not confined to viruses, however. Cases of cancer which have no viral involvement in humans have now been shown to be due either to the over-expression of individual cellular proto-oncogenes or to their alteration by mutation within the cellular genome. Hence these genes play an important role in the generation of human cancer.

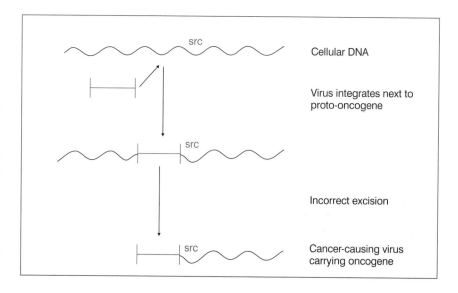

Figure 9.6

Model for the way in which a retrovirus could have picked up the cellular src gene by integration adjacent to the gene and subsequent incorrect excision.

Figure 9.7

A cellular proto-oncogene can
be converted into a cancer-
causing oncogene by increased
expression or by mutation.

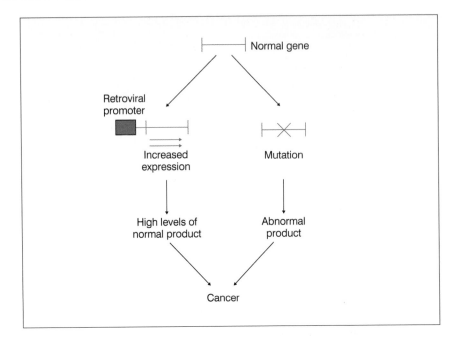

The potential risk of such proto-oncogenes causing cancer raises the question of why these genes have not been deleted during evolution. In fact, proto-oncogenes have been highly conserved in evolution, equivalents to mammalian and chicken oncogenes having been found not only in other vertebrates but also in invertebrates, such as *Drosophila*, and even in single-celled organisms, such as yeast.

The extraordinary evolutionary conservation of many of these genes, despite their potential danger, led to the suggestion that their products were essential for the processes regulating the growth of normal cells, and that their malregulation or mutation therefore results in abnormal growth and cancer. This idea has been confirmed abundantly as more and more proto-oncogenes have been characterised and shown to encode growth factors that stimulate the growth of normal cells, cellular receptors for growth factors and other cellular proteins involved in transmitting the growth signal within the cell, either by acting as a protein kinase enzyme or by binding GTP.

Ultimately the growth regulatory pathways controlled by oncogene products end in the nucleus, with the activation of genes whose corresponding proteins are required by the growing cell. It is not surprising, therefore, that several proto-oncogenes have been shown to encode transcription factors that regulate the expression of genes activated in growing cells (for reviews see Lewin, 1991; Forrest and Curran, 1992) (Table 9.2).

Hence, oncogenes present two aspects of importance from the point of view of the regulation of gene expression. First, since cancer is often caused by elevated expression of cellular oncogenes, the processes whereby this occurs are of interest, both from the point of view of the aetiology of cancer

Oncogene	Comments
erbA	Mutant form of the thyroid hormone receptor
ets	Binding site often found in association with API site
fos	Binds to API site as Fos-Jun dimer
jun	Can bind to API site alone as Jun-Jun dimer
mdm2	Inhibits gene activitation by p53
myb	—
myc	Requires Max protein to bind to DNA
rel	Member of the NFκB family
spi-1	Identical to PU.1 transcription factor

Table 9.2

Oncogenes and transcription factors

and for the light they throw on the mechanisms which regulate gene expression. This topic is discussed in Section 9.3. Secondly, the study of the transcription factors encoded by a few proto-oncogenes has led to a better understanding of the processes regulating gene expression in cells growing normally and in cancer cells, and this is discussed in Section 9.4.

9.3 ELEVATED EXPRESSION OF ONCOGENES

The products of cellular proto-oncogenes play a critical role in cellular growth control and, in many cases, are synthesised only at specific times and in small amounts. It is not surprising, therefore, that transformation into a cancer cell can result when these genes are expressed at high levels in particular situations.

The simplest example of such over-expression occurs in the case of retroviruses where, as we have already discussed, the oncogene comes under the influence of the strong promoter contained in the retroviral long terminal repeat (LTR) region and is hence expressed at a high level (Fig. 9.7).

A similar up-regulation due to the activity of a retroviral promoter is also seen in the case of avian leukosis virus (ALV) of chickens. Unlike the retroviruses described so far, however, this virus does not carry its own oncogene. Rather it transforms by integrating into cellular DNA next to a cellular oncogene, the c-*myc* gene (Fig. 9.8a) (for reviews of c-*myc*, see Grandori *et al.*, 2000, Eisenman, 2001). The expression of the c-*myc* gene is brought under the control of the strong promoter in the retroviral LTR, and it is hence expressed at levels 20–50 times higher than normal, producing transformation. This process is known as promoter insertion.

In other cases of this type, the ALV virus has been shown to have integrated downstream rather than upstream of the c-*myc* gene. Hence, the elevated expression of the c-*myc* gene in these cases cannot be due to promoter insertion. Rather, it involves the action of an enhancer element in the viral LTR which activates the c-*myc* gene's own promoter. As discussed in Section 7.3, enhancers, unlike promoters, can act in either orientation and at a distance, and hence the ALV enhancer can activate the *myc* promoter readily from a position downstream of the gene (Fig. 9.8b).

These cases are of interest from the point of view of gene regulation, indicating how viral regulatory systems can subvert cellular control processes.

Figure 9.8

Avian leukosis virus (ALV) can
increase expression of the *myc*
proto-oncogene either via
promoter insertion (a) or by the
action of its enhancer (b).

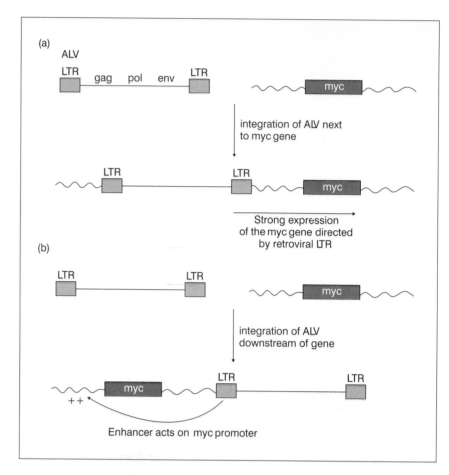

Of potentially greater interest, however, are the cases where up-regulation
of a cellular oncogene occurs through the alteration of internal cellular regulatory processes, rather than through viral intervention. The best-studied
example of this type concerns the increased expression of the c-*myc* oncogene which occurs in the transformation of B cells in cases of Burkitt's
lymphoma in humans or in the similar plasmacytomas that occur in mice.

When these tumours were studied, it was noted that they commonly
contained very specific chromosomal translocations which involved the
exchange of genetic material between chromosome 8 and chromosome 14
(Fig. 9.9). Most interestingly, the region of chromosome 8 involved includes
the c-*myc* gene, and the translocation results in the *myc* gene being moved to
chromosome 14, where it becomes located adjacent to the gene encoding
the immunoglobulin heavy chain. This translocation results in the increased
expression of the *myc* gene which is observed in the tumour cells.

Such translocation of specific oncogenes to the immunoglobulin locus
resulting in increased expression is not unique to the *myc* gene but has been
observed for a variety of other oncogenes in different B cell leukaemias

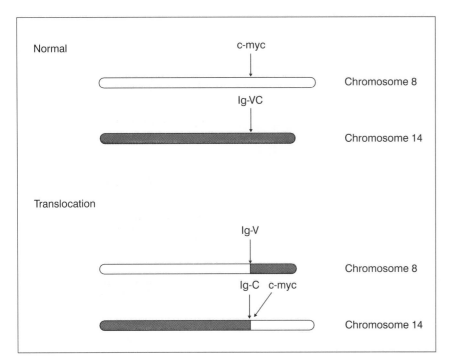

Figure 9.9

Translocation of the c-*myc* gene
from chromosome 8 to the
immunoglobulin heavy-chain
gene locus on chromosome 14
that occurs in cases of Burkitt's
lymphoma.

while similar translocations involving oncogene translocation to the T cell
receptor gene locus (which is highly expressed in T lymphocytes) have been
observed in T cell leukaemias (for reviews see Rabbitts, 1994; Latchman,
1996) (see also Section 9.4.3). The mechanisms responsible for the observed
up-regulation of oncogene expression have been best described, however, in
the case of the *myc* oncogene and these will now be discussed.

Detailed study of the processes mediating increased *myc* gene expression
has indicated that it is produced by different mechanisms in different
lymphomas, depending on the precise break points of the translocation
within the c-*myc* and immunoglobulin genes. In all cases studied, however,
the break point of the translocation occurs within the immunoglobulin gene,
resulting in a truncated gene lacking its promoter being linked to the c-*myc*
gene. This fact, together with the fact that the genes are always linked in a
head-to-head orientation (Fig. 9.10) indicates that the up-regulation of the
c-*myc* gene does not occur via a simple promoter insertion mechanism (see
Fig. 9.8a) in which it comes under the control of the immunoglobulin
promoter. In some cases, however, the B-cell-specific enhancer element,
which is located between the joining and constant regions of the
immunoglobulin genes (see Sections 2.4 and 7.3.2), is brought close to the
myc promoter by the translocation (Fig. 9.11). This enhancer element is
highly active in B cells and can activate the *myc* promoter in a manner analogous to the enhancer of ALV (see Fig. 9.8b).

Hence, in this case the c-*myc* gene is up-regulated by the action of the B-
cell-specific regulatory mechanisms of the immunoglobulin gene, the

Figure 9.10

Head-to-head orientation of the translocated *myc* gene and the immunoglobulin gene.

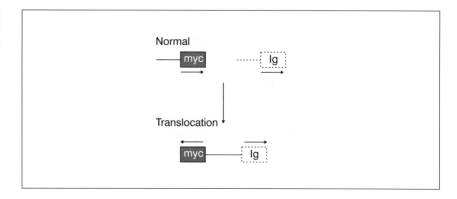

Figure 9.11

In some cases the enhancer of the immunoglobulin heavy-chain gene activates the *myc* gene promoter.

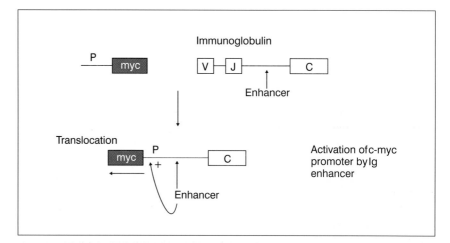

immunoglobulin enhancer activating the c-*myc* promoter rather than its own promoter, which has been removed by the translocation. In other cases, however, this does not appear to be the case, the break point of the translocation having removed both the immunoglobulin promoter and enhancer. This leaves the c-*myc* gene adjacent to the constant region of the immunoglobulin gene without any obvious B-cell-specific regulatory elements. In these cases it is likely that the up-regulation of the c-*myc* gene arises from its own truncation in the translocation. This results in the removal of negative regulatory elements that normally repress its expression, such as the upstream silencer element which normally represses the c-*myc* promoter (see Section 7.4.1 and Fig. 9.12).

More extensive truncation of the c-*myc* gene, involving the removal of transcribed sequences rather than upstream elements, has also been observed in some tumours. Frequently, this involves the removal of the first exon of the c-*myc* gene which does not contain any protein-coding information. This exon may thus fulfil a regulatory role by modulating the stability of the c-*myc* RNA or by affecting its translatability. Hence, its removal could enhance the level of c-*myc* protein by increasing the stability or the effi-

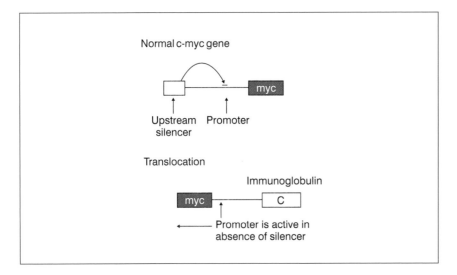

Figure 9.12

Activation of *myc* gene expression can be achieved by removal of the upstream silencer element.

ciency of translation of the c-*myc* RNA produced at a constant level of transcription. A similar increase in gene expression could be achieved by the removal of sequences within the first intervening sequence which inhibit the transcriptional elongation of the nascent c-*myc* transcript (see Section 4.3.2).

The increased expression of an oncogene produced by the removal of sequences that negatively regulate it is also seen in the case of the *lck* proto-oncogene, which encodes a tyrosine kinase related to the c-*src* gene product. In this case, activation of the oncogene in tumours is accompanied by the removal of sequences within its 5′ untranslated region, upstream of the start site of translation. The removal of these sequences results in a 50-fold increase in the initiation of translation of the *lck* mRNA into protein. Most interestingly, the region removed contains three AUG translation initiation codons which are located upstream of the correct initiation codon for production of the Lck protein (Fig. 9.13). The elimination of these codons results in increased translation initiation from the correct AUG, suggesting that initiation at the upstream codons inhibits correct initiation. This is exactly analogous to the regulation of translation of the GCN4 protein, which was discussed in Section 5.6.2.

It is likely that the processes regulating the expression of oncogenes such as c-*myc* and *lck*, which are revealed by studying their over-expression in tumours, also play a role in their normal pattern of regulation during cellular growth. Hence, their further study will throw light not only on the mechanisms of tumorigenesis but also on the processes regulating gene expression in normal cells.

Other examples of up-regulation of oncogene expression in tumorigenesis may occur, however, by mechanisms that are unique to the transformed cell. Thus, as discussed in Section 2.3, DNA amplification is relatively rare in normal cells. In tumours, however, it is observed frequently for specific

Figure 9.13

Increased translation of the *Ick* proto-oncogene mRNA can be achieved by deletion of AUG translational start sites upstream of the AUG that initiates the coding sequence of the Lck protein.

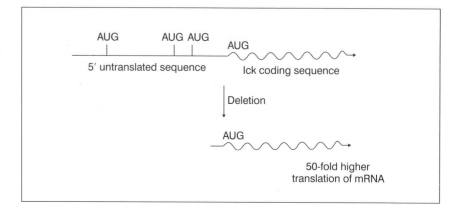

oncogenes and results in the presence of regions of amplified DNA which are visible in the microscope as homogeneously staining regions or as double minute chromosomes. Such amplification is especially common in human lung tumours and brain tumours, and frequently involves the c-*myc*-related genes, N-*myc* and L-*myc*. The expression of these genes in the tumour is increased dramatically due to the presence of up to 1000 copies of the gene in the tumour cell.

A variety of mechanisms, involving both the subversion of normal control processes or abnormal events occurring in the tumour cell, therefore results in the observed over-expression of oncogenes in tumour cells. When taken together with the production of abnormal oncogene products due to mutations, it is likely that these processes are involved in a wide variety of human cancers.

9.4 TRANSCRIPTION FACTORS AS ONCOGENES

As described in Section 9.2, the isolation of the cellular genes encoding particular oncogene products led to the realisation that they were involved in many of the processes regulating cellular growth. Ultimately the onset and continuation of cellular growth is likely to involve the activation of cellular genes that are not expressed in quiescent cells. It is not surprising, therefore, that several proto-oncogenes have been shown to encode transcription factors which regulate the transcription of genes activated in growing cells (Table 9.2) and several of these cases will be discussed (for reviews see Lewin, 1991; Forrest and Curran, 1992).

9.4.1 Fos, Jun, and AP1

The chicken retrovirus, avian sarcoma virus ASV17, contains an oncogene, v-*jun*, whose equivalent cellular proto-oncogene encodes a nuclearly located DNA binding protein. Sequence analysis of this protein revealed that it showed significant homology to the DNA binding domain of the yeast transcription factor GCN4, suggesting that it might bind to similar DNA sequences (Fig. 9.14). Interestingly, GCN4 itself had been shown previously

Figure 9.14

Comparison of the C-terminal amino acid sequences of the chicken Jun protein and the yeast transcription factor GCN4. Boxes indicate identical residues.

to bind to similar sequences to those bound by a factor, AP1 (activator protein 1), which had been detected in mammalian cell extracts by its DNA-binding activity (Fig. 9.15).

This relationship of Jun and AP1 to the sequence and binding activity, respectively, of GCN4 led to the suggestion that Jun might be related to AP1. This was confirmed by the findings that antibody to Jun reacted with purified AP1 preparations, and that Jun expressed in bacteria was capable of binding to AP1 binding sites in DNA. Moreover, Jun was capable of stimulating transcription from promoters containing AP1 binding sites but not from those which lacked these sites. Hence, the *jun* oncogene encodes a sequence-specific DNA binding protein capable of stimulating transcription of genes containing its binding site, which is identical to the AP1 binding site.

Although Jun undoubtedly binds to AP1-binding sites, preparations of AP1 purified on the basis of this ability contain several other proteins in addition to c-Jun. Several of these are encoded by genes related to *jun*, but another is the product of a different proto-oncogene, namely c-*fos*. Interestingly, however, although Fos is present in AP1 preparations, it does not bind to DNA when present alone but requires the product of the c-*jun* gene for DNA binding. Hence, in addition to its ability to bind to AP1 sites alone, Jun can also form a complex with Fos that binds to this site. As discussed in Section 8.2.4, this association takes place through the leucine zipper domains of the two proteins and results in a heterodimer complex which binds to the AP1 binding site with much greater affinity than the Jun homodimer.

Hence, both Fos and Jun, which were identified originally through their association with oncogenic retroviruses, are also cellular transcription factors (for reviews see Piechaczyk and Blanchard, 1994; Vogt, 1995; Karin

```
DNA binding site

GCN 4    5' T G A C/G T C A T 3'
  AP1    3' T G A  G  T C A G 3'
```

Figure 9.15

Relationship of the DNA binding sites for the yeast transcription factor GCN4 and the mammalian transcription factor AP1.

et al., 1997). Such a finding raises the question of the normal role of these factors and how they can cause cancer. In this regard it is of obvious interest that the AP1-binding site is involved in mediating the induction of genes that contain it in response to treatment with phorbol esters, which are also capable of promoting cancer. Thus, not only do many phorbol-ester-inducible genes contain AP1 binding sites (see Table 7.2) but, in addition, transfer of AP1 binding sites to a normally non-inducible gene renders that gene inducible by phorbol-esters. Increased levels of Jun and Fos are also observed in cells after treatment with phorbol-esters, indicating that these substances act by increasing the levels of Fos and Jun, which in turn cause increased transcription of other genes containing AP1-binding sites which mediate induction by the Fos–Jun complex.

Most interestingly, increased levels of Jun and Fos are also produced by treatment with serum or growth factors which stimulates the growth of quiescent cells. Hence, the transduction of the signal to grow, which begins with the growth factors and their cellular receptors and continues with intra-cytoplasmic signal transducers such as protein kinases and GTP-binding proteins, ends in the nucleus with the increased level of the transcription factors Jun and Fos (Fig. 9.16). These proteins will then activate the genes whose products are necessary for the process of growth itself.

Clearly, it is relatively easy to fit the oncogenic properties of Jun and Fos into this framework. Thus, if these proteins are normally produced in response to growth-inducing signals and activate growth, their continual abnormal synthesis will result in a cell which will be stimulated to grow

Figure 9.16

Growth factor stimulation of cells results in increased transcription of the c-*fos* and c-*jun* genes, which in turn stimulates transcription of genes which are activated by the Fos–Jun complex.

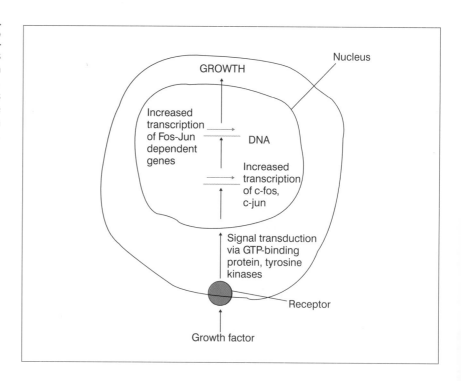

continually and will not respond to growth-regulating signals. Such continuous uncontrolled growth is characteristic of the cancer cell.

In agreement with this idea, mutations in the leucine zipper region of Fos, which abolish its ability to dimerise with Jun and induce genes containing AP1 sites, also abolish its ability to transform cells to a cancerous phenotype. Hence, the ability of Fos to cause cancer is directly linked to its ability to act as a transcription factor for genes containing the appropriate binding site.

Interestingly, AP1 sites in growth-regulated genes are often located close to binding sites for the Ets protein, another transcription factor which is encoded by a cellular proto-oncogene (see Table 9.2). Hence several different oncogenic transcription factors may co-operate to produce high-level transcription of specific genes in actively growing cells.

9.4.2 v-*erb*A and the thyroid hormone receptor

Unlike most other retroviruses, avian erythroblastosis virus (AEV) carries two cellular oncogenes, v-*erb*A and v-*erb*B. When c-*erb*A, the cellular equivalent of v-*erb*A, was cloned, it was shown to encode the cellular receptor that mediates the response to thyroid hormone.

As discussed in Sections 8.3.4 and 8.4.2, this receptor is a member of the superfamily of steroid–thyroid hormone receptors which, following binding of a particular hormone, induce the transcription of genes containing a binding site for the hormone–receptor complex. In the case of ErbA, the protein contains a region that can bind thyroid hormone. Following such binding, the hormone–receptor complex induces transcription of thyroid–hormone-responsive genes, such as those encoding growth hormone or the heavy chain of the myosin molecule (Fig. 9.17).

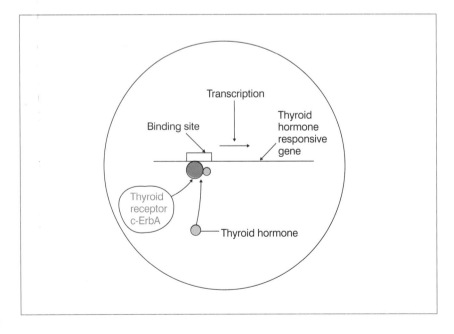

Figure 9.17

The c-*erb*A gene encodes the thyroid hormone receptor and activates transcription in response to thyroid hormone.

The finding that the cellular homologue of the v-*erb*A oncogene is a hormone-responsive transcription factor provides a further connection between oncogenes and cellular transcription factors. It raises the question, however, of the manner in which the transfer of the thyroid hormone receptor to a virus can result in transformation. To answer this question it is necessary to compare the protein encoded by the virus with its cellular counterpart. As shown in Fig. 9.18 the ErbA protein has the typical structure of a member of the steroid–thyroid hormone family (see also Fig. 8.15), containing both DNA binding and hormone-binding regions. The viral ErbA protein is generally similar except that it is fused to a portion of the retroviral gag protein at its N-terminus. It also contains a number of mutations in both the DNA binding and hormone-binding regions, as well as a small deletion in the hormone-binding domain.

Of these changes, it is the alterations in the hormone-binding domain which have the most significant effects on the function of the protein and which are thought to be critical for transformation. Thus, these changes abolish the ability of the protein to bind thyroid hormone and activate transcription. However, the protein retains the inhibitory domain which allows the Erb-A protein to repress gene transcription in the absence of thyroid hormone (see Section 8.3.4). Thus, the viral protein is functionally analogous to the alternatively spliced form of the c-*erb*A gene product discussed in Section 8.4.2, which lacks the hormone-binding domain and dominantly represses the ability of the hormone-binding receptor form to activate thyroid–hormone-responsive genes.

The idea that the non-hormone-binding viral ErbA protein might also be able to do this has been confirmed by studying the effect of this oncogene on thyroid–hormone-responsive genes. As expected, the v-*erb*A gene product was able to abolish the responsiveness of such genes to thyroid hormone by binding to the thyroid–hormone-response elements in their promoters and preventing activation by the cellular ErbA protein-thyroid hormone complex (Fig. 9.19).

Interestingly, however, this repression by v-*erb*A does not simply involve the passive blockage of binding by c-*erb*A following thyroid hormone treatment. Thus, mutations in the inhibitory domain described in Section 8.3.4

Figure 9.18

Relationship of the cellular ErbA protein and the viral protein. The black dots indicate single amino acid differences between the two proteins while the arrow indicates the region where nine amino acids are deleted in the viral protein.

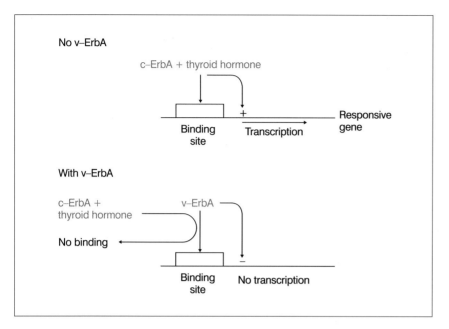

Figure 9.19

Inhibitory effect of the viral ErbA protein on gene activation by the cellular protein, in response to thyroid hormone. Note the similarity to the action of the α-2 form of the c-ErbA protein, illustrated in Fig. 8.42.

can abolish the oncogenic activity of v-*erb*A. These mutations do not affect the ability of v-*erb*A to bind to DNA but prevent it from recruiting the inhibitory co-repressor which is essential for its ability to actively repress transcription (see Section 8.3.4) (for a review see Perlmann and Vennstrom, 1995). Hence the ability to actively repress transcription is essential for transformation by v-*erb*A (Fig. 9.20).

The explanation of how such gene repression by viral ErbA can result in transformation is provided by the observation that the introduction of the viral gene into cells can repress transcription of the avian erythrocyte anion transporter gene. This gene is one of those which is switched on when chicken erythroblasts differentiate into erythrocytes. It has been known for some time that the viral ErbA protein can block this process, and it is now clear that this is achieved by blocking the induction of the genes needed for this to occur. In turn, such blockage of differentiation allows the cells to continue to proliferate. When this is combined with the introduction of the v-*erb*B gene, which encodes a truncated form of the epidermal growth factor receptor and renders cell growth independent of external growth factors, transformation results (Fig. 9.21).

Transformation caused by v-*erb*A thus represents an example of the activation of an oncogene by mutation resulting, in this case, in its losing the ability to activate transcription but retaining its ability to act as a dominant repressor of transcription. As discussed in Section 8.4.2, however, one alternatively spliced transcript of the c-*erb*A gene is also able to do this and, like the v-*erb*A gene product, cannot bind thyroid hormone. Hence, this repression of transcription by a non-hormone-binding form of the receptor is likely to be of importance in normal cells also.

Figure 9.20

The ability of v-*erb*A to transform cells to a cancerous phenotype requires it to be able to actively inhibit transcription via its inhibitory domain (dark blue) which acts by recruiting an inhibitory co-repressor (COR) (panel a). Thus mutations in the v-*erb*A inhibitory domain (white) which abolish its ability to bind the co-repressor also abolish its ability to transform cells even though it can still bind to DNA (panel b).

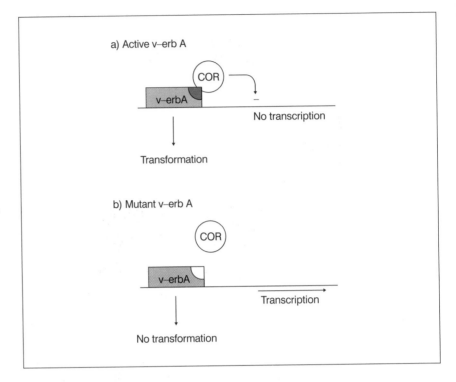

9.4.3 Other transcription-factor-related oncogenes

Although the *fos–jun* and *erb*A cases represent the best characterised examples of the connection between oncogenes and transcription factors, several other cellular oncogenes encode transcription factors (see Table 9.2; for a review see Lewin, 1991).

One of these, the Myc protein, has been studied intensively in view of its over-expression in many human tumours (see Section 9.3) (for reviews see Grandori *et al.*, 2000; Eisenman, 2001). The Myc protein clearly encodes a transcription factor which contains the leucine zipper motif characteristic of many transcription factors, including Fos and Jun (see Section 8.2.4), as well as a helix-loop-helix motif (see Section 8.2.4). Moreover, mutations in the leucine zipper region of the protein abolish its oncogenic ability to transform normal cells, suggesting that the ability to act as a transcription factor is essential for Myc-induced transformation.

Despite all this evidence, for many years the actual role of Myc in transcriptional control remained unclear. This was because it was not possible to demonstrate the binding of Myc to a specific DNA sequence in the manner that had been shown to occur for Jun and ErbA. This problem was resolved, however, by the finding that Myc has to heterodimerise with a second factor, Max, in order to bind to DNA (for reviews see Baudino and Cleveland, 2001; Grandori *et al.*, 2000). Hence, Myc resembles Fos in requiring another factor for sequence-specific binding. This finding indicates once again the importance of heterodimerisation in regulating the activity of tran-

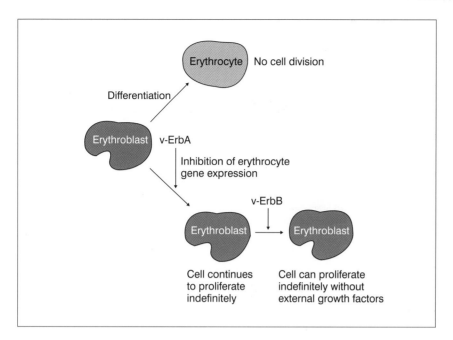

No cell division

Differentiation

Erythrocyte

Erythroblast   v-ErbA

Inhibition of erythrocyte
gene expression

v-ErbB

Erythroblast → Erythroblast

Cell continues
to proliferate
indefinitely

Cell can proliferate
indefinitely without
external growth factors

Figure 9.21

Inhibition of erythrocyte-specific
gene expression by the v-ErbA
protein prevents erythrocyte
differentiation and allows
transformation by the v-ErbB
protein.

scription factors and also illustrates how the function of a particular factor
can remain obscure simply because its partner has not yet been isolated.

Although our discussion of oncogenes as transcription factors has focused
on the cellular genes which are initially identified in RNA tumour viruses, a
number of oncogenes encoding transcription factors have also been identified on the basis of their involvement in the chromosomal translocations
which occur in human leukaemias (see Section 9.3). Thus, as well as involving previously characterised oncogenes such as the *myc* gene, these translocations can involve genes not previously shown to produce cancer when
over-expressed. For example, a member of the homeobox family of transcription factors (see Section 8.2.2), Hox-11, has been shown to be involved
in cases of acute childhood leukaemia where its expression is activated by
translocation to the T cell receptor gene locus.

As well as resulting in increased expression of a proto-oncogene by
translocation to a highly active immunoglobulin or T cell receptor gene
locus, such chromosomal translocations can also cause leukaemias by
producing fusion genes between two genes which were previously located
on separate chromosomes (for reviews see Rabbits, 1994; Latchman, 1996;
Look, 1997) (Fig. 9.22). Such fusion proteins presumably produce cancer
because the fusion protein has cancer-causing properties distinct from those
of either intact protein alone. As in the case of translocations causing over-
expression, such translocations can involve genes encoding transcription
factors previously identified within RNA tumour viruses or those which
have not previously been shown to have an oncogenic effect.

Thus, the c-*ets*-1 proto-oncogene originally identified in a chicken
retrovirus (see Table 9.1) which was discussed in Section 8.2.5 becomes

Figure 9.22

Activation of an oncogene through chromosomal translocation can occur if its expression is increased by translocation to the actively transcribed immunoglobulin or T cell receptor gene loci (panel a) or if a fusion gene encoding a novel protein with oncogenic properties is created (panel b).

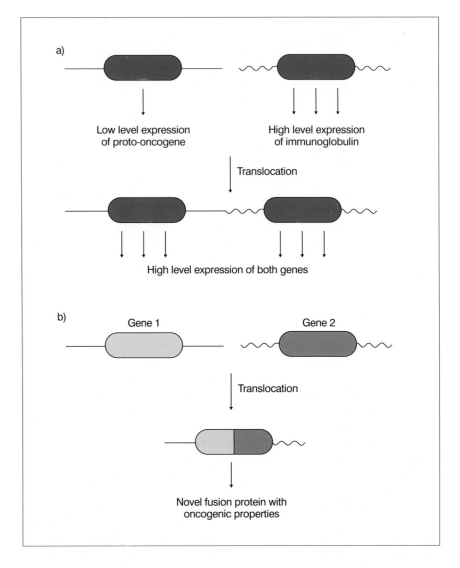

fused to the platelet-derived growth factor receptor gene in chronic myelomonocytic leukaemia. In contrast, in acute promyelocytic leukaemia (PML) the translocation involves the fusion of two transcription factor genes not characterised in oncogenic retroviruses, namely the retinoic acid receptor α gene (a member of the steroid/thyroid hormone receptor family discussed in Section 8.2.3) and the PML transcription factor, a zinc finger transcription factor which was originally characterised on the basis of its involvement in this disease.

Interestingly, the ELL gene which is involved in chromosomal translocations leading to fusion protein products in acute myeloid leukaemia is a transcription elongation factor (for a review see Conaway and Conaway, 1999). This indicates that oncogenes can encode factors involved in transcriptional elongation as well as transcriptional initiation (see Section 4.3.2 for discussion of transcriptional elongation and its control).

In addition to such cellular oncogenes identified in RNA tumour viruses or at sites of chromosomal translocations, it is worth noting that DNA viruses that can cause cancer also encode oncogenes capable of regulating cellular gene expression (for a review see Moran, 1993). In particular, both the large T oncogenes of the small DNA viruses, SV40 and polyoma, and the E1A protein of adenovirus are capable of affecting the transcription of specific cellular genes, and this ability is critical for the ability of these proteins to transform cells. Unlike the oncogenes of RNA viruses, however, the genes encoding these viral proteins do not appear to have specific cellular equivalents, and are likely to have evolved within the virus rather than having been picked up from the cellular genome.

The similar ability of oncogenes from DNA and RNA tumour viruses to affect cellular gene expression, despite their very different origins, suggests that such modulation of cellular gene expression is critical for the transforming ability of these viruses.

9.5 ANTI-ONCOGENES

9.5.1 Nature of anti-oncogenes

Following the discovery of cellular oncogenes it was rapidly shown that they encoded proteins which promoted cellular growth so that their activation by over-expression or by mutation resulted in abnormal growth leading to cancer. Subsequently, however, it became clear that cancer could also result from the deletion or mutational inactivation of another group of genes. This indicated that these genes encoded products which normally restrained cellular growth so that their inactivation would result in abnormal, unregulated growth. These genes were therefore named anti-oncogenes or tumour suppressor genes (for reviews see Knudson, 1993; Weinberg, 1993).

Clearly the inhibitory role of anti-oncogenes indicates that cancers will involve the deletion of these genes or the occurrence of mutations within them which inactivate their protein product, rather than their over-expression or mutational activation as occurs in the oncogenes (Fig. 9.23). Thus far a number of genes of this type have been defined on the basis of their mutation or deletion in specific tumour types (Table 9.3). Most interestingly, the three best-defined anti-oncogene products all encode transcription factors. Two of these, p53 and the Wilms' tumour gene product, appear to act by binding to target sites in the DNA of specific genes and regulating their expression while the third, the retinoblastoma gene product (Rb-1), acts predominantly via protein–protein interactions with other regulatory transcription factors. The p53 and Rb-1 proteins will therefore be discussed as examples of these two types (for a review of the Wilms' tumour gene product see Hastie, 2001).

9.5.2 p53

The p53 protein was originally identified as a 53 kDa protein which bound to the large T oncogene protein of the small DNA virus SV40. Subsequent

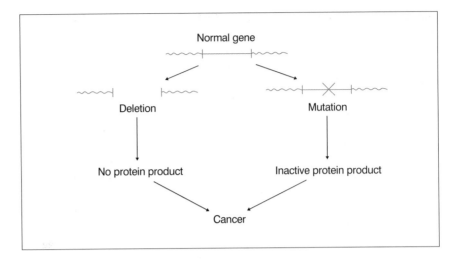

studies have shown that the gene encoding this protein is mutated in a very wide variety of human tumours, especially carcinomas (for reviews of p53 see Lohrum and Vousden, 2000; Vogelstein *et al.*, 2000). In normal cells, p53 is induced in response to DNA damage and its activation results in growth arrest and/or cell death. Most interestingly, inactivation of the p53 gene in mice does not prevent the normal development of the animal. Rather it results in an abnormally high rate of tumour formation which results in the early death of the animal. This has led to the idea that the p53 gene product normally acts to prevent cells with damaged DNA from proliferating by inducing their growth arrest or death. In the absence of p53 such cells proliferate and form tumours which occur at high frequency when p53 is inactivated by mutation.

The detailed characterisation of the p53 gene product has shown that it is a transcription factor capable of binding to a specific DNA sequence and activating the expression of specific genes. The mutations which occur in

Table 9.3

Anti-oncogenes and their functions

Anti-oncogene	Tumours in which gene is mutated	Nature of protein product
APC	Colon carcinoma	Cytoplasmic protein
BRCA-1	Breast cancer	DNA repair, transcription factor?
BRCA-2	Breast cancer	DNA repair, transcription factor?
DCC	Colon carcinoma	Cell adhesion molecule
NFI	Neurofibromatosis	Activator of Ras GTPase activity
NF2	Schwannomas, meningiomas	Cytoplasmic protein
p53	Sarcomas, breast carcinomas, leukaemia, etc	Transcription factor
RBI	Retinoblastoma, osteosarcoma, small lung cell carcinoma	Transcription factor
VHL	Phaeochromocytoma, kidney carcinoma	Transcriptional elongation regulator
WTI	Wilms' tumour	Transcription factor

human tumours result in a loss of the ability to bind to DNA, indicating that the ability of p53 to do this is crucial for its ability to control cellular growth and suppress cancer. The p53 protein thus functions, at least in part, by activating the expression of specific genes whose protein products act to inhibit cellular growth (Fig. 9.24i). Its inactivation by gene deletion (Fig. 9.24ii) or by mutation (Fig. 9.24.iii) thus leads to a failure to express these genes, leading to uncontrolled growth.

Interestingly, a failure of p53-mediated gene activation can also occur, even in the presence of functional p53 (Fig. 9.24.iv). Thus, many human soft tissue sarcomas contain intact p53 protein but have amplified the cellular *mdm2* oncogene. The MDM2 oncoprotein binds to p53 and masks its activation domain, so preventing it from activating transcription and hence resulting in cancer. This evidently represents an example of the "quenching" mechanism of transcriptional repression which was discussed in Section 8.3.4. In addition, it has been shown that the interaction of MDM2 with p53 also results in the rapid degradation of p53. This indicates a second mechanism by which MDM2 prevents p53 from acting and provides further evidence for the regulation of transcription activity by control of protein stability as discussed in Section 8.4.3.

The interaction of p53 with oncogenic proteins is not unique to the MDM2 oncoprotein. Thus, as noted above, it was the interaction of p53 with the

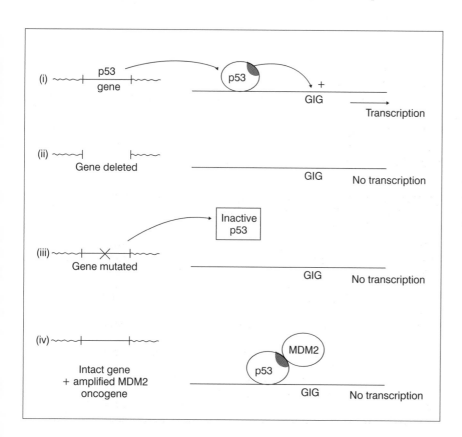

Figure 9.24

The functional p53 protein acts to stimulate the transcription of genes (GIG) whose protein products inhibit growth (i). This effect can be prevented, however, by the deletion of the p53 gene (ii) or by its inactivation by mutation (iii) as well as by the MDM2 oncoprotein which binds to p53 and masks its activation domain (iv). The activation domain of p53 is indicated by the coloured box.

SV40 large T oncoprotein which led to the original identification of the p53 protein. As in the case of the MDM2 protein, the interaction of p53 with either the large T protein or the transforming proteins of several other DNA viruses prevents p53 from activating its target genes. Such functional inactivation of p53 appears to play a critical role in the ability of these viruses to transform cells to a cancerous phenotype. When taken together with the action of the MDM2 protein this indicates that functional interactions between oncogene and anti-oncogene products are likely to be critical in the control of cellular growth. Hence changes in this balance due to the over-expression of specific oncogenes or the loss of anti-oncogenes will result in cancer.

As with many other transcription factors (see Section 8.4.3), the activity of p53 is also regulated by post-translational modifications. Thus, the phosphorylation of p53 at its N-terminus prevents its interaction with MDM2, thereby stabilising the protein whilst the addition of acetyl residues at its C-terminus enhances its DNA binding ability. Hence, as in the case of the histones discussed in Section 6.6, the activity of p53 is modulated by both phosphorylation and acetylation (Fig. 9.25).

The critical role of p53 in regulating the expression of genes encoding growth inhibitory proteins focuses attention on the nature of these growth inhibitory genes. A number of potential target genes for p53 have now been identified. One of these encodes a protein (p21) that acts as an inhibitor of cyclin-dependent kinases. Cyclin-dependent kinases are enzymes that stimulate cells to enter cell division. Hence, the identification of a p53-regulated gene as an inhibitor of these enzymes immediately suggested that p53 acts by stimulating the expression of this inhibitory factor, which in turn inhibits the cyclin-dependent kinases and therefore prevents cells replicating their DNA and undergoing cell division (Fig. 9.26).

Similarly, the observation that p53 also stimulates the expression of the bax gene, whose protein product stimulates programmed cell death or

Figure 9.25

The addition of phosphate residues at the N-terminus of the p53 protein reduces the interaction with MDM2 and improves its stability whilst the addition of acetyl residues at the C-terminus enhances its DNA binding ability.

apoptosis, supports a role for p53 in promoting the death of cells which have become abnormal and can no longer divide normally.

It is likely therefore that p53 acts as a sensor for damage to cellular DNA, for example, by irradiation, which could result in mutations occurring. When it is activated, growth arrest genes are stimulated and the cell ceases to divide so that it can repair the damage. If the damage to the cell is irreparable, however, genes inducing apoptosis are activated by p53 and the cell dies. Evidently, in the absence of p53, cells with DNA damage or mutations will continue to replicate and if the mutations activate specific oncogenes then a cancer will result (Fig. 9.27).

Hence, the p53 gene product plays a key role in regulating cell division, acting by regulating the expression of specific target genes. Its inactivation by mutations or by specific oncogene products is likely to play a critical role in the majority of human cancers. Although p53 was initially believed to be a unique protein, unrelated to any others, two proteins which are related to p53 have now been identified and have been named p63 and p73 on the basis of their molecular weights (for reviews see Lohrum and Vousden, 2000; Davis and Dowdy, 2001). Neither p63 or p73 appears to be commonly mutated in human cancer but they do appear to be required for normal development. Thus, loss or mutation of p63 leads to the human disease known as EEC syndrome (ectrodactyly, ectodermal dysplasia and cleft lip) in which patients have limb defects and facial clefts (Celli *et al.*, 1999).

9.5.3 The retinoblastoma protein

The retinoblastoma gene (Rb-1) was the first anti-oncogene to be identified on the basis that its inactivation results in the formation of eye tumours known as retinoblastomas (for a review of Rb see Harbour and Dean, 2000). As with p53, the Rb-1 protein acts as an anti-oncogene by regulating the expression of specific target genes. Unlike p53, however, it appears to act

Figure 9.26

The p53 protein stimulates the transcription of the gene encoding the 21 kDa inhibitor of cyclic dependent kinases (CDK). By inhibiting the activity of the kinases, the 21 kDa protein prevents DNA synthesis and thus cell division.

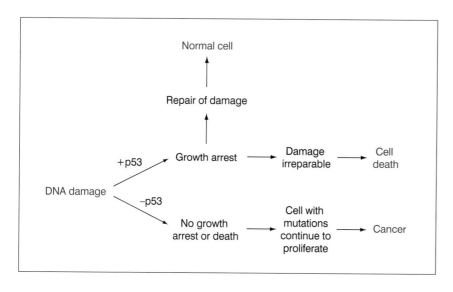

Figure 9.27

In normal cells, p53 is activated by damage to DNA. The cell is then induced to growth arrest, allowing the damage to be repaired. If the damage cannot be repaired, p53 activates cell death and the damaged cell dies. In the absence of p53, the damaged cell continues to proliferate and, if it has mutations in oncogenes, will form a cancer.

primarily via protein–protein interactions with other transcription factors.

In particular, RB-1 has been shown to interact with the cellular transcription factor E2F. E2F normally stimulates the transcription of several growth-promoting genes such as the cellular oncogenes c-*myc* and c-*myb* (see Section 9.4.3) and the genes encoding DNA polymerase α and thymidine kinase. The interaction of Rb-1 with E2F does not affect the ability of E2F to bind to its target sites in the DNA but prevents it stimulating transcription. This is due to Rb-1 binding to the transcriptional activation domain of E2F and blocking its activity via a quenching type mechanism (see Section 8.3.4). In addition, however, it has been shown that Rb-1 can bind to both histone deacetylases (Luo *et al.*, 1998) and histone methylases (Ringrose and Paro, 2001) and thereby promote histone deacetylation and methylation. As discussed in Section 6.6 both these modifications promote a more closed chromatin structure incompatible with transcription.

Hence, Rb-1 can block the action of E2F both by inhibiting its ability to activate transcription and via promoting a closed chromatin structure (Fig. 9.28). Hence Rb-1 acts as an anti-oncogene by preventing the transcription of several growth-promoting genes, including oncogenes such as c-*myc* which themselves encode transcription factors.

During the normal cell cycle of dividing cells, the Rb-1 protein becomes phosphorylated and this prevents its interacting with E2F, allowing the E2F factor to activate the growth-promoting genes whose protein products are necessary for cell cycle progression (Fig. 9.29a). Such an effect can also be achieved by the deletion of the Rb-1 gene or its inactivation by mutation which prevents the production of a functional Rb-1 protein (Fig. 9.29b). Similarly, it is also possible for such an absence of functional Rb-1 protein to arise from transcriptional inactivation of the Rb-1 gene which is otherwise intact. Thus 6 out of 77 retinoblastomas were found to have a heavily methylated Rb-1 gene which would result in a failure of its transcription (see Section 6.5) even though the gene itself was theoretically capable of encoding a functional protein. In addition, as with the p53 protein, the Rb-1 protein can be inactivated by protein–protein interaction with the oncogene products of DNA viruses. In this case, however, the association of the viral protein with Rb-1 dissociates the Rb-1-E2F complex, releasing free E2F which can then activate gene expression (Fig. 9.29c).

Figure 9.28

Following binding of Rb to the DNA bound E2F transcription factor, transcription is inhibited since Rb both blocks the ability of E2F to activate transcription and promotes the organisation of a tightly packed chromatin structure (wavy line) which does not allow transcription to occur.

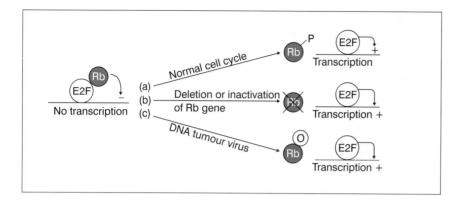

Figure 9.29

The retinoblastoma protein (Rb) binds to the E2F transcription factor and represses transcription. This inhibition can be relieved allowing E2F to activate transcription when: (a) the Rb protein is phosphorylated which occurs during the normal cell cycle and prevents its interacting with E2F; (b) the gene encoding Rb is deleted or inactivated by mutation; or (c) the Rb protein binds to the product of a DNA tumour virus oncogene (o) which releases it from E2F.

Thus, the Rb-1 protein plays a critical role in cellular growth regulation, modulating the expression of specific oncogenes and acting as a target for the transforming oncoproteins of specific viruses. Interestingly, Rb-1 has also been shown to interact with the RNA polymerase I transcription factor UBF (see Section 3.2.1) and thereby repress the transcription of the genes encoding ribosomal RNA. In addition, it can also repress the transcription of the 5S rRNA and tRNA genes by RNA polymerase III (see Section 3.2.2), probably by interacting with TFIIIB (for a review see White, 1997). Since transcription of the RNAs involved in protein synthesis is evidently necessary for cellular growth, Rb-1 can act as a global inhibitor of cellular growth by preventing the transcription of all the genes transcribed by RNA polymerases I and III as well as those of specific genes essential for growth by RNA polymerase II (Fig. 9.30).

Interestingly, Rb-1 itself is inhibited by association with the Id2 transcription factor, a member of the inhibiting class of helix-loop-helix proteins which was discussed in Section 8.2.4. In tumour cells over-expressing the N-*myc* oncogene transcription factor (see Section 9.3), the Id2 gene is transcriptionally activated by N-*myc*. The excess Id2 binds to Rb-1 and inactivates it, allowing the tumour to grow (Lasorella *et al.*, 2000). This therefore represents another example of oncogenes and anti-oncogenes interacting antagonistically to regulate cellular growth with the over-expression of an oncogene protein resulting in the inactivation of an anti-oncogene protein.

Together with p53 therefore Rb-1 plays a key role in restraining cellular growth. Indeed, many parallels exist between these two proteins. Thus, as well as being regulated by tumour virus proteins, Rb-1-like p53 (see Section 9.5.2) is inhibited by interaction with the MDM2 protein. Moreover, acetylation of Rb-1 enhances its interaction with MDM2 (Chan *et al.*, 2001) indicating that both p53 and Rb-1 can be regulated via phosphorylation and acetylation.

In addition, there is evidence that the p53 and Rb-1 pathways interact with one another. Thus, the phosphorylation of Rb-1 which regulates its activity is carried out by the cyclin-dependent kinases. Hence, the action of p53 which stimulates the p21 inhibitor of these kinases will reduce their activity and hence maintain Rb-1 in its non-phosphorylated growth inhibitory form (Fig. 9.31).

Figure 9.30

By binding to the polymerase transcription factor UBF, the polymerase II transcription factor E2F and the polymerase III transcription factor TFIIIB, Rb-1 can repress transcription of the ribosomal RNA genes by RNA polymerase I, the transcription of E2F-dependent genes by RNA polymerase II and the transcription of the tRNA and 5SrRNA genes by RNA polymerase III.

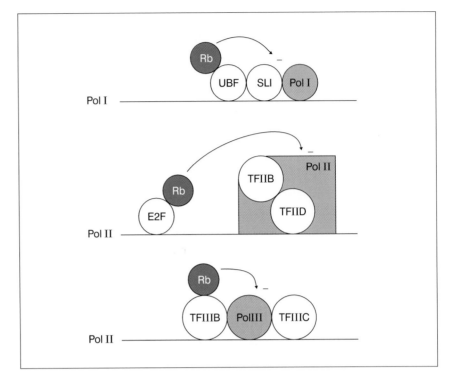

9.5.4 Other anti-oncogene transcription factors

In addition to p53, Rb-1 and the Wilms' tumour gene product which are well-defined transcription factors, it is probable that other anti-oncogenes are also transcription factors. Thus, the products of the BRCA-1 and BRCA-2 genes which are mutated in many cases of breast cancer appear to act primarily by regulating the rate at which damaged DNA is repaired. They also contain regions which can act as activation domains (see Section 8.3.2) and it has therefore been suggested that they also act as transcriptional regulators (for a review see Marx, 1997).

The vital role of p53, Rb-1 and the other anti-oncogene products, indicates that, although discovered later than the oncogenes, the anti-oncogenes are likely to play as critical a role in regulating cellular growth in general and the pattern of gene expression in particular. Hence, the precise rate of cellular growth is likely to be controlled by the balance between interacting oncogene and anti-oncogene production with cancer resulting from a change in this balance by activation or over-expression of oncogenes or inactivation of anti-oncogenes.

9.6 CONCLUSIONS

The key role of gene regulation in the control of normal development and functioning of the adult organism is indicated by the very large number of

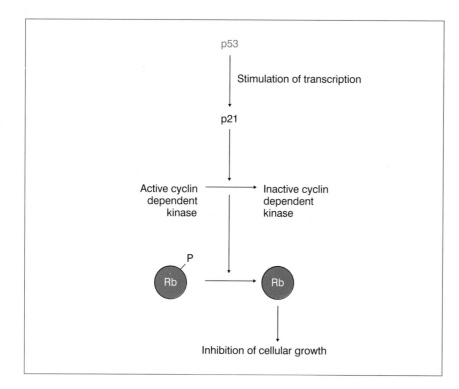

p53

Stimulation of transcription

p21

Active cyclin dependent kinase → Inactive cyclin dependent kinase

P

Rb → Rb

Inhibition of cellular growth

Figure 9.31

Activation of the gene encoding the p21 inhibitor of cyclin-dependent kinases by p53 results in the inhibition of these kinases which in turn maintains Rb-1 in its non-phosphorylated growth inhibitory form.

diseases which involve the mutation of genes encoding transcription factors or other proteins involved in gene regulation. However, the disease which most frequently results from alterations in gene regulation is cancer. Indeed, the study of the processes whereby increased expression or mutation of certain cellular genes or the deletion or inactivation of other genes can cause cancer has increased greatly our knowledge of the process of transformation, whereby normal cells become cancerous. Similarly, the recognition that the products of these cellular genes play a critical role in the growth regulation of normal cells has allowed insights obtained from studies of their activities in tumour cells to be applied to the study of normal cellular growth control.

Such a reciprocal exchange of information is well illustrated in the case of the oncogenes and anti-oncogenes that encode cellular transcription factors. Thus, the study of the Rb-1 and p53 proteins has greatly enhanced our knowledge of the processes regulating the transcription of genes whose protein products enhance or inhibit cellular growth. Similarly, the isolation of the *fos* and *jun* genes in tumorigenic retroviruses has aided the study of the effects of growth factors on normal cells, while the recognition that the v-*erb*A gene product is a truncated form of the thyroid hormone receptor has allowed the elucidation of its role in transformation via the inhibition of erythroid differentiation.

These two cases also illustrate the two mechanisms by which cellular genes can become oncogenic – namely mutation or over-expression. Thus, in the case of the v-*erb*A gene, mutations have rendered the protein different from the corresponding c-*erb*A gene from which it was derived. These alter the properties of the protein so it cannot bind thyroid hormone, and it behaves as a dominant repressor of transcription. Similar activation by mutation is also seen in oncogenes whose products function at other stages of the growth-signalling process. Notable examples include the truncation of the epidermal growth factor receptor in the v-*erb*B oncogene, and the loss of GTP hydrolysing ability in the *ras* oncogene, both of which result in proteins "frozen" into a positive growth-signalling form.

Interestingly, in a situation where an oncogene product and an anti-oncogene product interact, cancer can result from mutations which enhance the activity of the oncogenic protein or decrease the activity of the anti-oncogene protein. This is seen in the case of the adenomatous polyposis coli (APC) anti-oncogene protein (see Table 9.3) which is not a transcription factor and the β-catenin oncogene protein which plays a key role in cell–cell adhesion but also acts as a transcription factor (for a review see Peifer, 1997). These two factors normally interact and this interaction results in the export of β-catenin to the cytoplasm and its rapid degradation (Rosin-Arbesfeld *et al.*, 2000; Fig. 9.32a). This prevents it from moving to the nucleus and interacting with the LEF-1 transcription factor which was discussed in Section 7.3.3 and stimulating its ability to activate transcription (for reviews see Hunter, 1997; Nusse, 1997). In normal cells, in response to specific signals, β-catenin is stabilised and can stimulate the ability of LEF-1 to activate transcription resulting in cellular proliferation (Fig. 9.32b). However, such stabilisation can also be achieved either by the inactivation of APC by mutation or by mutation of β-catenin itself resulting in its enhanced stability (Fig. 9.32c).

Hence, normal cellular growth is controlled by interaction of the anti-oncogene product APC and the oncogene product β-catenin and can be stimulated by inactivating mutations in the anti-oncogenic partner or activating mutations in the oncogenic partner. In addition to illustrating another example of oncogene activation by mutation, β-catenin also illustrates a unique dual role, acting both as a factor critical for cell-to-cell adhesion and as an oncogenic transcription factor capable of regulating gene activity. Indeed, it is likely that this dual role allows it to transmit signals from cell adhesion components to the nucleus, resulting in changes in gene expression in response to extracellular events.

In contrast to the situation with oncogenes encoding proteins which pre-exist in an inactive form, and can be activated by mutation, oncogenes whose products are made only in response to a particular growth signal and whose activity then mediates cellular growth can cause cancer simply by the normal product being made at an inappropriate time. Thus, in the case of Fos or Jun, which are synthesised in normal cells in response to treatment with growth-promoting phorbol-esters or growth factors, their continuous synthesis is sufficient to transform the cell. A similar case is that of the *sis* oncogene, which encodes the cellular gene for platelet-derived growth factor and also causes cancer when over-expressed by a retrovirus.

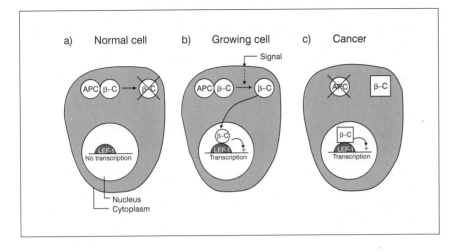

Figure 9.32

Interaction of the oncogene product β-catenin (β-C) and the anti-oncogene product APC (APC). In normal cells APC and β-catenin are associated in the cytoplasm and this results in the rapid degradation of β-catenin (panel a). In response to specific stimuli, however, β-catenin is stabilised and it can move to the nucleus where it interacts with the LEF-1 transcription factor (dark blue) and stimulates its ability to activate its target genes (panel b). This effect can also be achieved in cancer by the inactivation of APC by mutation or the activating mutation of β-catenin which converts it to a form (square) which is not susceptible to degradation. In either case the β-catenin moves to the nucleus and interacts with LEF-1 stimulating its activity (panel c).

Such cases of high-level expression of a normal oncogene product causing cancer also occur in a number of cases without any evidence of retroviral involvement. These examples of alterations in cellular regulatory processes producing increased expression of particular genes obviously provide another aspect to the connection between cancer and gene regulation. Thus, information on the origin of the translocations of the c-*myc* gene in Burkitt's lymphoma is obviously important for the study of cancer aetiology. Similarly, the fact that such translocations increase expression in some cases by removing elements that normally inhibit c-*myc* expression, allows the characterisation of such negative elements and their role in regulating c-*myc* expression in normal cells.

Hence, the study of cellular oncogenes and anti-oncogenes has contributed greatly to our knowledge both of cancer and of cellular growth-regulatory processes, and is likely to continue to do so in the future.

REFERENCES

Baudino, T.A. and Cleveland, J.L. (2001). The Max network gone mad. *Molecular and Cellular Biology* **21**, 691–702.

Baylin, S.B. and Herman, J.G. (2000). DNA hypermethylation in tumorigenesis: epigenetics joins genetics. *Trends in Genetics* **16**, 168–174.

Broach, J.R. and Levine, A.J. (1997). Oncogenes and cell proliferation. *Current Opinion in Genetics and Development* **7**, 1–6.

Celli, J., Duijf, P., Hamel, B.C.J., Bamshad, M., Kramer, B., Smits, A.P.T., Newbury-Ecob, R., Hennekam, R.C.M., Van Buggenhout, G., Van Haeringen, A., Woods, C.G., Van Essen, A.J., de Waal, R., Vriend, G., Haber, D.A., Yang, A., Mckeon, F., Brunner, H.G. and Van Bokhoven, H. (1999). Heterozygous germline mutations in the p53 homologue p63 are the cause of EEC syndrome. *Cell* **99**, 143–153.

Chan, H.M., Krstic-Demonacos, M., Smith, L., Demonacos, C. and La Thangue, N.B. (2001). Acetylation control of the retinoblastoma tumour-suppressor protein. *Nature Cell Biology* **3**, 667–674.

Conaway, J.W. and Conaway, R.C. (1999). Transcription elongation and human disease. *Annual Reviews of Biochemistry* **68**, 301–319.

D'Arcangelo, G. and Curran, T. (1995). Smart transcription factors. *Nature* **376**, 292–293.

Davis, P.K. and Dowdy, S.F. (2001). p73. *International Journal of Biochemistry and Cell Biology* **33**, 935–939.

Eisenman, R.N. (2001). Deconstructing Myc. *Genes and Development* **15**, 2023–2030.

Engelkamp, D. and Van Heyningen, V. (1996). Transcription factors in disease. *Current Opinion in Genetics and Development* **6**, 334–342.

Forrest, D. and Curran, T. (1992). Crossed signals: oncogenic transcription factors. *Current Opinion in Genetics and Development* **2**, 19–27.

Gibbons, R.J., Pickets, D.S., Villard, L. and Higgs, D.R. (1995). Mutations in a putative global transcriptional regulator cause X-linked mental retardation with α-thalassaemia (ATR-X syndrome). *Cell* **80**, 837–845.

Grandchamp, B., Picat, C., Mignotte, V., Wilson, J.H.P., Te Velde, K., Sandkuyl, L., Romeo, P.H., Goossers, M. and Nordmann, Y. (1989). Tissue-specific splicing mutation in acute intermittent porphyria. *Proceedings of the National Academy of Sciences USA* **86**, 661–664.

Grandori, C., Cowley, S.M., James, L.P. and Eisenman, R.N. (2000). The Myc/Max/Mad network and the transcriptional control of cell behaviour. *Annual Review of Cell and Developmental Biology* **16**, 653–699.

Harbour, J.W. and Dean, D.C. (2000). The Rb/E2F pathway: expanding roles and emerging paradigms. *Genes and Development* **14**, 2393–2409.

Hastie, N.D. (2001). Life, sex, and WT1 isoforms – three amino acids can make all the difference. *Cell* **106**, 391–394.

Hunter, T. (1997). Oncoprotein networks. *Cell* **88**, 333–346.

Karin, M., Liu, Z.-G. and Zandi, E. (1997). AP-1 function and regulation. *Current Opinion in Cell Biology* **9**, 240–246.

Kersten, S., Desvergne, B. and Wahli, W. (2000). Roles of PPARs in health and disease. *Nature* **405**, 421–424.

Knudson, A.G. (1993). Anti-oncogenes and human cancer. *Proceedings of the National Academy of Sciences USA* **90**, 10914–10921.

Lasorella, A., Noseda, M., Beyna, M. and Lavarone, A. (2000). Id2 is a retinoblastoma protein target and mediates signalling by Myc oncoproteins. *Nature* **407**, 592–598.

Latchman, D.S. (1996). Transcription factor mutations and disease. *New England Journal of Medicine* **334**, 28–33.

Lewin, B. (1991). Oncogenic conversion by regulatory changes in transcription factors. *Cell* **64**, 303–312.

Liu, J., Akoulitchev, S., Weber, A., Ge, H., Chuikov, S., Libutti, D., Wang, X.W., Conaway, J.W., Harris, C.C., Conaway, R.C., Reinberg, D. and Levens, D. (2001). Defective interplay of activators and repressors with TFIIH in Xeroderma Pigmentosum. *Cell* **104**, 353–363.

Lohrum, M.A.E. and Vousden, K.H. (2000). Regulation and function of the p53-related proteins: same family, different rules. *Trends in Cell Biology* **10**, 197–202.

Look, A.T. (1997). Oncogenic transcription factors in the human acute leukaemias. *Science* **278**, 1059–1064.

Luo, R.X., Postigo, A.A. and Dean, D.C. (1998). Rb interacts with histone deacetylase to repress transcription. *Cell* **92**, 463–473.

Marx, J. (1997). Possible function found for breast cancer genes. *Science* **276**, 531–532.

Moran, E. (1993). DNA tumour virus transforming proteins and the cell cycle. *Current Opinion in Genetics and Development* **3**, 63–70.

Nucifora, F.C. Jr., Sasaki, M., Peters, M.F., Huang, H., Cooper, J.K., Yamada, M., Takahashi, H., Tsuji, S., Troncoso, J., Dawson, V.L., Dawson, T.M. and Ross, C.A. (2001). Interference by Huntingtin and Atrophin-1 with CBP-mediated transcription leading to cellular toxicity. *Science* **291**, 2423–2428.

Nusse, R. (1997). A versatile transcriptional effector of wingless signalling. *Cell* **89**, 321–323.

Peifer, M. (1997). β-catenin as oncogene: the smoking gun. *Science* **275**, 1752–1753.

Pellizzoni, L., Kataoka, N., Charroux, B. and Dreyfuss, G. (1998). A novel function for SMN, the spinal muscular atrophy disease gene product, in pre-mRNA splicing. *Cell* **95**, 615–624.

Perlmann, T. and Vennstrom, B. (1995). The sound of silence. *Nature* **377**, 287.

Piechaczyk, M. and Blanchard, J.M. (1994). C-fos proto-oncogene regulation and function. *Critical Reviews in Oncology and Haematology* **17**, 93–131.

Rabbits, T.H. (1994). Chromosomal translocations in human cancer. *Nature* **372**, 143–149.

Ringrose, L. and Paro, R. (2001). Cycling silence. *Nature* **412**, 493–494.

Rosin-Arbesfeld, R., Townsley, F. and Bienz, M. (2000). The APC tumour suppressor has a nuclear export function. *Nature* **406**, 1009–1012.

Schwartz, M.W. and Kahn, S.E. (1999). Insulin resistance and obesity. *Nature* **402**, 860–861.

Takeya, T. and Hanafusa, H. (1985). Structure and sequence of the cellular gene homologous to the RSV-*src* gene and the mechanism for generating transforming virus. *Cell* **32**, 881–890.

Trottier, Y., Dewys, D. and Mandel, J.L. (1993). Fragile X syndrome: an expanding story. *Current Biology* **3**, 783–786.

Vogelstein, B., Lane, D. and Levine, A.J. (2000). Surfing the p53 network. *Nature* **408**, 307–310.

Vogt, P.K. (1995). The story of jun. *Archives in Biochemistry and Biophysics* **316**, 1–4.

Weinberg, R.A. (1993). Tumour suppressor genes. *Neuron* **11**, 191–196.

White, R.J. (1997). Regulation of RNA polymerase I and III by the retinoblastoma protein: a mechanism for growth control. *Trends in Biochemical Sciences* **22**, 77–80.

10 Conclusions and future prospects

The extraordinary rate of progress in the study of gene regulation can be gauged from the fact that, by 1980, no transcriptional regulatory protein or its DNA binding site in a regulated promoter had been defined. The tremendous advances since that time were evident by the time the first edition of this book was published. Thus, at that time (1990), a relatively clear picture of the action of individual factors which regulate gene expression was available. For example, in the case of gene induction by glucocorticoid hormone it was known that these agents act by activating specific receptor proteins (Section 8.2.3), that such activated receptors bind to specific DNA sequences upstream of the target gene (Section 7.2.3), displacing a nucleosome (Section 6.7.2) and that a region in the receptor protein then interacts with a factor bound to the TATA box to activate transcription (Section 8.3.3).

Clearly, therefore, the process by which a single agent can activate a specific transcription factor and thereby modulate gene expression was reasonably well understood in outline. However, in the period from 1990 to the publication of the second edition of this book in 1995, it became increasingly clear that the activation of a single transcription factor by a single agent cannot be assessed in isolation. Rather, the activity of a specific factor will depend upon its interaction with other transcription factors as well as with other proteins. Thus, for example, glucocorticoid hormone activates transcription by promoting the dissociation of the glucocorticoid receptor from the inhibitory hsp90 protein which otherwise anchors it in the cytoplasm and prevents it from moving to the nucleus and activating transcription (see Section 8.4.3).

A similar example of protein–protein interactions regulating transcription factor activity but in the opposite direction is observed in the Fos and Jun proteins which regulate gene expression in response to phorbol-ester treatment (Section 9.4.1). Thus, as noted in Section 8.2.4, the Fos protein cannot bind to DNA on its own but can do so only after forming a heterodimer with the Jun protein. This heterodimer then binds to DNA and activates transcription. Hence, in this case, the interaction with another protein allows Fos to activate transcription whereas in the case of the glucocorticoid receptor the interaction with hsp90 has the opposite effect.

Such types of protein–protein interactions regulating transcription factor activation are not confined to the regulation of gene expression by short-term inducers but are also involved in the regulation of tissue-specific gene expression. Thus, as discussed in Section 8.2.4, the MyoD transcription factor can bind to the promoters of muscle-specific genes and activate their expression, thereby causing the differentiation of a fibroblast cell line into muscle cells. The production of differentiated skeletal muscle cells in this situation requires, however, that the cells are cultured in the absence of growth factors. In the presence of growth factor, the cells contain high levels of the inhibitory factor Id which heterodimerises with MyoD and prevents it binding to DNA and activating transcription. The production of the skeletal muscle phenotype therefore requires not only high levels of the activating MyoD protein but also a fall in the levels of the inhibitory Id protein which is induced by growth factor removal.

It is clear, therefore, that in families of transcription factors which are capable of dimerisation by means of specific motifs such as the leucine zipper or helix-loop-helix (see Section 8.2.4), such dimerisation with another member of the family can result in activation (Fos–Jun) or repression (MyoD–Id).

The Id example also illustrates another theme which has become of increasing importance in gene regulation, namely the importance of specific inhibition as opposed to activation of gene expression (see Section 8.3.4). As well as transcription factors, such as the Id factor, which function only as inhibitors, cases also exist where the same factor can also function as a direct activator or repressor, depending on the circumstances. This is seen in the case of the thyroid hormone receptor which, unlike the glucocorticoid receptor, binds to its DNA target site even in the absence of hormone. As discussed in Section 8.3.4, this factor can directly inhibit promoter activity in the absence of thyroid hormone whereas in the presence of thyroid hormone it undergoes a conformational change which allows it to activate gene expression.

This case of the thyroid hormone receptor illustrates two of the major themes to have emerged between the publication of the second edition of this work and the present day, namely the role of co-repressor – co-activator molecules and the influence of transcription factors on chromatin structure. Thus, as discussed in Section 8.3.4, the inhibitory domain of the thyroid hormone receptor acts indirectly by recruiting an inhibitory molecule known as the nuclear receptor co-repressor which actually produces the inhibitory effect. Although this factor may interact directly with the basal transcriptional complex, it has also been shown to itself recruit another molecule which has the ability to acetylate histones (see Section 6.6.1). As such histone acetylation is associated with a tightly packed chromatin structure incompatible with transcription (see Section 6.6.1), the inhibitory effect of the thyroid hormone receptor appears to be mediated, at least in part, by the ability of its co-repressor to recruit a molecule which can organise the chromatin into a non-transcribable form.

As discussed in Section 8.3.4, the addition of thyroid hormone results in a conformational change in the thyroid hormone receptor which results in the release of its co-repressor and allows the receptor to activate transcription. Interestingly, as with transcriptional repression by the receptor, such activation requires a co-factor and involves alterations in histone acetylation. Thus, following exposure to thyroid hormone, the thyroid hormone receptor can bind the CBP co-activator. This factor was originally identified as a co-activator which binds to the CREB transcription factor but was subsequently shown to be involved in transcriptional activation by nuclear receptors such as the glucocorticoid receptor and the thyroid hormone receptor (see Section 8.3.3).

As discussed in Section 8.3.3, CBP is able to act as a co-activator by interacting with components of the basal transcriptional complex to stimulate its activity. In addition, however, it also has histone deacetylase activity (see Section 6.6.1), indicating that by deacetylating histones it can reverse the effects of the nuclear receptor co-repressor complex and convert the chromatin into a form where transcription can occur. Hence, the addition of thyroid hormone converts the thyroid hormone receptor from a form which can bind an inhibitory co-repressor complex with histone acetylase activity to a form where it can bind a stimulatory co-activator with histone deacetylase activity.

This example thus indicates the critical importance of co-activators and co-repressors as well as providing a critical link between the transcription factors discussed in Chapter 8 and the chromatin structure changes discussed in Chapter 6. Indeed, it has become increasingly clear that modification of histones by transcriptional regulators is a key event in the modulation of chromatin structure, leading to transcriptional activation or repression. As discussed in Section 6.6, such histone modifications include not only acetylation but also methylation and phosphorylation, with the existence of one modification at a particular amino acid, influencing positively or negatively the occurrence of another modification at another amino acid. This has led to the idea of a "histone code" in which a particular pattern of modification on the histones in a particular region of DNA promotes or inhibits the changes necessary for transcription to occur by affecting the recruitment of chromatin remodelling complexes and DNA binding transcriptional activators or repressors.

Interestingly, a number of co-activator/co-repressor molecules appear to interact respectively with a number of different activating or inhibitory transcription factors. In the case of the CBP co-activator which is used by transcription factors activated by different signalling pathways, this has been shown to result in interactions between these pathways. Thus, both the glucocorticoid receptor, which is activated by steroids, and the AP1 (Fos–Jun) complex, which is activated by phorbol esters, require the CBP co-activator in order to mediate this stimulatory effect on gene transcription. Because CBP levels in the cell are relatively low, when both pathways are activated simultaneously, they will compete for the limited amount of CBP and neither pathway will be able to activate transcription. Hence, the requirement of both of these pathways for CBP results in a mutual inhibition of gene activation by each of these factors.

This example illustrates how the effect of one signalling pathway can be affected by the activation of another signalling pathway. Hence, in the absence of glucocorticoid hormone, the glucocorticoid receptor will be anchored in the cytoplasm by hsp90 and the Fos–Jun complex will be able to interact with CBP and activate gene expression in response to growth factors or phorbol esters. However, when the glucocorticoid receptor is freed from hsp90 by the presence of glucocorticoid, it can interact with CBP removing it from Fos–Jun and preventing them activating gene expression. This is seen in the case of the gene encoding the collagenase enzyme which is activated by the Fos–Jun complex in response to phorbol esters and can produce severe tissue destruction in an inflamed area. As expected on the basis of the above model, the activation of this gene by phorbol esters is inhibited by treatment with glucocorticoid hormone, accounting for the anti-inflammatory effort of this steroid hormone. Conversely, the activation of glucocorticoid-responsive genes by hormone treatment will be inhibited by the presence of high levels of Fos and Jun induced by growth factor or phorbol-ester treatment since they will compete for the limited amount of available CBP.

These examples indicate therefore that the effect of a particular factor on gene expression will depend on the nature of the other factors present in the cell and the state of activation of specific signalling pathways. Moreover, it is clear that our simple picture of the glucocorticoid receptor acting in isolation has to be modified to consider its inhibition by interaction with hsp90 as well as its need to recruit other factors in order to stimulate transcription such as CBP and, as discussed in Section 6.7.2, the SWI–SNF complex which also reorganises chromatin structure.

Over the last few years it has become clear that the regulation of a single gene in an intact cell is vastly more complex than could be predicted from the study of an isolated transcription factor interacting with its binding site. At first sight it may appear that recent progress in this area has merely complicated the issue. However, the process of gene regulation must not only produce all the different types of cells in the body but also ensure that each cell type is produced in the right place and at the right time during development.

It is inevitable therefore that gene regulation must be a highly complex process with numerous factors interacting with each other and their respective DNA binding sites to activate or inhibit the expression of specific genes in specific cell types. Interestingly, the CBP factor which, as discussed above, plays a critical role in regulating gene activation in response to specific signalling pathways also appears to be involved in gene regulation during development. Thus, as discussed in Section 9.1, mutations in the CBP gene in humans result in Rubinstein–Taybi syndrome which is characterised by abnormal development, leading to mental retardation and physical abnormalities. Interestingly, this disease occurs when only one of the two copies of the CBP gene is mutated with the other being capable of producing functional CBP protein and individuals with two mutant CBP genes have never been identified. This indicates that CBP is likely to be of such importance that its complete loss by inactivation of both CBP genes is incompatible with survival. Moreover, as in the signalling processes described above, the precise level of CBP in a cell is critical for proper development and a single functional gene evidently cannot produce sufficient CBP for normal development to occur.

The central role of CBP in many different signalling pathways and in development, together with the critical role of the basal transcription factor TBP in transcription by all three RNA polymerases (see Section 3.2.4), has often made gene regulation in the last few years appear to be "the study of

CBP and TBP". Of course, despite the central role of these factors, it is far more complex than that. Nonetheless, the progress in understanding gene regulation in terms of factor–factor interactions, changes in chromatin structure, etc., which has taken place in the last few years offers hope that an understanding of the ultimate problem of mammalian development in terms of differential gene expression can eventually be achieved.

INDEX

Bold page numbers refer to illustrations, *italics* to tables